Communications
in Computer and Information Science   **774**

Commenced Publication in 2007
Founding and Former Series Editors:
Alfredo Cuzzocrea, Xiaoyong Du, Orhun Kara, Ting Liu, Dominik Ślęzak,
and Xiaokang Yang

More information about this series at http://www.springer.com/series/7899

Xueqi Cheng · Weiying Ma
Huan Liu · Huawei Shen
Shizheng Feng · Xing Xie (Eds.)

Social Media Processing

6th National Conference, SMP 2017
Beijing, China, September 14–17, 2017
Proceedings

 Springer

Editors
Xueqi Cheng
Institute of Computing Technology
Chinese Academy of Sciences
Beijing
China

Huawei Shen
Institute of Computing Technology
Chinese Academy of Sciences
Beijing
China

Weiying Ma
Beijing Jinri Toutiao Technology Co. Ltd
Beijing
China

Shizheng Feng
Renmin University of China
Beijing
China

Huan Liu
Arizona State University
Tempe, AZ
USA

Xing Xie
Microsoft Asia Research
Beijing
China

ISSN 1865-0929   ISSN 1865-0937 (electronic)
Communications in Computer and Information Science
ISBN 978-981-10-6804-1   ISBN 978-981-10-6805-8   (eBook)
https://doi.org/10.1007/978-981-10-6805-8

Library of Congress Control Number: 2017956083

Preface

Social media greatly facilitate the creation, delivery, and consumption of online content. Huge amounts of user-generated content are spread in social media platforms, offering us an unprecedented opportunity to study human society, including collective behavior modeling, user profiling, sentiment analysis, information propagation, network analysis, and many other tasks. Meanwhile, social media processing requires the integration of computer science and social science. The 6th National Conference on Social Media Processing (SMP 2017) was held in Beijing, China, in 2017, with the purpose of promoting original research in mining social media and applications, bringing together experts from related fields such as natural language processing, data mining, information retrieval, and social science, and providing a leading forum in which to exchange research ideas and results in emergent social media processing problems.

In this year, we received 140 valid submissions in total, a new record for SMP conferences. These submissions were divided into seven tracks according to their topics, including SMP-KDD (knowledge discovery from data), SMP-NET (social networks), SMP-NLP (natural language processing), SMP-SENT (sentiment analysis), SMP-SOC (computational social science), SMP-CCC (computational communication), and SMP-PRO (user profiling). Each track had two area chairs who were responsible for giving meta-reviews and decision suggestions for each paper in the track. Each paper was peer reviewed by at least three members of the Program Committee (PC) composed of international experts in natural language processing, data mining, information retrieval, complex network, and social sciences.

The PC members together with the area chairs worked hard to select papers through a rigorous review process and via extensive discussions. The competition was very tough; only 28 English papers were accepted. The conference also featured invited talks from outstanding researchers in social media processing and related areas: Tong Zhang (Tencent AI Lab), Shuicai Shi (TRS corporation), Tieyan Liu (Microsoft Research of Asia), Jie Tang (Tsinghua University), Wei Wang (University of California, Los Angeles), Tao Zhou (University of Electronic Science and Technology of China), Bing Qin (Harbin Institute of Technology), and Tianguang Meng (Tsinghua University).

Without the support of several funding agencies and industrial partners, the successful organization of SMP 2017 would not have been possible. We would also like to express our gratitude to the Steering Committee of the special group of Social Media Processing of the Chinese Information Processing Society for all their advice and the

Organizing Committee for their dedicated efforts. Last but not least, we sincerely thank all the authors, presenters, and attendees who jointly contributed to the success of SMP 2017.

September 2017

Xueqi Cheng
Weiying Ma
Huan Liu
Huawei Shen
Shizheng Feng
Xing Xie

Organization

Conference General Chairs

Xueqi Cheng — Institute of Computing Technology, Chinese Academy of Sciences, China
Weiying Ma — Beijing Jinri Toutiao Technology Co., Ltd, China
Huan Liu — Arizona State University, USA

Program Committee Chairs

Huawei Shen — Institute of Computing Technology, Chinese Academy of Sciences, China
Shizheng Feng — Renmin University of China, China
Xing Xie — Microsoft Research Asia, China

Area Chairs

Minlie Huang — Tsinghua University, China
Duyu Tang — Microsoft Research Asia, China
Tianguang Meng — Tsinghua University, China
Xin Zhao — Renmin University of China, China
Lun Zhang — Beijing Normal University, China
Guojie Song — Peking University, China
Linyuan Lv — Hangzhou Normal University, China
Chuan Shi — Beijing University of Posts and Telecommunications, China
Chong Feng — Beijing Institute of Technology, China
Zhongyu Wei — Fudan University, China
Guoqing Zhang — Beijing Uniwisdom Technology Co., Ltd, China
Ruifeng Xu — Harbin Institute of Technology (Shenzhen), China
Yang Yang — Zhejiang University, China
Chengzhi Zhang — Nanjing University of Science and Technology, China

Local Chair

Wengtao Ouyang — Institute of Computing Technology, Chinese Academy of Sciences, China

Competition Chairs

Shengyi Jiang — Guangdong University of Foreign Studies, China
Huaiyu Wan — Beijing Jiaotong University, China
Weinan Zhang — Harbin Institute of Technology, China

Publication Chair

Tieyun Qian   Wuhan University, China

Publicity Chair

Chenliang Li   Wuhan University, China

Data Chair

Qi Zhang   Fudan University, China

Forum Chair

Zhiyuan Liu   Tsinghua University, China

Best Paper Award Chair

Jun Ma   Shandong University, China

Sponsor Chair

Ting Liu   Harbin Institute of Technology, China

Website Maintenance

Yongqi Wang   Institute of Computing Technology,
                        Chinese Academy of Sciences, China
Cheng Yang   Tsinghua University, China

Program Committee

Tingting He   Central China Normal University, China
Shengyi Jiang   Guangdong University of Foreign Studies, China
Jun Ma   Shandong University, China
Bing Qin   Harbin Institute of Technology, China
Daling Wang   Northeastern University, China
Suge Wang   Shanxi University, China
Xiaoke Xu   Dalian Minzu University, China
Zhengtao Yu   Kunming University of Science and Technology, China
Ming Zhang   Peking University, China
Jun Zhao   Institute of Automation, Chinese Academy of Sciences, China
Zhumin Chen   Shandong University, China
Peng Cui   Tsinghua University, China
Jibing Gong   Yanshan University, China
Qilong Han   Harbin Engineering University, China

Contents

Knowledge Discovery for Data

Locality-Sensitive Hashing for Finding Nearest Neighbors
in Probability Distributions.. 3
  Yi-Kun Tang, Xian-Ling Mao, Yi-Jing Hao, Cheng Xu,
  and Heyan Huang

Supervised Hashing for Multi-labeled Data with Order-Preserving Feature ... 16
  Dan Wang, Heyan Huang, Hua-Kang Lin, and Xian-Ling Mao

Inferring User Profile Using Microblog Content and Friendship Network.... 29
  Zhishan Zhao, Jiachen Du, Qinghong Gao, Lin Gui, and Ruifeng Xu

EEG: Knowledge Base for Event Evolutionary Principles and Patterns 40
  Zhongyang Li, Sendong Zhao, Xiao Ding, and Ting Liu

Prediction of Cascade Structure and Outbreaks Recurrence in Microblogs ... 53
  Zhenhua Huang, Zhenyu Wang, Yingbo Zhu, Chengqi Yi,
  and Tingxuan Su

Exploring Effective Methods for On-line Societal Risk Classification
and Feature Mining.. 65
  Nuo Xu and Xijin Tang

A Markov Chain Monte Carlo Approach for Source Detection in Networks ... 77
  Le Zhang, Tianyuan Jin, Tong Xu, Biao Chang, Zhefeng Wang,
  and Enhong Chen

Natural Language Processing

Neural Chinese Word Segmentation as Sequence to Sequence Translation ... 91
  Xuewen Shi, Heyan Huang, Ping Jian, Yuhang Guo, Xiaochi Wei,
  and Yi-Kun Tang

Attention-Based Memory Network for Sentence-Level Question Answering ... 104
  Pei Liu, Chunhong Zhang, Weiming Zhang, Zhiqiang Zhan,
  and Benhui Zhuang

Entity Set Expansion on Social Media: A Study for Newly-Presented
Entity Classes... 116
  He Zhao, Chong Feng, Zhunchen Luo, and Yuxia Pei

An Effective Approach of Sentence Compression Based on "Re-read"
Mechanism and Bayesian Combination Model   129
   Zhonglei Lu, Wenfen Liu, Yanfang Zhou, Xuexian Hu, and Binyu Wang

Terminology Translation Error Identification and Correction.   141
   Mengyi Liu, Jian Tang, Yu Hong, and Jianmin Yao

Opinion Target Understanding in Event-Level Sentiment Analysis   153
   Suyang Zhu, Shoushan Li, and Guodong Zhou

Topic Enhanced Word Vectors for Documents Representation   166
   Dayu Li, Yang Li, and Suge Wang

Text Mining and Sentiment Analysis

Social Annotation for Query Expansion Learning from Multiple
Expansion Strategies   181
   Yuan Lin, Bo Xu, Luying Li, Hongfei Lin, and Kan Xu

Supervised Domain Adaptation for Sentiment Regression.   193
   Jian Xu, Hao Yin, Shoushan Li, and Guodong Zhou

Dependency-Attention-Based LSTM for Target-Dependent
Sentiment Analysis   206
   Xinbo Wang and Guang Chen

A Novel Fuzzy Logic Model for Multi-label Fine-Grained
Emotion Retrieval   218
   Chu Wang, Daling Wang, Shi Feng, and Yifei Zhang

Deep Transfer Learning for Social Media Cross-Domain
Sentiment Classification.   232
   Chuanjun Zhao, Suge Wang, and Deyu Li

Local Contexts Are Effective for Neural Aspect Extraction   244
   Jianhua Yuan, Yanyan Zhao, Bing Qin, and Ting Liu

Context Enhanced Word Vectors for Sentiment Analysis   256
   Zhe Ye and Fang Li

Social Network Analysis and Social Computing

Divergence or Convergence: Interaction Between News Media Frames
and Public Frames in Online Discussion Forum in China.   271
   Lun Zhang

A Unified Framework of Lightweight Local Community Detection
for Different Node Similarity Measurement 283
    Jinglian Liu, Daling Wang, Weiji Zhao, Shi Feng, and Yifei Zhang

Estimating the Origin of Diffusion in Complex Networks
with Limited Observations 296
    Shuaishuai Xu, Yinzuo Zhou, and Zike Zhang

Exploring the Country Co-occurrence Network in the Twittersphere
at an International Economic Event 308
    Xinzhi Zhang

The 2016 US Presidential Election and Its Chinese Audience 319
    Jiahua Yue, Yuke Li, and James Sundquist

Understanding the Pulse of the Online Video Viewing Behavior
on Smart TVs. 331
    Tao Lian, Zhumin Chen, Yujie Lin, and Jun Ma

Hierarchical Community Detection Based on Multi Degrees of Distance
Space and Submodularity Optimization 343
    Shu Zhao, Chengjin Yu, and Yanping Zhang

Author Index 355

Knowledge Discovery for Data

Locality-Sensitive Hashing for Finding Nearest Neighbors in Probability Distributions

Yi-Kun Tang, Xian-Ling Mao$^{(\boxtimes)}$, Yi-Jing Hao, Cheng Xu, and Heyan Huang

School of Computer Science and Technology, Beijing Institute of Technology,
Haidian District, Beijing 100081, China
{tangyk,maoxl,hyj,1120141839,hhy63}@bit.edu.cn

Abstract. In the past ten years, new powerful algorithms based on efficient data structures have been proposed to solve the problem of Approximate Nearest Neighbors search (ANN). To find the nearest neighbors in probability-distribution-type data, the existing Locality Sensitive Hashing (LSH) algorithms for vector-type data can be directly used to solve it. However, these methods do not consider the special properties of probability distributions. In this paper, based on the special properties of probability distributions, we present a novel LSH scheme adapted to angular distance for ANN search in high-dimensional probability distributions. We define the specific hashing functions, and prove their local-sensitivity. Also, we propose a Sequential Interleaving algorithm based on the "Unbalance Effect" of Euclidean and angular metrics for probability distributions. Finally, we compare, through experiments, our methods with the state-of-the-art LSH algorithms in the context of ANN on six public image databases. The results prove the proposed algorithms can provide far better accuracy in the context of ANN than baselines.

Keywords: Approximate Nearest Neighbors · Locality Sensitive Hashing · Arccos-distance

1 Introduction

In the past decade, we have witnessed an explosive growth of data on the Internet. Billions of data are publicly available on the Web, and it brings both challenges and opportunities to traditional algorithms developed on small to median scale data sets. Particularly, nearest neighbor search has become a key ingredient in many large-scale machine learning and computer vision tasks. In big data applications, it is typically time-consuming or impossible to return the exact nearest neighbors to the given queries. In fact, approximate nearest neighbors (ANN) [13,24] are enough to achieve satisfactory performance in many applications, such as the image retrieval task in search engines. Moreover, ANN search is usually more efficient than exact nearest neighbor search to solve large-scale problems. Hence, ANN search has attracted more and more attention in this big data era [24].

© Springer Nature Singapore Pte Ltd. 2017
X. Cheng et al. (Eds.): SMP 2017, CCIS 774, pp. 3–15, 2017.
https://doi.org/10.1007/978-981-10-6805-8_1

Because of its low storage cost and fast retrieval speed, hashing is one of the popular solutions for ANN search [2,23]. The hashing techniques used for ANN search are usually called similarity-preserving hashing or Locality Sensitive Hashing (LSH), and its basic idea is to transform the data points from the original feature space into a binary-code Hamming space, where the similarity in the original space is preserved. More specifically, the Hamming distance between the binary codes of two points should be small if these two points are similar in the original space. Otherwise, the Hamming distance should be as large as possible. With the binary-code representation, the storage cost can be substantially reduced and the query speed can be dramatically improved for ANN search [13,26]. For example, if we encode each point with 256 bits, we can store a data set of 1 million points with only 32M memory.

Many hashing methods have been proposed by researchers from different research communities. The existing hashing methods can be mainly divided into two categories [17,26]: data-independent methods and data-dependent methods. Data-dependent hashing methods [18,19,23,29], which are also called learning to hash (LH) methods, whose hash functions are learned from the training data; Data-independent hashing methods use simple random projections which are independent of the training data for hash functions. Both Data-independent and Data-dependent hashing methods have an important property that points with high similarity will have high probability to be mapped to the same hashcodes. Compared with the data-dependent methods, data-independent methods need longer codes to achieve satisfactory performance [9], which will be less efficient due to the higher storage and computational cost; However, the data-dependent hashing algorithms are not dynamic, in contrast to the data-independent hashing methods, which allow dynamically updates to the point set. In this paper, we focus on the data-independent hashing methods.

A well-defined distance is crucial in data-independent hashing methods. Most of the popular distances are subject to the metric axioms, i.e., non-negativity, symmetry and triangular inequality. In existing LSH families, there are different distances and corresponding LSH for different types of data. For vector-type data, l_p distance is often used to develop a LSH method, e.g. p-stable LSH [6], Leech lattice LSH [1], Spherical LSH [27], and Beyond LSH [2]; Also, angle-based metric (arccos) is popular for vector-type data, e.g. Random Projection (PR) [1,4], Super-bit LSH [16], Kernel LSH [20], Concomitant LSH [7], and Hyperplane hashing [14]; Chi-squared Distance [10] and Bregman divergence [25] have also been used as similarity function for vector-type data. For set-type data, Jaccard Coefficient based LSH include Min-hash [3], K-min Sketch [22], Min-max hash [15], B-bit minwise hashing [21], and Sim-min-hash [30]. More details refer to the paper [28] for a brief survey.

So far as we know, few works focus on the LSH algorithms for probability-distribution-type data. The reason may be: (i) The distance metrics for probability distributions, e.g. KL-divergence and JSD, are not the well-defined distance metrics, which do not satisfy triangle inequality, and is hard to develop similarity-preserved hashing; (ii) The performance of metrics for vector-type data is a pretty good approximation to, sometimes even better than, the one of JSD or KL (see Table 1). In practice, it is common to use Euclidean (l_2) and Angular (arccos) metrics for LSH to measure the distance between two probability distributions, i.e. take the probability distributions as general vectors, and do not consider the specific properties of probability distributions. Through the practical experiments, from Table 1, the state-of-the-art angle-based algorithm (Random Projection, RP), performs worse than Euclidean-based method (l_2). However, through brute-force search (linear search, ls), RP should be better than l_2, which means the performance of current LSH for angular metric still can be improved greatly. Meanwhile, we found the performances for RP and l_2 are unstable, i.e. the performance varies with different hash bits.

Compared with vector-type data, the special properties of probability-distribution-type data is that all probability distributions are only located in $\frac{1}{2^d}$ areas in d-dimensional space, and satisfies the conditions of Non-negative and Sum-equal-one. In this paper, we proposed a angle-based LSH algorithm, called pbRP, which considers the inner properties of probability-distribution-type data.

The remainder of this paper is organized as follows. We first present some preliminaries, and then propose a similarity-preserving hashing method, pbRP. Furthermore, a sequential interleaving algorithm (si) is introduced. Finally, empirical evaluations on various benchmarks are presented.

2   Preliminary

LSH was first introduced by Indyk and Motwani in [13] for the Hamming metric. They defined the requirements on hash function families to be considered as locality-sensitive (Locality-Sensitive Hashing functions). In this section, we present an overview of the LSH scheme. The intuition behind LSH is to use hash functions to map points into buckets, such that nearby objects are more likely to map into the same buckets than objects that are farther away.

Let S be the domain of the objects and D the distance measure between objects.

Definition 1 (D1). *A function family $H = h : S \rightarrow U$ is called (r_1, r_2, p_1, p_2)-sensitive, with $r_1 < r_2$ and $p_1 > p_2$, for D if for any p, q $\in$ S*

- *if $D(q, p) \leq r_1$ then $P_H[h(q) = h(p)] \geq p_1$,*
- *if $D(q, p) > r_2$ then $P_H[h(q) = h(p)] \leq p_2$.*

Intuitively, the definition states that nearby objects (those within distance r_1) are more likely to collide ($p_1 > p_2$) than objects that are far apart (those with a distance greater than r_2). To decrease the probability of false detection p_2, several functions are concatenated: for a given integer N, let us define a new function family $G = g : S \rightarrow U^N$ such that $g(p) = (h_1(p), ..., h_N(p))$, where $h_i \in H$. Thus, the probability of good detection p_1 decreases too. Several functions g are used to compensate the decrease in p_1. For a given integer U, choose g_1, ..., g_U from G, independently and uniformly at random, each one defining a new hash table, in order to get U hash tables.

3   LSH for Probability Distributions

Assume that K^{+d} is the point set, where each point is a probability distribution with d-dimensions. First, we investigate a special attribution of probability-distribution-type data, as following theorems.

Theorem 1 (Range). *Given any two probability distribution p_1 and p_2, p_1, $p_2 \in K^{+d}$, the angle between p_1 and p_2 is θ. $\theta \in [0, \frac{\pi}{2}]$.*

Proof. p_1 and p_2 are probability distributions, thus $\sum_{i=1}^{d} p_{1i} = 1$, $\sum_{i=1}^{d} p_{2i} = 1$, $p_{1i} \geq 0$ and $p_{2i} \geq 0$, $i = 1, 2, ..., d$. The CauchySchwarz inequality [8] states that for all vectors x and y of an inner product space it is true that $\langle x, y \rangle \leq \|x\| \cdot \|y\|$, where $\langle \cdot, \cdot \rangle$ is the inner product, and $\|x\|$ referring to the norms of the vector x. Moreover, the two sides are equal if and only if $\frac{x_1}{y_1} = \frac{x_2}{y_2} = \cdot = \frac{x_n}{y_n}$, if any of the vectors' magnitude is not zero. Thus, $\cos\theta = \frac{\langle p_1, p_2 \rangle}{\|p_1\| \cdot \|p_2\|} \leq \frac{\|p_1\| \cdot \|p_2\|}{\|p_1\| \cdot \|p_2\|} = 1$; Meanwhile, because $\|p_1\| = \|p_2\| = 1$, it is clear that we have equality when $\frac{p_{11}}{p_{21}} = \frac{p_{12}}{p_{22}} = \cdot = \frac{p_{1d}}{p_{2d}}$, and in this case $p_1 = p_2$. On the other hand, $\cos\theta = \frac{\langle p_1, p_2 \rangle}{\|p_1\| \cdot \|p_2\|} = \frac{\sum_{i=1}^{d} p_{1i} \cdot p_{2i}}{\|p_1\| \cdot \|p_2\|} \geq 0$, because of $p_{1i} \cdot p_{2i} \geq 0$; Meanwhile, two sides are equal if and only if $p_{1i} \cdot p_{2i} = 0$, i.e., p_1 is a probability distribution orthogonal to the probability distribution p_2. So far, we have proved $0 \leq \cos\theta \leq 1$, as we know, function $\cos\theta$ decreases monotonically between 0 and π, thus, $\theta \in [0, \frac{\pi}{2}]$, and the conclusion holds.

Given a collection of probability distributions in K^{+d}, we consider the family of hash functions defined as follows: We choose a random probability distribution r from the d-dimensional Dirichlet distribution (i.e. r is drawn from symmetry Dirichlet distribution, and the parameter $\alpha = 1$). Corresponding to this probability distribution r, we define a hash function h_r as follows:

$$h_r(u) = \begin{cases} 1 \text{ if } \theta(u, r) \leq \frac{\pi}{4}, \\ 0 \text{ if } \theta(u, r) > \frac{\pi}{4}. \end{cases} \tag{1}$$

Furthermore, we have following theorem:

Theorem 2 (T2). $Pr[h_r(u) \neq h_r(v)] = \frac{2\theta(u,v)}{\pi}$

Proof. A restatement of the lemma is that the probability the random hyperplane separates the two probability distributions is directly proportional to the angle between the two probability distributions; That is, it is proportional to the angle $\theta = \arccos(u, v)$, i.e. $\theta(u, v)$. By symmetry, $Pr[h_r(u) \neq h_r(v)] = 2Pr[\theta(u, r) \leq \frac{\pi}{4}, \theta(v, r) > \frac{\pi}{4}]$. The set $\{r : \theta(u, r) \leq \frac{\pi}{4}, \theta(v, r) > \frac{\pi}{4}\}$ corresponds to the intersection of two half-spaces whose dihedral angle is precisely θ;

By using Theorem 1, $\theta(\boldsymbol{u}, \boldsymbol{v}) \in [0, \frac{\pi}{2}]$, and $\boldsymbol{u}, \boldsymbol{v} \in K^{+d}$, its intersection with a n-dimensional unit sphere is a spherical digon of angle $\theta(\boldsymbol{u}, \boldsymbol{v})$ and, by symmetry of the sphere, thus has measure equal to $\frac{\theta(\boldsymbol{u},\boldsymbol{v})}{\pi}$ times the measure of the full sphere. In other words, $Pr[\theta(\boldsymbol{u}, \boldsymbol{r}) \leq \frac{\pi}{4}, \theta(\boldsymbol{v}, \boldsymbol{r}) > \frac{\pi}{4}] = \frac{\theta(\boldsymbol{u},\boldsymbol{v})}{\pi}$. Thus, $Pr[h_{\boldsymbol{r}}(\boldsymbol{u}) \neq h_{\boldsymbol{r}}(\boldsymbol{v})] = 2Pr[\theta(\boldsymbol{u}, \boldsymbol{r}) \leq \frac{\pi}{4}, \theta(\boldsymbol{v}, \boldsymbol{r}) > \frac{\pi}{4}] = \frac{2\theta(\boldsymbol{u},\boldsymbol{v})}{\pi}$, and the lemma follows.

Then, for probability distributions $\boldsymbol{u}$ and $\boldsymbol{v}$,

$$Pr[h_{\boldsymbol{r}}(\boldsymbol{u}) = h_{\boldsymbol{r}}(\boldsymbol{v})] = 1 - \frac{2\theta(\boldsymbol{u}, \boldsymbol{v})}{\pi} \tag{2}$$

Finally, we demonstrate that the original LSH scheme (Definition D1) still holds for this family H.

Theorem 3 (pbRP sensitivity). *The pbRP hash function family H, defined in Eq. (1), is $(\theta_1, \theta_2, p_1, p_2)$-sensitive.*

Proof. Let us define p as the probability distribution of the hash functions to be locality-sensitive: Through Eq. (2), p decreases monotonically with respect to angle distance $\theta(\boldsymbol{u}, \boldsymbol{v})$. Reminding that angles $\theta_1 < \theta_2$, if we set $p_1 = p(\theta_1)$ and $p_2 = p(\theta_2)$, for any two probability distributions $\boldsymbol{u}$ and $\boldsymbol{v}$, if $\theta(\boldsymbol{u}, \boldsymbol{v}) \leq \theta_1$ then $P_H[h(\boldsymbol{u}) = h(\boldsymbol{v})] \geq p_1$; If $\theta(\boldsymbol{u}, \boldsymbol{v}) > \theta_2$ then $P_H[h(\boldsymbol{u}) = h(\boldsymbol{v})] \leq p_2$. It is evident that $p_2 < p_1$. This concludes the proof of the Theorem: The proposed hash family H is $(\theta_1, \theta_2, p_1, p_2)$-sensitive.

4   Sequential Interleaving

It is easy to prove the following theorem:

Theorem 4 (Unbalance Effect). *Given a probability distribution $\boldsymbol{p}$, assume that K^{+d} is the set of all probability distributions, the sets $S_1 = \{\boldsymbol{p}_i | \arccos(\frac{\boldsymbol{p}_i \cdot \boldsymbol{p}}{\|\boldsymbol{p}_i\| \, \|\boldsymbol{p}\|}) \leq \theta, \boldsymbol{p}_i \in K^{+d}\}$ and $R_1 = \{\boldsymbol{p}_i | \arccos(\frac{\boldsymbol{p}_i \cdot \boldsymbol{p}}{\|\boldsymbol{p}_i\| \, \|\boldsymbol{p}\|}) = \theta, \boldsymbol{p}_i \in K^{+d}\}$, and assume that $\boldsymbol{p}_0$ is a probability distribution, which satisfies $\boldsymbol{p}_0 = argmin_{\boldsymbol{p}_i \in R_1} \|\boldsymbol{p}_i - \boldsymbol{p}\|$, and the set $S_2 = \{\boldsymbol{p}_i | \, \|\boldsymbol{p}_i - \boldsymbol{p}\| \leq \|\boldsymbol{p}_0 - \boldsymbol{p}\|, \boldsymbol{p}_i \in K^{+d}\}$, then $S_2 \subseteq S_1$. Similarly, assume that $\boldsymbol{p}_1$ is a probability distribution, which satisfies $\boldsymbol{p}_1 = argmax_{\boldsymbol{p}_i \in R_1} \|\boldsymbol{p}_i - \boldsymbol{p}\|$, and the set $S_2 = \{\boldsymbol{p}_i | \, \|\boldsymbol{p}_i - \boldsymbol{p}\| \leq \|\boldsymbol{p}_1 - \boldsymbol{p}\|, \boldsymbol{p}_i \in K^{+d}\}$, then $S_2 \supseteq S_1$.*

Remark: The point $\boldsymbol{p}_0$ and $\boldsymbol{p}_1$ locates in the bound of S_1 and S_2, however, the set for Euclidean-based methods is subset or supset of the one for angle-based methods, which means that the scope of finding ANN points for angle-based methods is different with the one for Euclidean-based methods. Thus, we can combine the strength of two types of methods to obtain better ANN performance.

Given two ranking lists generated by different methods, to obtain benefit from the both results, we consider an interleaving process on these two ranking lists to obtain a better ranking list. Lots of works have been done to interleave two ranking lists of documents in information retrieval area [5,11,12]. We borrow the interleaving idea in information retrieval area to handle this problem, and propose a novel interleaving algorithm, called Sequential Interleaving, to obtain a better ranking list by merging two ranking lists.

As a retrieved list, the ranking position of a data point in each list generated by an algorithm can be treated as a reflection of "confidence level", which is how "good" the algorithm thinks the data point is similar to the given query. Intuitively, the confidence level $CL(p, L_r)$ for the point p of the position r in the ranking list L, should be an inverse function form of the ranking position r. To define the function form of $CL(p, L_r)$, we inspect the metrics used to evaluating a retrieved list in information retrieval. DCG is a ranking-aware metric which can effectively evaluate how relevant a ranking list is for the query. Its widely-used binary value form [7] is defined as:

$$DCG(L) = \sum_{r=1}^{len(L)} \frac{rel_{L_r}}{log_2(r+1)} \tag{3}$$

where rel_{L_r} is the relevance of the document ranked at r in the ranking list L. In this formula, DCG reflects total confidence level of all documents in the ranking list, thus we can define $CL(p, L_r)$ as:

$$CL(p, L_r) = \begin{cases} \frac{1}{log_2(r+1)} & p \in L, \\ 0 & p \notin L. \end{cases} \tag{4}$$

where p is a data point, and r is the ranking postion of p in the ranking list L.

Given a query, there are two ranking lists of top-k similar data points, and the goal is to combine the two ranking lists to obtain a better top-k ranking list. We first obtain the union of the two ranking lists L_1 and L_2, then for each data point p in the union, to compute the total confidence level of p by the following simple formula:

$$TotalCL(p, L_1, L_2) = \alpha \sum_{L_{1_i}=p} CL(p, L_{1_i}) + (1-\alpha) \sum_{L_{2_j}=p} CL(p, L_{2_j}) \tag{5}$$

where $L_{1_i} = p$ denotes the data point p is at the position i in the ranking list L_1, and α is a prior weighting factor. Finally, we sort all the data points by using $TotalCL(p, L_1, L_2)$, and then give the top-k similar interleaved data points as final ranking result. The proposed algorithm is called Sequential Interleaving, described in Algorithm 1. We combine the ranking lists of pbRP and l_2 by proposed Sequential Interleaving method to obtain a ranking list, and abbreviated as si.

Algorithm 1. Sequential Interleaving

1: INPUT: Ranking Lists L_1 and L_2
2: Probs = union(L_1, L_2)
3: **for** $\forall$ p $\in$ Probs **do**
4:    CL[p] = TotalCL(p, L_1, L_2) (Formula (5))
5: **end for**
6: SortedProbs = SortByCL(Probs, CL)
7: return SortedProbs

---

5 Experiments

5.1 Data Sets and Evaluation Protocols

Six publicly available image datasets, namely CIFAR10, CIFAR100-20, CIFAR100-100, Local-Patch, MNIST and COVTYPE, are used to compare the proposed approach against state-of-the-art methods. **CIFAR10** dataset consists of 60K 32×32 colour images in 10 classes. Every images is represented by a 512-dimensional GIST feature vector. CIFAR-100 is just like the CIFAR-10, except that it has 20 "coarse" and 100 "fine" superclasses, denoted as **CIFAR100-20** and **CIFAR100-100**. **Local-Patch** contains roughly 300K 32×32 image patches from photos of Trevi Fountain (Rome), Notre Dame (Paris) and Half Dome (Yosemite). For each image patch, we compute a 128-d SIFT vector as the holistic descriptor. **MNIST** consists of a total of 70000 hand-written digit samples, each with 780 features. **COVTYPE** is a common benchmark featuring 54 dimensions (Blackard and Dean, 1999). The feature vectors of all datasets need to be reformed by L1 normalization, and be transformed into probability distributions.

In this paper, the most representative methods, Random Projection (PR) [1,28] and p-stable LSH (l_2) [6,28], are chosen to evaluate the effectiveness of the proposed methods. PR is based on angular distance for vectors, while p-stable LSH is based on l_2 distance for vectors.

All the experimental results are averaged over 10 random training/test partitions. For each partition, we randomly select 1000 images with their tags as queries, and the remaining images and tags as reference database images. We use mean Average Precision (mAP), p@N and Precision-Recall curves to illustrate performances of different methods.

All experiments are conducted on our workstation with Intel(R) Xeon(R) CPU X7560@2.27 GHz and 32G memory.

Parameter α. Figure 1 shows the effect of prior weighting factor in Eq. 5 at code size of 128 bits on six datasets. As one can see, Sequential Interleaving algorithm can reach the best performance when $\alpha = 0.65$. Similar trends have been observed at code size 8, 16, 32, 64 and 256 bits, which are not presented here because of limited space. In the following experiments, we set parameter $\alpha = 0.65$.

Table 1. mAP on six datasets. The best mAP is shown in bold face. "ls" means linear search.

#bits	Local-Patch					CIFAR100-100				
	RP	l_2	pbRP	jsd	kl	RP	l_2	pbRB	jsd	kl
8	0.0159	0.3739	**0.3852**	-	-	0.0161	0.0331	**0.0361**	-	-
16	0.0006	0.3714	**0.3852**	-	-	0.0009	0.0331	**0.0361**	-	-
32	0.0000	0.3642	**0.3852**	-	-	0.0003	0.0331	**0.0361**	-	-
64	0.0001	0.2731	**0.3852**	-	-	0.0004	0.0331	**0.0361**	-	-
128	0.0000	0.2103	**0.3852**	-	-	0.0004	0.0331	**0.0361**	-	-
256	0.0000	0.2146	**0.3852**	-	-	0.0003	0.0331	**0.0361**	-	-
ls	0.3852	0.3844	0.3852	0.4044	0.4022	0.0361	0.0331	0.0361	0.0364	0.0345

#bits	CIFAR100-20					CIFAR10				
	RP	l_2	pbRP	jsd	kl	RP	l_2	pbRP	jsd	kl
8	0.0086	0.0801	**0.0823**	-	-	0.0416	0.1661	**0.1745**	-	-
16	0.0020	0.0801	**0.0823**	-	-	0.0037	0.1661	**0.1745**	-	-
32	0.0001	0.0801	**0.0823**	-	-	0.0003	0.1661	**0.1745**	-	-
64	0.0002	0.0801	**0.0823**	-	-	0.0003	0.1661	**0.1745**	-	-
128	0.0001	0.0801	**0.0823**	-	-	0.0005	0.1661	**0.1745**	-	-
256	0.0002	0.0801	**0.0823**	-	-	0.0001	0.1661	**0.1745**	-	-
ls	0.0823	0.0801	0.0823	0.0824	0.0807	0.1745	0.1661	0.1745	0.1721	0.1684

#bits	MNIST					COVTYPE				
	RP	l_2	pbRP	jsd	kl	RP	l_2	pbRP	jsd	kl
8	0.0321	0.3743	**0.4305**	-	-	0.0559	0.0784	**0.4455**	-	-
16	0.0149	0.3743	**0.4305**	-	-	0.0329	0.0539	**0.4455**	-	-
32	0.0007	0.3743	**0.4305**	-	-	0.0066	0.0265	**0.4455**	-	-
64	0.0013	0.3741	**0.4305**	-	-	0.0068	0.0031	**0.4455**	-	-
128	0.0009	0.3739	**0.4305**	-	-	0.0052	0.0008	**0.4455**	-	-
256	0.0003	0.3654	**0.4305**	-	-	0.0084	0.0001	**0.4455**	-	-
ls	0.4305	0.3743	0.4305	0.4310	0.4271	0.4455	0.4441	0.4455	0.4446	0.3728

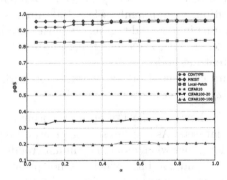

Fig. 1. Impact of parameter α in Sequential Interleaving over six public datasets at code size of 128 bits.

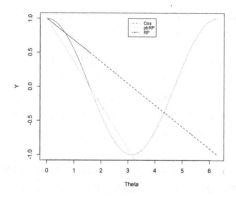

Fig. 2. For small angles (not too close to orthogonal), RP is a pretty good approximation to $\cos(\theta)$; For all angles ($[0, \pi/2]$), pbRP is better approximation to $\cos(\theta)$ than RP.

5.2   Results

The mAP values for different methods with different code sizes on six datasets are shown in Table 1. The value of each entry in the tables is the mAP of a combination of a method under a specific code size. The best mAP among RP, pbRP, l_2 and si under the same setting is shown in bold face. From the table, we can make several observations: (1) For all datasets, pbRP performs better than RP and l_2, which shows that the proposed method is effective. Especially, pbRP and RP are both based on angular metric, however, pbRP can greatly improve the performance against RP. The reasons maybe (i) pbRP restricts the scope of the random projection in $1/2^d$ the whole high-dimensional space, d is the dimension of a probability distribution, which avoid the invalid random projection; (ii) From Fig. 2, for small angles (not too close to orthogonal), RP is a pretty good approximation to $\cos(\theta)$; For all angles ($[0, \pi/2]$), pbRP is better approximation to $\cos(\theta)$ than RP. (2) For all datasets, the performance of pbRP is very stable, and reach the performance of brute-force search (linear search, ls); While the performance of RP and l_2 are not stable, varies with different hash bits. For example, for l_2 over MNIST dataset, mAP at 128 bits is 0.3739, and mAP at 256 is 0.3654. (3) By brute-force search, the performance of metrics (arccos and l_2) for vector-type data is a pretty good approximation to, sometimes even better than, the one of JSD or KL.

The p@5 values for different methods with different code sizes on six datasets are shown in Table 2, and we can obtain similar conclusions with mAP values in Table 1. In Table 2, we also observe: For all datasets, si performs better than

Table 2. p@5 on six public datasets. The best p@5 among RP, pbRP, l_2 and si under the same setting is shown in bold face. "ls" means linear search.

#Bits	Local-Patch				CIFAR100-100			
	RP	l_2	pbRP	si	RP	l_2	pbRP	si
8	0.739	0.858	0.848	**0.859**	0.124	0.194	0.198	**0.211**
16	0.658	0.857	0.848	**0.859**	0.076	0.194	0.198	**0.211**
32	0.633	0.845	0.848	**0.851**	0.044	0.194	0.198	**0.211**
64	0.622	0.847	0.848	0.845	0.052	0.194	0.198	**0.211**
128	0.715	0.828	0.848	0.834	0.046	0.194	0.198	**0.211**
256	0.696	0.825	0.848	**0.851**	0.047	0.194	0.198	**0.211**
ls	0.848	0.857	0.848	-	0.198	0.194	0.198	-

#Bits	CIFAR100-20				CIFAR10			
	RP	l_2	pbRP	si	RP	l_2	pbRP	si
8	0.276	0.322	0.342	**0.351**	0.448	0.478	0.493	**0.509**
16	0.189	0.322	0.342	**0.351**	0.387	0.478	0.493	**0.509**
32	0.101	0.322	0.342	**0.351**	0.239	0.478	0.493	**0.509**
64	0.133	0.322	0.342	**0.351**	0.264	0.478	0.493	**0.509**
128	0.131	0.322	0.342	**0.351**	0.234	0.478	0.493	**0.509**
256	0.146	0.322	0.342	**0.351**	0.22	0.478	0.493	**0.509**
ls	0.342	0.322	0.342	-	0.493	0.478	0.493	-

#Bits	MNIST				COVTYPE			
	RP	l_2	pbRP	si	RP	l_2	pbRP	si
8	0.855	0.956	**0.962**	0.961	0.953	0.949	0.952	**0.953**
16	0.699	0.956	**0.962**	0.961	0.944	0.949	0.952	**0.953**
32	0.781	0.956	**0.962**	0.961	0.926	0.941	0.952	0.944
64	0.825	0.956	**0.962**	0.961	0.932	0.924	0.952	0.936
128	0.862	0.956	**0.962**	0.961	0.94	0.891	0.952	**0.953**
256	0.821	0.956	**0.962**	0.961	0.923	0.918	0.952	0.938
ls	0.962	0.956	0.962	-	0.952	0.952	0.952	-

pbRP and l_2 under most settings, which shows that the Sequential Interleaving algorithm is effective. Moreover, p@10, p@15 and p@20 values have similar trends.

Figure 3 shows the precision-recall curves on six datasets. Once again, we can easily find that proposed method pbRP significantly outperforms other state-of-the-art methods under most settings.

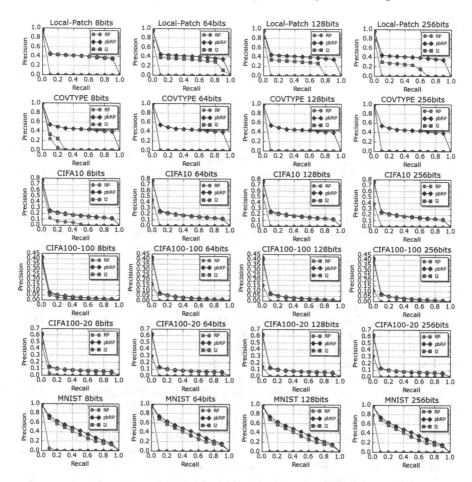

Fig. 3. Precision-recall curve on six public data sets.

6 Conclusion

Traditional LSH methods focus on vector-type data. In this paper, we investigate the practicability of hashing methods for probability-distribution-type data, and propose a novel angle-based hashing method and a Sequential Interleaving algorithm. The experiments show that proposed algorithms are more effective than the state-of-the-art baselines.

Acknowledgments. This work was supported by 863 Program (2015AA015404), 973 Program (2013CB329303), China National Science Foundation (61402036, 60973083, 61273363), Beijing Advanced Innovation Center for Imaging Technology (BAICIT-2016007).

References

1. Andoni, A., Indyk, P.: Near-optimal hashing algorithms for approximate nearest neighbor in high dimensions. In: 47th Annual IEEE Symposium on Foundations of Computer Science, FOCS 2006, pp. 459–468. IEEE (2006)
2. Andoni, A., Indyk, P., Nguyen, H.L., Razenshteyn, I.: Beyond locality-sensitive hashing. In: Proceedings of the Twenty-Fifth Annual ACM-SIAM Symposium on Discrete Algorithms, pp. 1018–1028. SIAM (2014)
3. Broder, A.Z., Glassman, S.C., Manasse, M.S., Zweig, G.: Syntactic clustering of the web. Comput. Netw. ISDN Syst. **29**(8), 1157–1166 (1997)
4. Charikar, M.S.: Similarity estimation techniques from rounding algorithms. In: Proceedings of the Thiry-Fourth Annual ACM Symposium on Theory of Computing, pp. 380–388. ACM (2002)
5. Chukllin, A., Schuth, A., Zhou, K., De Rijke, M.: A comparative analysis of interleaving methods for aggregated search. ACM Trans. Inf. Syst **33**(2), 5 (2015)
6. Datar, M., Immorlica, N., Indyk, P., Mirrokni, V.S.: Locality-sensitive hashing scheme based on p-stable distributions. In: Proceedings of the Twentieth Annual Symposium on Computational Geometry, pp. 253–262. ACM (2004)
7. Eshghi, K., Rajaram, S.: Locality sensitive hash functions based on concomitant rank order statistics. In: Proceedings of the 14th ACM SIGKDD International Conference on Knowledge Discovery and Data Mining, pp. 221–229. ACM (2008)
8. Gilbert, S.: Linear algebra and its applications, Thomson, Brooks/Cole, Belmont, CA. Technical report (2006). ISBN 0-030-10567-6
9. Gong, Y., Lazebnik, S.: Comparing data-dependent and data-independent embeddings for classification and ranking of internet images. In: 2011 IEEE Conference on Computer Vision and Pattern Recognition (CVPR), pp. 2633–2640. IEEE (2011)
10. Gorisse, D., Cord, M., Precioso, F.: Locality-sensitive hashing for chi2 distance. IEEE Trans. Pattern Anal. Mach. Intell. **34**(2), 402–409 (2012)
11. Hofmann, K., Whiteson, S., de Rijke, M.: A probabilistic method for inferring preferences from clicks. In: Proceedings of the 20th ACM International Conference on Information and Knowledge Management, pp. 249–258. ACM (2011)
12. Hofmann, K., Whiteson, S., de Rijke, M.: Estimating interleaved comparison outcomes from historical click data. In: Proceedings of the 21st ACM International Conference on Information and Knowledge Management, pp. 1779–1783. ACM (2012)
13. Indyk, P., Motwani, R.: Approximate nearest neighbors: towards removing the curse of dimensionality. In: Proceedings of the Thirtieth Annual ACM Symposium on Theory of Computing, pp. 604–613. ACM (1998)
14. Jain, P., Vijayanarasimhan, S., Grauman, K.: Hashing hyperplane queries to near points with applications to large-scale active learning. In: Advances in Neural Information Processing Systems, pp. 928–936 (2010)
15. Ji, J., Li, J., Yan, S., Tian, Q., Zhang, B.: Min-max hash for Jaccard similarity. In: 2013 IEEE 13th International Conference on Data Mining (ICDM), pp. 301–309. IEEE (2013)
16. Ji, J., Li, J., Yan, S., Zhang, B., Tian, Q.: Super-bit locality-sensitive hashing. In: Advances in Neural Information Processing Systems, pp. 108–116 (2012)
17. Jiang, Q.Y., Li, W.J.: Scalable graph hashing with feature transformation. In: IJCAI (2015)
18. Kong, W., Li, W.J.: Isotropic hashing. In: Advances in Neural Information Processing Systems, pp. 1646–1654 (2012)

19. Kong, W., Li, W.J., Guo, M.: Manhattan hashing for large-scale image retrieval. In: Proceedings of the 35th International ACM SIGIR Conference on Research and Development in Information Retrieval, pp. 45–54. ACM (2012)
20. Kulis, B., Grauman, K.: Kernelized locality-sensitive hashing. IEEE Trans. Pattern Anal. Mach. Intell. **34**(6), 1092–1104 (2012)
21. Li, P., Konig, A., Gui, W.: B-bit minwise hashing for estimating three-way similarities. In: Advances in Neural Information Processing Systems, pp. 1387–1395 (2010)
22. Li, P., Owen, A., Zhang, C.H.: One permutation hashing. In: Advances in Neural Information Processing Systems, pp. 3113–3121 (2012)
23. Liu, W., Mu, C., Kumar, S., Chang, S.F.: Discrete graph hashing. In: Advances in Neural Information Processing Systems, pp. 3419–3427 (2014)
24. Liu, Y., Cui, J., Huang, Z., Li, H., Shen, H.T.: SK-LSH: an efficient index structure for approximate nearest neighbor search. Proc. VLDB Endowment **7**(9), 745–756 (2014)
25. Mu, Y., Yan, S.: Non-metric locality-sensitive hashing. In: AAAI (2010)
26. ODonnell, R., Wu, Y., Zhou, Y.: Optimal lower bounds for locality-sensitive hashing (except when q is tiny). ACM Trans. Comput. Theor. (TOCT) **6**(1), 5 (2014)
27. Terasawa, K., Tanaka, Y.: Spherical LSH for approximate nearest neighbor search on unit hypersphere. In: Dehne, F., Sack, J.-R., Zeh, N. (eds.) WADS 2007. LNCS, vol. 4619, pp. 27–38. Springer, Heidelberg (2007). doi:10.1007/978-3-540-73951-7_4
28. Wang, J., Shen, H.T., Song, J., Ji, J.: Hashing for similarity search: a survey. arXiv preprint arXiv:1408.2927 (2014)
29. Zhang, T., Qi, G.J., Tang, J., Wang, J.: Sparse composite quantization. In: Proceedings of the IEEE Conference on Computer Vision and Pattern Recognition, pp. 4548–4556 (2015)
30. Zhao, W.L., Jégou, H., Gravier, G.: Sim-Min-Hash: an efficient matching technique for linking large image collections. In: Proceedings of the 21st ACM International Conference on Multimedia, pp. 577–580. ACM (2013)

Supervised Hashing for Multi-labeled Data with Order-Preserving Feature

Dan Wang, Heyan Huang[(⊠)], Hua-Kang Lin, and Xian-Ling Mao

Beijing Institute of Technology, Beijing, China
{wangdan12856,hhy63,1120141916,maoxl}@bit.edu.cn

Abstract. Approximate Nearest Neighbors (ANN) Search has attracted much attention in recent years. Hashing is a promising way for ANN which has been widely used in large-scale image retrieval tasks. However, most of the existing hashing methods are designed for single-labeled data. On multi-labeled data, those hashing methods take two images as similar if they share at least one common label. But this way cannot preserve the order relations in multi-labeled data. Meanwhile, most hashing methods are based on hand-crafted features which are costing. To solve the two problems above, we proposed a novel supervised hashing method to perform hash codes learning for multi-labeled data. In particular, we firstly extract the order-preserving data features through deep convolutional neural network. Secondly, the order-preserving features would be used for learning hash codes. Extensive experiments on two real-world public datasets show that the proposed method outperforms state-of-the-art baselines in the image retrieval tasks.

Keywords: Order-preserving feature · Supervised hashing · Multi-labeled data

1 Introduction

Approximate nearest neighbor (ANN) [1,2] search plays a fundamental role in machine learning and related areas, such as image retrieval, pattern recognition and computer vision [3,22,25].

Hashing is an effective technology to solve ANN problems. It has received more and more attention in the big data era because of its fast retrieval speed and low storage cost. Hashing maps the data points from the original feature space into Hamming space to generate compact binary/hash codes. The more similar two points are in the original feature space, the smaller their hamming distance is. On the contrary, when two data points are dissimilar, a high hamming distance is expected between their hash codes.

Generally speaking, hashing methods can be divided into data-independent methods and data-dependent methods. Compared with data-independent hashing methods like Locality Sensitive Hashing (LSH) [5], data-dependent hashing

© Springer Nature Singapore Pte Ltd. 2017
X. Cheng et al. (Eds.): SMP 2017, CCIS 774, pp. 16–28, 2017.
https://doi.org/10.1007/978-981-10-6805-8_2

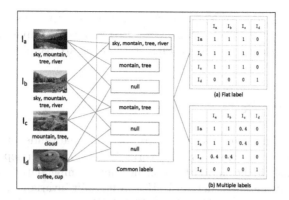

Fig. 1. Example of multi-labeled data. The left panel shows four images and their labels. The middle panel presents the common labels between images. The traditional similarities are listed in the upper-right panel (a): two images are similar when they share at least one label. The bottom-right panel (b) is the similarities considering the numbers of common labels. The more common labels two images have, the more similar they are.

methods can achieve comparable or better accuracy with shorter codes by utilizing the train data [6, 16, 17, 23]. Hence, data-dependent methods have become more popular than data-independent methods.

However, most of the existing hashing methods are designed for single-labeled data which cannot be generalized for multi-labeled data directly, such as ImageNet[1], CIFAR-100[2], IAPRTC-12[3], MIRFLICKER[4], and NUS-WIDE[5]. In fact, most existing hashing methods take two images as similar if they share at least one label as shown in Fig. 1(a). However, it is a natural assumption that two images are more similar if they share more common labels. A multi-labeled example has been shown in Fig. 1, I_a has labels "$sky, mountain, tree, river$", I_b has labels "$sky, mountain, tree, river$", and I_c has labels "$mountain, tree, cloud$". The three images are considered to have great similarity because they share the common labels "$mountain, tree$". However, their similarity should be in different degrees according to the number of common labels. Since I_a and I_b has more common labels than I_a and I_c, I_a and I_b are more similar.

Moreover, most hashing methods are based on hand-crafted features such as Gist, Sift and color moments. Those hand-crafted features are costing but not optimal. Recently, deep learning has been widely used for tasks on images. It can be also used to extract the data features.

To solve the above two issues, we proposed a novel supervised hashing method for multi-labeled data. Specifically, we designed a novel feature extraction

---

[1] http://www.image-net.org/.

[2] https://www.cs.toronto.edu/~kriz/cifar.html.

[3] http://imageclef.org/SIAPRdata.

[4] http://press.liacs.nl/mirflickr/mirdownload.html.

[5] http://lms.comp.nus.edu.sg/research/NUS-WIDE.htm.

algorithm to obtain the "order-preserving" features by using deep convolutional neural network. Extensive experiments on two real-world public datasets show that the proposed method outperforms state-of-the-art baselines in the image retrieval task.

2 Related Work

The existing hashing methods can be grouped into data-independent methods and data-dependent methods. Early works focus on data-independent methods to learn hash functions without using any training data. Representative data-independent methods include LSH [1,5,9], shift-invariant kernels hashing (SIKH) [19], and lots of extensions [4,11,12,19]. Different from data-independent methods, data-dependent methods learn hash functions from training data. Using the same length of hash codes or smaller ones, the data-dependent methods can achieve comparable or even better performance comparing to data-independent methods.

Furthermore, existing data-dependent methods can be further divided into three categories: unsupervised methods, supervised methods and semi-supervised methods. Unsupervised hashing tries to preserve the Euclidean similarity between the training points, while supervised hashing [15,18,23] and semi-supervised try to preserve the semantic similarity constructed from the semantic labels of the training points.

Unsupervised methods use unlabeled data to learn hash functions and try to keep the neighborhood relations of data in the original space. Representative unsupervised hashing methods include K-means hashing (KMH) [7], Iterative Quantization (ITQ) [6], Spherical Hashing (SH) [8], Discrete Graph Hashing (DGH) [16] and Asymmetric Innerproduct Binary Coding (AIBC) [21], etc. Supervised hashing approaches utilize the label information to build the similarity matrix of training data and further to learn the hash functions. Notable methods in this category include Kernel-Based Supervised Hashing (KSH) [17], Latent Factor Hashing (LFH) [23] and Column Sampling Based Discrete Supervised Hashing (COSDISH) [10], etc. Semi-supervised methods use both labeled and unlabeled data to train hashing functions. Usually, supervised hashing methods achieve higher accuracy than unsupervised methods due to the semantic gap problem. Hence, many recent works focus on supervised methods.

Recently, deep hashing methods have been proposed to perform feature learning and hash codes learning simultaneously. Those methods usually make use of convolutional neural networks and force the output to be hash codes. After iterative training, they can get similarity-preserving hash codes. Deep Pairwise Supervised Hashing (DPSH) [14] and Deep Supervised Ranking Hashing (DSRH) [24] are the representatives of deep hashing.

3   Our Method

3.1   Framework

Figure 2 shows the overall framework of our method. It consists of two part: order-preserving feature learning and hash codes learning.

- **Order-preserving Feature Learning.** As shown in Fig. 2(a), this part contains one input layer, five convolutional layers, two full-connected layers, and one output layer. The raw image will be reshaped to 224×224 as input. The 1st convolutional layer contains 64 filters whose size is 5×5 and the stride is four pixels. We use 256 filters in the 2nd–5th convolutional layers and the stride is one pixel. The max-pooling operator size is set as 2×2 for the 1st, 4th and 5th convolutional layers. Following the CNN part, two full-connected layers contain 4,096 hidden units follow for each. The activation function for all hidden layers is the rectification linear unit (RELU). The last layer is the output whose length is 512. The Gaussian activation function is used for the output layer.
- **Hash Codes Learning.** As shown in Fig. 2(b), the data features generated from the order-preserving feature learning part would become the input of hashing codes learning part. In this part, a supervised hashing algorithm based on maximum likelihood function is designed to optimize hash codes.

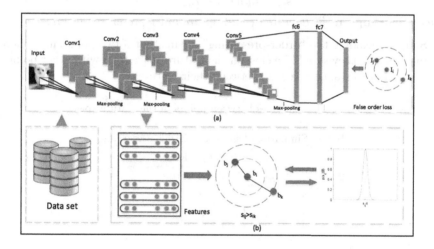

Fig. 2. The framework of our method. Part (a) shows the deep architecture of neural network that produces the order-preserving feature by taking raw images as input. Part (b) show the architecture which generates hash codes for multi-labeled images.

3.2   Similarity Definition

Given a dataset $\{\mathbf{X}, \mathbf{L}\}$, $\mathbf{X} = \{\mathbf{x_i}\}_{i=1}^{N}$ with N points, $\mathbf{x_i}$ is the pixels matrix of the ith image. $\mathbf{L}$ is a set consisting of $l_1, l_2, \cdots, l_N$ in which l_i is the label set of the point $\mathbf{x_i}$.

In the light of multi-labeled data, the intuitional assumption is that the more common labels two labels share, the higher their similarity is. Thus, the similarity between two points can be defined as the proportion of the common labels in the whole labels:

$$s_{ij} = 2 \times \frac{|l_i \cap l_j|}{|l_i \cup l_j|} - 1, \tag{1}$$

where l_i is the labels of the point $\mathbf{x_i}$. $|l|$ is the size of set l. According to the above equation, $\mathbf{S} = \{s_{ij}\} \in \mathbb{R}^{N \times N}$ can be calculated. $s_{ij} \in [-1, 1]$ represents the degree of similarity between the points $\mathbf{x_i}$ and $\mathbf{x_j}$. The bigger s_{ij} is, the more similar points $\mathbf{x_i}$ and $\mathbf{x_j}$ are.

3.3   Order-Preserving Feature Learning

For a given data $\mathbf{x_i}$, the features $\mathbf{u_i} \in \mathbb{R}^{512}$ can be get by passing it to the convolutional neural network. Assume there are three items $\mathbf{x_i}, \mathbf{x_j}$ and $\mathbf{x_k}$, their output feature are $\mathbf{u_i}, \mathbf{u_j}$ and $\mathbf{u_k}$ respectively. We define the "Order-preserving feature" as below:

Definition 1. *For any point* $\mathbf{x_i}$, *if there are two other points* $\mathbf{x_j}$ *and* $\mathbf{x_k}$ *with the similarities* s_{ij} *and* s_{ik} *respectively, such that, their features* $\mathbf{u_i}, \mathbf{u_j}$ *and* $\mathbf{u_k}$ *satisfy:*

$$(s_{ij} - s_{ik})(E_{ij} - E_{ik}) < 0,$$

where $E_{ij} = |\mathbf{u_i} - \mathbf{u_j}|_F^2$, *then* $\mathbf{u_i}, \mathbf{u_j}$ *and* $\mathbf{u_k}$ *are order-preserving features.*

Simply speaking, the "order-preserving" means that if two points have more common labels, they should be close in the feature space than with other points. There are four distance relations between their features:

Table 1. The distances between points in the feature space.

Index	Similarity	Distance	Order-preserving
1	$s_{ij} > s_{ik}$	$E_{ij} > E_{ik}$	False
2	$s_{ij} > s_{ik}$	$E_{ij} < E_{ik}$	True
3	$s_{ij} < s_{ik}$	$E_{ij} > E_{ik}$	True
4	$s_{ij} < s_{ik}$	$E_{ij} < E_{ik}$	False

In Table 1, $s_{ij} > s_{ik}$ represents that point $\mathbf{x_i}$ is more similar with point $\mathbf{x_j}$ than with point $\mathbf{x_k}$. In this case, $\mathbf{u_i}$ should be closer to $\mathbf{u_j}$ than $\mathbf{u_k}$ as shown in the 2nd and 3rd lines. Considering the two wrong cases in the 1st and 4th lines, the loss function of order-preserving feature can be defined as below:

$$J_{OF} = \frac{1}{N^3} \sum_{i=1}^{N} \sum_{j=1}^{N} \sum_{k=1}^{N} max((E_{ij} - E_{ik})(s_{ij} - s_{ik}), 0) \tag{2}$$

where $max(\cdot, \cdot)$ function returns the bigger value.

Algorithm 1. Order-preserving Feature Learning

Require: Dataset $\{\mathbf{X}, \mathbf{L}\}$, minibatch size (128 default), learning rate (0.001 initi-
ated), max iterations (100 default)
Ensure: Data features $\mathbf{U}$
1: calculate simiarity $\mathbf{S}$ according to Eqn.(1);
2: **repeat**
3:     Randomly sample a minibatch of points from $\mathbf{X}$, name as $\mathbf{X_{batch}}$;
4:     Calculate features for $\mathbf{X_{batch}}$ by forward propagation, name as $\mathbf{U_{batch}}$;
5:     For $\mathbf{u_i}$ in $\mathbf{U_{batch}}$:
6:     Compute derivatives according to Eqn.(3);
7:     Update the network parameters by utilizing back propagation;
8:     Endfor
9: **until** max iteration number;
10: Put all data into network to generate data features.

Optimization. To solve the optimization problems listed above, we employed
the stochastic gradient descent (SGD) to minimize the objective function. In
Eq. (2), for any items $\mathbf{x_i}, \mathbf{x_j}$ and $\mathbf{x_k}$, if

$$(E_{ij} - E_{ik})(s_{ij} - s_{ik}) > 0,$$

the derivatives of Eq. (2) with respect to output vectors $\mathbf{u_i}$ are given by:

$$\frac{\partial J_{OF}}{\partial \mathbf{u_i}} = 2(s_{ij} - s_{ik})(\mathbf{u_k} - \mathbf{u_j}) \tag{3}$$

These derivative values can be fed into the underlying CNN via back-
propagation algorithm to update the parameters. Algorithm 1 summarizes the
process of the order-preserving feature learning part.

3.4   Hashing for Multi-labeled Data

The target of hashing for multi-labeled data is to learn the binary codes $\mathbf{b_i} \in \{-1, 1\}^d$ for each point $\mathbf{x_i}$, where d is the code length. The inner product between
two points $\mathbf{b_i}$ and $\mathbf{b_j}$ can be written as below:

$$\theta_{ij} = \mathbf{b_i}^T \mathbf{b_j} \tag{4}$$

Suppose $\mathbf{B} = \{\mathbf{b_i}\}_{i=1}^N$ is the hash codes for all data, then we can define the
likelihood for multi-labeled data as Eq. (5):

$$p(s_{ij}|\mathbf{B}) = e^{-(\theta_{ij} - s_{ij}d)^2}. \tag{5}$$

The likelihood in Eq. (5) forces the θ_{ij} to be an appropriate value and further
adjust the hash codes. It significantly differs from the likelihood on singled-
labeled data as shown in Fig. 3.

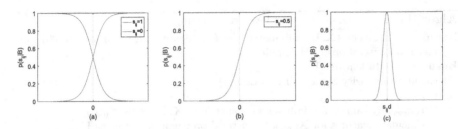

Fig. 3. (a) Traditionally, the likelihood is defined as $p(s_{ij}|\mathbf{B}) = a_{ij}^{s_{ij}}(1 - a_{ij})^{1-s_{ij}}$ s.t. $s_{ij} \in \{0, 1\}$ where $a_{ij} = 1/(1 + e^{-0.5\theta_{ij}})$. (b) For partially similar points, traditional methods treat them as $s_{ij} = 0$ if they share at least one label. This way will weaken the inherent semantic difference and conflict with the characteristics of multi-labeled data. (c) Our likelihood pushes the inner product between two points close to $s_{ij}d$.

By taking *log* function on Eq. (5), the objective function of hashing for multi-labeled data would be:

$$
\begin{aligned}
J_{hash} &= log\, p(\mathbf{B}|\mathbf{S}) \\
&= log\, p(\mathbf{S}|\mathbf{B})p(\mathbf{B}) \\
&= - \sum_{s_{ij} \in S} (\theta_{ij} - s_{ij}d)^2 + log\, p(\mathbf{B}).
\end{aligned}
\tag{6}
$$

Optimization. Since $\mathbf{B}$ is discrete, Eq. (6) is not a convex to optimize. Utilizing the techniques in [23], $\mathbf{B}$ should be relaxed to be $\mathbf{H}$ and $\mathbf{B} = sgn(\mathbf{H})$. We put a normal distribution on $p(\mathbf{H})$. The Eq. (6) should finally become:

$$
J_{hash} = - \sum_{s_{ij} \in S} (\theta_{ij} - s_{ij}d)^2 - \frac{1}{2\lambda}\|\mathbf{H}\|_F^2 + c,
\tag{7}
$$

where λ is the hyper-parameter. The gradient vector and the Hessian matrix of the objective function J_{hash} with respect to $\mathbf{h_i}$ can be derived as:

$$
\mathbf{D^1} = \frac{\partial J}{\partial \mathbf{h_i}^T} = -2 \sum_{j:s_{ij} \in \mathbf{S}} (\mathbf{h_i}^T\mathbf{h_j} - s_{ij}d)\mathbf{h_j}^T - 2 \sum_{j:s_{ji} \in \mathbf{S}} (\mathbf{h_j}^T\mathbf{h_i} - s_{ji}d)\mathbf{h_j}^T - \frac{1}{\lambda}\mathbf{h_i}^T,
$$

$$
\mathbf{D^2} = \frac{\partial^2 J}{\partial \mathbf{h_i}^T\mathbf{h_i}} = -2 \sum_{j:s_{ij} \in \mathbf{S}} \mathbf{h_j}\mathbf{h_j}^T - 2 \sum_{j:s_{ji} \in \mathbf{S}} \mathbf{h_j}\mathbf{h_j}^T - \frac{1}{\lambda}\mathbf{I}.
$$

It can be found that the Hessian matrix is a negative definite matrix. So, this surrogate learning algorithm can be viewed as a generalization of the expectation maximization (EM) algorithm.

– **E-step:** update $\mathbf{h_i}$ with the following rule: $\mathbf{h_i}(t+1) = \mathbf{h_i}(t) - \mathbf{D^1}(t)^T\mathbf{D^2}(t)^{-1}$.
– **M-step:** Compute J_{hash} in Eq. (7) using the new $\mathbf{h_i}$.

The updating propose can be controlled by the maximum allowed number of iteration T. The initial values of $\mathbf{H}$ can be obtained through PCA on the order-preserving features $\mathbf{U}$.

4 Experiments

4.1 Datasets and Evaluation Metrics

We evaluate the proposed method on two public datasets of multi-labeled images:

- **CIFAR-100** has 60,000 images in total. We randomly selected 10,000 images as the test query set, and the rest images are used as training samples. Since the labels in CIFAR-100 is hierarchical, we ignored the hierarchy and taken each image as double-labeled.
- **IAPRTC-12** includes the 20,000 segmented images. It is a multi-labeled dataset in which each image has been manually segmented and the resultant regions have been annotated according to a predefined vocabulary of labels. We randomly sampled 2,000 query images as testset and used the rest as the training set. The labels of images in this dataset are hierarchical and multiple. In experiments, we uses the last label of each hierarchy as the label of images. So, each data has 4.1239 labels in average. Since few images share the same label set, the similarity is small but higher than dissimilarity. It makes few positive examples in the training data. To balance the positive and negative examples, the similarity has been strengthened by using $a(s_{ij}) = 1/(1 + e^{-5(1+s_{ij})/2})$.

We measured the performance of methods by Average Cumulative Gain (ACG), Normalized Discounted Cumulative Gain (NDCG) [20], Mean Average Precision (MAP) and Weighted mAP [13].

4.2 Experimental Settings

We implemented the proposed method based on the open-source MatConvnet[6] framework. The batch size is 128. In training, the weights of the layers are initialized by the pre-trained imagenet-matconvnet-vgg-m[7] model. The learning rate is initialized to be 0.001. After every 10 epochs on the training data, the learning rate is adjusted to one third of the current value. λ is set as 1.

4.3 Results on Accuracy

In our experiments, some state-of-the-art hashing methods such as DPSH, COS-DISH, LFH, KMH, and ITQ are involved as comparison. The first three methods are supervised and the rest two are unsupervised methods. For all supervised baselines, two images are similar if they share at least one common label, which is set as their authors do. We resize all images to be 224×224 pixels and directly use the raw images as input for DPSH. The rest hashing methods use 512-D gist features as input.

---

[6] http://www.vlfeat.org/matconvnet/.
[7] http://www.vlfeat.org/matconvnet/pretrained/.

Table 2 shows the comparison results w.r.t. MAP, Weighted mAP, ACG@100, and NDCG@100 on CIFAR-100. As shown in this table, our method achieves the best performance at most cases. But in 32 bits, our method works not good as COSDISH. It is because when the code length is short, the interval information of multi-labeled data cannot be expressed well. The shorter hash codes may lose the information contained in multi-labeled data. With code length grows, the performance of our method increases quickly. Taking NDCG@100 as example, the effectness at 64 bits and 128 bits increases by 17.04%, 8.80% respect to the value of 32 bits and 64 bits respectively. Moreover, DPSH works well on average in short code lengths. But our method can achieve comparable results with longer code length as shown in Table 3. Compared with single-labeled data, the information in the multi-labeled data works in the form of the different degrees of similarity. Thus, for multi-labeled data, long hash codes should be generated.

The experimental results on IAPRTC-12 are shown in Table 4. Compared with all baselines, our method takes down the best effectness at most criterias. Similar with CIFAR-100, the performance increases quickly with the length of hash codes growing.

Table 2. MAP, Weighted mAP, ACG@100, and NDCG@100 values on CIFAR-100 varying different code lengths.

Method	MAP			Weighted mAP		
	32	64	128	32	64	128
COSDISH	0.2571	0.2327	0.2690	0.3085	0.2792	0.3227
LFH	0.2086	0.2251	0.2508	0.2503	0.2701	0.3010
KMH	0.0700	0.0701	0.0680	0.0881	0.0889	0.0862
ITQ	0.0734	0.0763	0.0789	0.0928	0.0972	0.1010
DPSH	**0.4367**	0.4504	0.4595	**0.5384**	0.5593	0.5757
Ours	0.2113	**0.4684**	**0.5840**	0.2537	**0.5639**	**0.7007**
Method	ACG@100			NDCG@100		
	32	64	128	32	64	128
COSDISH	0.2128	0.1794	0.2257	0.5112	0.4939	0.4888
LFH	0.1513	0.1658	0.1884	0.4721	0.4848	0.4967
KMH	0.1801	0.2017	0.2095	0.4075	0.4097	0.4053
ITQ	0.1900	0.2162	0.2370	0.4110	0.4172	0.4222
DPSH	**0.6905**	**0.7313**	**0.7714**	**0.7007**	**0.7103**	0.7171
Ours	0.2152	0.5101	0.6430	0.4663	0.6367	**0.7247**

4.4 Results of Order-Preserving Features

We further compare the proposed order-preserving feature with hand-craft features (512-D gist feature). We used the order-preserving feature as the input for

Table 3. MAP, Weighted mAP, ACG@100, and NDCG@100 values on CIFAR-100 at 160 bits and 256 bits.

Methods	MAP		Weighted mAP		ACG@100		NDCG@100	
	160	256	160	256	160	256	160	256
DPSH	0.4726	0.5057	0.5935	0.6379	**0.7938**	**0.8452**	0.7226	0.7365
Ours	**0.6304**	**0.6709**	**0.7558**	**0.8057**	0.7190	0.7518	**0.7526**	**0.7762**

Table 4. MAP, Weighted MAP, ACG@100, and NDCG@100 values on IAPRTC-12 varying different code lengths.

Method	MAP			Weighted MAP		
	32	64	128	32	64	128
COSDISH	0.4208	0.4549	0.4753	0.5982	0.6546	0.6912
LFH	0.4557	0.4741	0.4793	0.6464	0.6807	0.6914
KMH	0.3350	0.3297	0.3237	0.4629	0.4549	0.4450
ITQ	0.3508	0.3539	0.3559	0.4898	0.4959	0.5005
DPSH	**0.5183**	0.5268	0.5327	**0.7642**	0.7777	0.7907
Ours	0.5116	**0.5501**	**0.5659**	0.7504	**0.8163**	**0.8555**
Method	ACG@100			NDCG@100		
	32	64	128	32	64	128
COSDISH	0.6270	0.7827	0.8361	0.5326	0.5575	0.5676
LFH	0.8188	0.8662	0.8802	0.5527	0.5583	0.5717
KMH	0.5999	0.6000	0.5876	0.4847	0.4792	0.4724
ITQ	0.6281	0.6497	0.6649	0.5020	0.5071	0.5107
DPSH	**1.0982**	1.1458	1.1789	**0.6278**	**0.6356**	0.6404
Ours	1.0940	**1.1868**	**1.2875**	0.6048	0.6288	**0.6462**

baselines. To seperate these new versions from existed baselines, we renamed the new versions by adding "+CNN", such as "COSDISH+CNN".

Figure 4 illustrates the curves of MAP, Weighted mAP, ACG@100, and NDCG@100 on CIFAR-100. The dotted lines show the results using order-preserving features. The solid lines represent the results using hand-crafted features. It can be found that by using the order-preserving feature, all baselines achieve higher values at most cases. It proved that hand-crafted features cannot capture the inherent characteristics of images.

Figure 5 illustrates the curves of MAP, Weighted mAP, ACG@100, and NDCG@100 on IAPRTC-12. Similar with the curves on CIFAR-100, the order-preserving features bring in better performance. The order-preserving features are more suitable than hand-crafted features on image retrieval tasks.

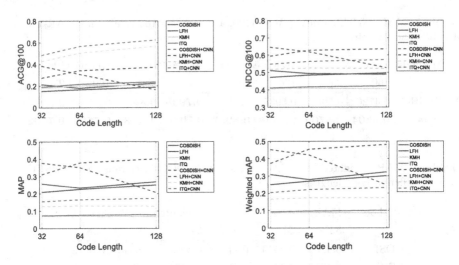

Fig. 4. Curves of MAP, Weighted mAP, ACG@100, and NDCG@100 on CIFAR-100 varying different code lengths after introducing order-preserving features.

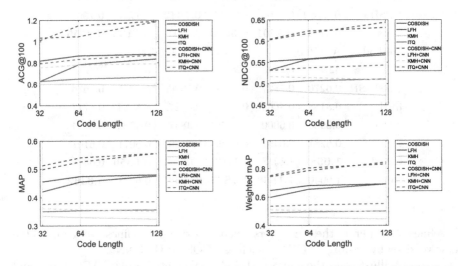

Fig. 5. Curves of MAP, Weighted mAP, ACG@100, and NDCG@100 on IAPRTC-12 varying different code lengths after introducing order-preserving features.

5   Conclusion

In this paper we have proposed a supervised hashing algorithm for multi-labeled images. The CNN model is used to learn order-preserving feature representations. We used some classic ranking metrics to evaluate the performance of our method. The experiment result shows that the characteristics of multi-labeled data are important to increase the effectiveness of feature representation and hash functions learning.

Acknowledgement. This work was supported by 863 Program (2015AA015404), 973 Program (2013CB329303), China National Science Foundation (61402036, 60973083, 61273363), Beijing Advanced Innovation Center for Imaging Technology (BAICIT-2016007).

References

1. Andoni, A., Indyk, P.: Near-optimal hashing algorithms for approximate nearest neighbor in high dimensions. In: Foundations of Computer Science Annual Symposium, vol. 51, no. 1, pp. 459–468 (2006)
2. Arya, S., Mount, D.M., Netanyahu, N.S., Silverman, R., Wu, A.Y.: An optimal algorithm for approximate nearest neighbor searching fixed dimensions. J. ACM (JACM) **45**(6), 891–923 (1998)
3. Beis, J.S., Lowe, D.G.: Shape indexing using approximate nearest-neighbour search in high-dimensional spaces. In: 1997 IEEE Computer Society Conference on Proceedings of Computer Vision and Pattern Recognition, pp. 1000–1006. IEEE (1997)
4. Datar, M., Immorlica, N., Indyk, P., Mirrokni, V.S.: Locality-sensitive hashing scheme based on p-stable distributions. In: Proceedings of the Twentieth Annual Symposium on Computational Geometry, pp. 253–262. ACM (2004)
5. Gionis, A., Indyk, P., Motwani, R.: Similarity search in high dimensions via hashing. In: International Conference on Very Large Data Bases, pp. 518–529 (2000)
6. Gong, Y., Lazebnik, S.: Iterative quantization: a procrustean approach to learning binary codes. In: IEEE Conference on Computer Vision and Pattern Recognition, pp. 817–824 (2011)
7. He, K., Wen, F., Sun, J.: K-means hashing: an affinity-preserving quantization method for learning binary compact codes. In: IEEE Conference on Computer Vision and Pattern Recognition, pp. 2938–2945 (2013)
8. Heo, J.P., Lee, Y., He, J., Chang, S.F., Yoon, S.E.: Spherical hashing. In: 2012 IEEE Conference on Computer Vision and Pattern Recognition (CVPR), pp. 2957–2964, June 2012
9. Indyk, P., Motwani, R.: Approximate nearest neighbors: towards removing the curse of dimensionality. In: Proceedings of the Thirtieth Annual ACM Symposium on Theory of Computing, pp. 604–613. ACM (1998)
10. Kang, W.C., Li, W.J., Zhou, Z.H.: Column sampling based discrete supervised hashing. In: AAAI Conference on Artificial Intelligence (2016)
11. Kulis, B., Grauman, K.: Kernelized locality-sensitive hashing for scalable image search. In: 2009 IEEE 12th International Conference on Computer Vision, pp. 2130–2137. IEEE (2009)
12. Kulis, B., Jain, P., Grauman, K.: Fast similarity search for learned metrics. IEEE Trans. Pattern Anal. Mach. Intell. **31**(12), 2143–2157 (2009)
13. Lai, H., Yan, P., Shu, X., Wei, Y., Yan, S.: Instance-aware hashing for multi-label image retrieval. IEEE Trans. Image Proces. **25**(6), 2469–2479 (2016). https://doi.org/10.1109/TIP.2016.2545300
14. Li, W.J., Wang, S., Kang, W.: Feature learning based deep supervised hashing with pairwise labels. In: Proceedings of the Twenty-Fifth International Joint Conference on Artificial Intelligence, IJCAI 2016, New York, NY, USA, 9–15 July 2016, pp. 1711–1717 (2016)

15. Lin, G., Shen, C., Shi, Q., van den Hengel, A., Suter, D.: Fast supervised hashing with decision trees for high-dimensional data. In: 2014 IEEE Conference on Computer Vision and Pattern Recognition, pp. 1971–1978, June 2014
16. Liu, W., Mu, C., Kumar, S., Chang, S.F.: Discrete graph hashing. Adv. Neural Inf. Process. Syst. **4**, 3419–3427 (2014)
17. Liu, W., Wang, J., Ji, R., Jiang, Y.G., Chang, S.F.: Supervised hashing with kernels. In: 2012 IEEE Conference on Computer Vision and Pattern Recognition (CVPR), pp. 2074–2081, June 2012
18. Norouzi, M., Blei, D.M.: Minimal loss hashing for compact binary codes. In: Proceedings of the 28th International Conference on Machine Learning (ICML 2011), pp. 353–360 (2011)
19. Raginsky, M., Lazebnik, S.: Locality-sensitive binary codes from shift-invariant kernels. In: Advances in Neural Information Processing Systems 22: Proceedings of a Meeting Held 7–10 December 2009, Vancouver, British Columbia, Canada, pp. 1509–1517 (2009)
20. Rvelin, K., Kekäläinen, J.: IR evaluation methods for retrieving highly relevant documents. In: International ACM SIGIR Conference on Research and Development in Information Retrieval, pp. 41–48 (2000)
21. Shen, F., Liu, W., Zhang, S., Yang, Y.: Learning binary codes for maximum inner product search. In: IEEE International Conference on Computer Vision (2015)
22. Wang, J., Yang, J., Yu, K., Lv, F., Huang, T., Gong, Y.: Locality-constrained linear coding for image classification. In: 2010 IEEE Conference on Computer Vision and Pattern Recognition (CVPR), pp. 3360–3367. IEEE (2010)
23. Zhang, P., Zhang, W., Li, W.J., Guo, M.: Supervised hashing with latent factor models. In: International ACM SIGIR Conference on Research and Development in Information Retrieval, pp. 173–182 (2014)
24. Zhao, F., Huang, Y., Wang, L., Tan, T.: Deep semantic ranking based hashing for multi-label image retrieval, pp. 1556–1564 (2015). IEEE Computer Society
25. Zheng, L., Wang, S., Tian, L., He, F., Liu, Z., Tian, Q.: Query-adaptive late fusion for image search and person re-identification. In: Proceedings of the IEEE Conference on Computer Vision and Pattern Recognition, pp. 1741–1750 (2015)

Inferring User Profile Using Microblog Content and Friendship Network

Zhishan Zhao[1], Jiachen Du[1], Qinghong Gao[1], Lin Gui[2], and Ruifeng Xu[1,3(✉)]

[1] School of Computer Science and Technology, Harbin Institute of Technology
Shenzhen Graduate School, Shenzhen, China
zhishan777@gmail.com, dujiachen199165@gmail.com,
gaoqinghong1994@gmail.com, xuruifeng@hit.edu.cn
[2] School of Mathematics and Computer Science, Fuzhou University, Fuzhou, China
guilin.nlp@gmail.com
[3] Guangdong Provincial Engineering Technology Research Center for Data Science,
Guangzhou, China

Abstract. With the rapid development of microblogs in recent years, accurate prediction of microblog user profiles is valuable for marketing, personalized recommendation, and legal investigation. Microblog users post rich contents everyday and build a complex friendship network with "following" behaviors. Both of user-generated content and friendship network are crucial for user profiling. In this work, we propose a neural-network based model for user profiling. It takes advantages of both user-generated content and friendship network with attentional multi-scale convolutional neural networks and graph embeddings. We evaluate our model on SMP CUP 2016 dataset whose task is to infer age, gender and region of microblog users. The experiment results show that utilizing information from user generated content and friend network, our method obtains the state-of-the-art performance on all of three sub-tasks.

Keywords: Social network analysis · User profiling · Neural networks

1 Introduction

User profiles in social networks are important source of valuable information for variable fields such as marketing, sociology and advertising et al. The overwhelming popularity of microblogs creates an opportunity for users to display some aspects of themselves. While many users do not explicitly list all their personal information in their online profile, their user-generated content often contains strong evidence to suggest many types of user attributes, such as gender, age, and region. Meanwhile, the friendship network that build by users with "following" behaviors also contains useful information about user attributes. Many attributes of users have been investigated based on microblogs data, such as gender [1–3,22], age [4], political polarity [5,6], or profession [7].

However, there are two facts making inferring user profile challenging. (1) User-generated content is complex. Each user may generate many short

X. Cheng et al. (Eds.): SMP 2017, CCIS 774, pp. 29–39, 2017.
https://doi.org/10.1007/978-981-10-6805-8_3

messages with diverse topics. And those messages are from multiple sources, e.g. write by user himself, repost from other users, or shared from another platform. Hence, how to focus on messages that contain stronger evidence of one attribute of users is important for accurately user profiling. (2) Friendship network contains useful information about user attributes. How to incorporate user-generated content and friendship network structure information appropriately is still a difficult problem.

To address these challenges, we propose a unified neural-network based model for user profiling. The model is based multi-scale Convolutional Neural Networks whose aim is to capture local and temporal information of user-generated text. To better focus on important phrases in text, self-attention mechanism is applied for text representation. Besides, network embedding, which aims at learning low-dimensional vector representation of users in friendship network, is concatenated to capture the hidden characteristics of users. We evaluate our method on SMP CUP 2016 dataset that collected from Sina Weibo. The experiment result shows that with incorporating content and friendship network structure information, our method obtains the state-of-the-art performance on inferring use's gender, age and region tasks.

2   Related Work

Inferring attributes of microblog users is a growing area of interest, much work has been done on the problem of infer various attributes of a user. Several of the recent works were focused on predicting ethnicity [5,8], age [9,10], gender [11,12], interests [13], tags [14–16] etc.

Research in those area have primarily focused on the construction of more sophisticated feature-based classifiers. For instance, [8] developed models to predict the gender of Twitter users using Support Vector Machines (SVM), with emoticons features, Web abbreviation features, and word unigrams and bigrams features extracted from the concatenation of each user's tweets. [23] developed a hierarchical Bayesian model for predicting the ethnicity and gender of Facebook users from Nigeria using letter n-grams from user names, as well as word n-grams from user content as features. Besides, social network structure also be considered in some research. [17–20] extract features for inferring user gender, including blog words, words' POS tag, word classes, content word classes, results of dictionary based content analysis, POS unigram, and personality types to capture stylistic behavior of authors' writings etc. [21] proposed a cascading topic model for classifying user region, which jointly identifies words with high regional affinity, geographically coherent linguistic regions, and the relations between regional and topic variation. [24] proposed a content-enhanced network embedding (CENE) method, which is capable of jointly leveraging the social network structure and the content information in an unsupervised way.

In this paper, we focus on explore an efficient method to incorporating user-generated content and friendship network structure information supervised, which has been less studied by previous work.

3   Model Architecture

Our model structure is shown in Fig. 1. Each message is represented by a message embedding through convolution and max-pooling layers. Then we obtain a user's content embedding from weighted sum all messages embedding of the user. The weight value of each message embedding is counted by self-attention mechanism. Content embedding and network embedding are concatenated as user embedding, and finally fed to output layer.

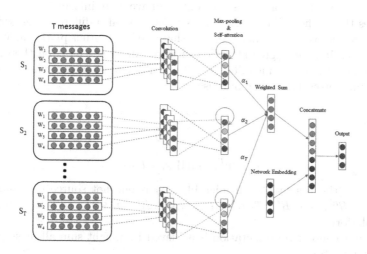

Fig. 1. Model architecture

3.1   User Context Representation

Multi-scale CNNs for Message Embedding. CNNs [25] have achieved remarkable results on computer vision [26] and have also been shown to be effective for various NLP tasks [27]. Architectures employed for NLP applications differ in that they typically involve temporal rather than spatial convolutions.

In our CNNs model, it takes a message, as a sequence of words $x = [x_1, x_2, \cdots, x_n]$ where each word is represented as a d-dimensional vector, and returns a vector S which represents local and temporal information about the sequence. A narrow convolution is applied between x and a kernel $W \in R^{kd}$ of width k.

$$h_i = f(W x_{i:i+k-1} + b) \tag{1}$$

where $x_{i:j}$ is the concatenation of the word sequence from i-th word to j-th word, f is nonlinear function and b is bias. In this paper, we use different kernel sizes to obtain multiple local contextual feature maps. Then take the max-over-time pooling operation [28] to capture the most important features for each feature map. The vector output from the max-over-time pooling layer is the representation of the message.

Self-attention. From the above CNNs method, several message representation vectors are obtained for one user. We can simply average them to get user content representation:

$$c(u) = \frac{1}{T} \sum_{i=1}^{T} s_i \tag{2}$$

where T is message number of the user, $c(u)$ is user content embedding.

However, each message representation vector could reflect different aspects of the user. For example, some message contents have more information about user attributes than others. The source of each message also different, e.g. write by user himself, repost from other users, or shared from another platform. We may want to pay different degree attention on those messages basis on their source and content. Hence, we introduce a self-attention mechanism, which assign a weight value for each message of a user.

$$\alpha = \frac{exp(e_i)}{\sum_{j=1}^{T} e_j} \tag{3}$$

$$e_i = v^T tanh(W s_i + U h_i) \tag{4}$$

Where α_i is the weight of i-th microblog, s_i is one hot source representation vector, $v \in R^{n'}$, $W \in R^{n' \times m}$, $U \in R^{n' \times n}$, $s_i \in R^m$, $h_i \in R^n$, m is the number of source platforms.

The user context representation is achieved by weight sum all message representation vectors.

$$c(u) = \sum_{i=1}^{T} \alpha_i h_i \tag{5}$$

3.2 User Network Representation

The user friendship network in microblog, which treats each user as a node and treats "following" relationship as an edge between two nodes, hidden huge information about users. For mining useful information from friendship network, network embedding, which aims at learning low-dimensional vector representation of users in networks, has been proposed as a critical technique for network analysis tasks.

In this work, we focus on embeddings of friendship networks. A friendship network is defined as $G = (V, E)$, where V is the set of vertices, each representing a microblog user and E is the set of edges between the users, each representing a followed relationship between two users. Each edge $e \in E$ is an ordered pair $e = (u, v)$, which represent user u is followed by v.

In this paper, we use the loss function described in [29], which preserve both the *first-order* and *second-order* proximities between users. The *first-order* proximity refers to the local pairwise proximity between the vertices in the network.

To model the *first-order* proximity, for each edge (i, j), joint probability between vertex v_i and v_j is defined as follows:

$$p_1(v_i, v_j) = \frac{1}{1 + exp(-u_i \cdot u_j)} \tag{6}$$

$$O_1 = - \sum_{(i,j) \in E} \log p_i(v_i, v_j) \tag{7}$$

where $\boldsymbol{u_i} \in R^d$ is low-dimensional vector representation if vector v_i.

The *second-order* proximity assumes that vertices sharing many connections to other vertices are similar to each other. It specifies the following objective function for each edge (i, j):

$$O_2 = - \log \sigma(u_j'^T \cdot u_i) + \sum_{i=1}^{K} E_{v_n \sim P_n(v)} [\log \sigma(-u_n'^T \cdot u_i)] \tag{8}$$

where $\sigma(x) = \frac{1}{1 + exp(-x)}$ is the sigmoid function. u_i is the representation of v_i when it is treated as a vertex while $u_i'^T$ is the representation of v_i when it is treated as a specific "context". The first term models the observed edges, the second term models the negative edges drawn from the noise distribution and K is the number of negative edges. Following [29], We set $P_n(v) \propto d_v^{\frac{3}{4}}$, where d_v is the out-degree of vertex v.

4 Experiment

4.1 Datasets

The dataset used in this paper is released by SMP CUP 2016[1], which collected from Sina Weibo. The competition contains three tasks: predicting user's gender (male/female), age ($-1979/1980$–$1989/1990+$) and region (Northeast China/North China/Central China/East China/Northwest China/Southwest China/South China/Overseas). We use **Accuracy** as evaluation metrics for three tasks, same as evaluation metrics used in the competition. The statistics of dataset are shown in Table 1.

In our experiment, LTP[2] was used for word segmentation. For gender and age tasks, we did not perform any dataset preprocessing, apart from replacing all digits with a zero and deleting URLs in the user's microblogs. For region task, we construct Gazetteer features[3] by using geography knowledge and Sina Weibo location information.

---

[1] https://biendata.com/competition/smpcup2016/.

[2] http://pyltp.readthedocs.io/zh_CN/latest/.

[3] As training data is insufficient, the model is difficult to learn how to map a specific location that shows in user-generated content to its belonging region. Hence, we construct a region dictionary using geography knowledge and Sina Weibo location information to help our model find the relation between location and region.

Table 1. Detail of dataset

Dataset	Training data	Validation data	Testing data
#friendship network nodes	2,565,000	1,000	1,000
#friendship network edges	550,000,000	150,000	130,000
#users with content	44,000	980	1,240
#users with label	3,200	980	1,240

4.2 Pretrained Embeddings

We use pretrained word embeddings and network embedding to initialize our embedding lookup table. Word embedding are pretrained using Glove[4] with a minimum word frequency cutoff of 5, and a window size of 8. The dimension of word embedding is set to 100. Network embedding are pretrained using LINE[5] with first order embedding dimension 150 and second order embedding dimension 150, negative edges number is set to 5. For users who are not shown in friendship network, we initialize all zeros as their network embeddings.

4.3 Training

In order to train efficiently and augment training data, instead of using all messages of users, we random select $T = 20$ messages of users at every training epoch. For all tasks, nonlinear function is rectified linear units (RELU), learning rate is 0.001, dropout rate is 0.2, l2 regularization is $5E - 5$, mini-batch size is 50. All parameters in our network are initialized by randomly sampling from uniform distribution in $[-0.1, 0.1]$. CNN filter window sizes are set to 1, 3, 5 with 64 feature maps each. The number of hidden units in the self-attention mechanism n' is 64.

4.4 Experiment Analysis

In this section, we first report the results of our model on the SMP CUP 2016 dataset in comparison to some baseline methods. Then we explore the effectiveness of our proposed model architecture.

4.4.1 Baselines

To illustrate the performance boost of our proposed model, we compare our model with some baseline methods.

- **BOW:** Bag of words (BOW) is commonly used in methods of document classification. In this model, a text (such as a sentence or a document) is represented as the bag of its words, disregarding grammar and even word

---

[4] https://nlp.stanford.edu/projects/glove/.
[5] https://github.com/tangjianpku/LINE.

order. In our experiment, we concatenate all messages of one user as a document and using chi-square to select 3000 feature words. Then SVD is used for dimensionality reduction.

- **Paragraph Vector:** Paragraph Vector [30] is an unsupervised algorithm that learns fixed-length feature representations from variable-length pieces of texts. The vector representations are learned to predict the surrounding words in contexts sampled from the paragraph. In our experiment, all messages of one user are concatenate and treated as a paragraph. The dimension of paragraph vector is set to 300.
- **Top teams of SMP CUP 2016:** We also compare our model with top 3 teams of SMP CUP 2016: HLT-HITSZ, DUTIR-TONE, 卢泓宇. All of these teams have done a lot of feature engineering for every specific task. Except team HLT-HITSZ, none of other teams use friendship network structure information.

4.4.2 Experiment Result

Table 2 shows comparison result of our model to all the baseline models mentioned above. As we can see, our model performs consistently and significantly better than BOW and Paragraph Vector baselines on three tasks. The results indicate that our proposed method can better predict microblog user's profile. Without handcraft features, our model's result competitive with top teams' results of SMP CUP 2016. In order to further improve accuracy and robustness, we also compute an ensemble model by averaging 10 weak models that trained independently on 10 sampled training subsets according to the bagging algorithm. Through the model averaging process, our model obtains state-of-the-art performance on three tasks. It is interesting to see that our model and HLT-HITSZ, which both use friendship network information, significantly outperform other methods on region task.

To discover the contribution of each component to model performance, we conduct an ablation study by removing one component at a time in our network architecture. Specifically, we have tested the performance of the model without

Table 2. Results of our proposed model against baseline methods

Baselines	Gender (Accuracy)	Age (Accuracy)	Region (Accuracy)
BOW	85.42	62.01	65.50
Paragraph Vector	82.33	58.67	61.62
HLT-HITSZ (ensemble)	88.30	64.80	72.70
DUTIR-TONE (ensemble)	89.27	67.98	69.76
卢泓宇	88.31	65.73	68.06
Our model (single)	88.69	67.43	72.61
Our model (ensemble)	**89.75**	**68.14**	**73.42**

Table 3. An ablation study on SMP CUP 2016 dataset

Model	Gender	Age	Region
Full architecture	**88.69**	**67.43**	**72.61**
Without self-attention	88.01 (−0.68)	66.27 (−1.16)	72.14 (−0.47)
Without network embedding	87.07 (−1.62)	64.33 (−3.10)	66.38 (−6.23)

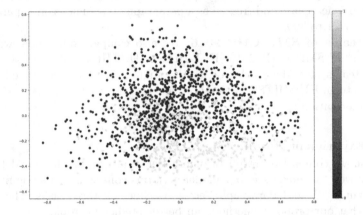

Fig. 2. 2-dimensional t-SNE visualization of the network embedding for user gender inferring task. Color of a node indicates the gender of the user (Color figure online)

the self-attention mechanism or network embedding. From the results presented in Table 3, we observe that the self-attention mechanism impacts considerably the performance of all three tasks. The network embedding, which can capture relationship between microblog users, gives biggest performance gains for all three tasks, especially for region task.

We also use t-SNE [31] to obtain a 2-dimensional visualization of the user network embedding for user gender inferring task, shown in Fig. 2. We can clearly see that nodes with same colors are distributed closer, which gives credible evidence of the good performance of user network embedding.

5   Case Study

To gain insights about the self-attention mechanism's behavior at a more detailed level, we randomly select a user in SMP CUP 2016 dataset. For each message of the user, attention value was extracted from gender task and we sample seven messages attention value for analysis, as outlined in Table 4 .

The first message is shared from the platform "快盘", which is outside of microblog platform, has the smallest attention value among the seven messages. The result verify that our self-attention mechanism has the abilitcm other platform and pay little attention on them. All the second to seventh messages are generated by the user but not the same topic. The terms "姐" and "裙子" in the

Table 4. The analysis of the effectiveness of self-attention mechanism on gender task

Source	content	Attention Value
快盘	# 金山快盘 # 签或不签到，快盘就在这里，静静等您，来快盘签到，或让快盘住进你的心里，默然欢笑，寂静欢喜！【手机体验地址：体验地址：】原图	0.00077
Android 客户端	姐要减肥！中码的裙子还太小。	0.12851
Android 客户端	凯文老师可以推荐几款可爱的新娘服装吗？	0.18436
Android 客户端	最近就一个字，累	0.00841
Android 客户端	猜猜这是那里？原图	0.00142
Android 客户端	第一名都还没有公布，大奖就已经被工作人员拎着了	0.00374
Android 客户端	昨天去孤儿院和敬老院了！从来不知道他们就在我身边离得那么近	0.00952

second message and "可爱的" and "新娘" in third message give a strong hint that the gender of the user is female, which guide self-attention mechanism to assigns a higher attention value to those two messages. On the other side, the fourth to seventh messages have little evidence about the user's gender, which led to be assigned a smaller attention value by self-attention mechanism.

6 Conclusion

In this paper, we have presented a new unified neural network based method for microblog user profiles inference. Multi-scale CNNs and self-attention mechanism are used to capture useful features in user-generated content. Network embedding is used to extract user characteristics hidden in friendship network. With incorporate content features and network embedding, our method obtains the state-of-the-art performance on SMP CUP 2016 dataset. For future work, we will extend our methods to involves more diverse contents such as avatars and user names.

Acknowledgments. This work was supported by the National Natural Science Foundation of China 61370165, U1636103, 61632011, Shenzhen Foundational Research Funding JCYJ20150625142543470, JCYJ20170307150024907 and Guangdong Provincial Engineering Technology Research Center for Data Science 2016KF09.

References

1. Ciot, M., Sonderegger, M., Ruths, D.: Gender inference of twitter users in nonEnglish contexts. In: Proceedings of EMNLP, pp. 18–21 (2013)
2. Wendy, L., Derek, R.: What's in a name? Using first names as features for gender inference in twitter. In: AAAI Spring Symposium Series (2013)

3. Liu, W., Zamal, F.A., Ruths, D.: Using social media to infer gender composition of commuter populations. In: Proceedings of the International Conference on Weblogs and Social Media (2102)
4. Rao, D., Yarowsky, D.: Detecting latent user properties in social media. In: Proceedings of the NIPS MLSN Workshop (2010)
5. Pennacchiotti, M., Popescu, A.M.: A machine learning approach to twitter user classification. In: Proceedings of ICWSM (2011)
6. Conover, M.D., Ratkiewicz, J., Francisco, M., et al.: Political polarization on twitter. In: Proceedings of ICWSM (2011)
7. Tu, C., Liu, Z., Sun, M.: PRISM: Profession Identification in Social Media with personal information and community structure. In: Proceedings of Social Media Processing (2015)
8. Rao, D., Yarowsky, D., Shreevats, A., Gupta, M.: Classifying latent user attributes in twitter. In: Proceedings of the 2nd International Workshop on Search and Mining User-Generated Contents, pp. 37–44 (2010)
9. Rosenthal, S., McKeown, K.: Age prediction in blogs: a study of style, content, and online behavior in pre-and post-social media generations. In: Proceedings of the 49th Annual Meeting of the Association for Computational Linguistics, Human Language Technologies, vol. 1, pp. 763–772 (2011)
10. Nguyen, D., Smith, N.A., Rosé, C.P.: Author age prediction from text using linear regression. In: Proceedings of the 5th ACL-HLT Workshop on Language Technology for Cultural Heritage, pp. 115–123 (2011)
11. Burger, J.D., Henderson, J., Kim, G., Zarrella, G.: Discriminating gender on twitter. In: Proceedings of the Conference on Empirical Methods in Natural Language Processing, pp. 1301–1309 (2011)
12. Al Zamal, F., Liu, W., Ruths, D.: Homophily and latent attribute inference: inferring latent attributes of twitter users from neighbors. In: Proceedings of ICWSM (2012)
13. Lim, K.H., Datta, A.: Finding twitter communities with common interests using following links of celebrities. In: Proceedings of the 3rd International Workshop on Modeling Social Media, pp. 25–32 (2012)
14. Tu, C., Liu, Z., Sun, M.: Inferring correspondences from multiple sources for microblog user tags. In: Huang, H., Liu, T., Zhang, H.-P., Tang, J. (eds.) SMP 2014. CCIS, vol. 489, pp. 1–12. Springer, Heidelberg (2014). doi:10.1007/978-3-662-45558-6_1
15. Gui, L., Xu, R, He, Y., Lu, Q., Wei, Z.: Intersubjectivity and Sentiment: from Language to Knowledge. In: Proceedings of 25th International Joint Conference on Artificial Intelligence (IJCAI) (2016)
16. Gui, L., Zhou, Y., Xu, R., He, Y., Lu, Q.: Learning representations from heterogeneous network for sentiment classification of product reviews. In: Proceedings of Knowledge-Based Systems, pp. 34–45 (2017)
17. Yan, X., Yan, L.: Gender classification of weblog authors. In: Proceedings of the Association for the Advancement of Artificial Intelligence. Computational Approaches to Analyzing Weblogs (2006)
18. Tuv, E., Borisov, A., Runger, G., Torkkola, K.: Feature selection with ensembles, artificial variables, and redundancy elimination. Proc. J. Mach. Learn. Res. 10, 1341–1366 (2009)
19. Houvardas, J., Stamatatos, E.: N-gram feature selection for authorship identification. In: Proceedings of the 12th International Conference on Artificial Intelligence: Methodology, Systems, Applications, pp. 77–86 (2006)

20. Schler, J., Koppel, M., Argamon, S., Pennebaker, J.: Effects of age and gender on blogging. In: Proceedings of the Association for the Advancement of Artificial Intelligence Spring Symposium Computational Approaches to Analyzing Weblogs (2006)

21. Eisenstein, J., O'Connor, B., Smith, N.A., et al.: A latent variable model for geographic lexical variation. In: Proceedings of Empirical Methods in Natural Language Processing. Association for Computational Linguistics, pp. 1277–1287 (2010)

22. Mukherjee, A., Liu, B.: Improving gender classification of blog authors. In: Proceedings of the 2010 Conference on Empirical Methods in Natural Language Processing, Cambridge, MA. Association for Computational Linguistics, October 2010

23. Rao, D., Fink, C., Oates, T.: Hierarchical Bayesian models for latent attribute detection in social media. In: Proceedings of the 5th International Conference in Weblogs and Social Media (2011)

24. Sun, X., Guo, J., Ding, X., Liu, T.: A general framework for content-enhanced network representation learning. arXiv preprint (2016)

25. LeCun, Y., Boser, B., Denker, J.S., Henderson, D., Howard, R.E., Hubbard, W., Jackel, L.D.: Handwritten digit recognition with a backpropagation network. In: Proceedings of NIPS (1989)

26. Krizhevsky, A., Sutskever, I., Hinton, G.: ImageNet classification with deep convolutional neural networks. In: Proceedings of NIPS (2012)

27. Kim, Y.: Convolutional neural networks for sentence classification. In: Proceedings of the 2014 Conference on Empirical Methods in Natural Language Processing, pp. 1746–1751 (2014)

28. Collobert, R., Weston, J., Bottou, L., et al.: Natural language processing (almost) from scratch. J. Mach. Learn. Res. **12**(8), 2493–2537 (2011)

29. Tang, J., Qu, M., Wang, M., et al.: LINE: Large-scale Information Network Embedding. In: Proceedings of the 24th International Conference on World Wide Web, pp. 1067–1077 (2015)

30. Le, Q.V., Mikolov, T.: Distributed representations of sentences and documents. In: Proceedings of the 31st International Conference on Machine Learning (2014)

31. van der Laurens, M., Hinton, G.: Visualizing data using t-SNE. Proc. J. Mach. Learn. Res. **9**, 2579–2605 (2008)

EEG: Knowledge Base for Event Evolutionary Principles and Patterns

Zhongyang Li, Sendong Zhao, Xiao Ding, and Ting Liu[✉]

Research Center for Social Computing and Information Retrieval,
Harbin Institute of Technology, Harbin, China
{zyli,sdzhao,xding,tliu}@ir.hit.edu.cn

Abstract. The evolution and development of events has its underlying principles, leading to events happened sequentially. Therefore, the discovery of such evolutionary patterns between events are of great value for event prediction, decision-making and scenario design of dialog system. In this paper, we propose **Event Evolutionary Graph** (EEG), which reveals evolutionary patterns and development logics between events. Specifically, we propose to construct EEG by recognizing the sequential relation between events and the direction of each sequential relation. For sequential relation and direction recognition, we explore the effectiveness of 4 categories of features: count-based, ratio-based, context-based and association-based features for correctly identifying sequential relations and corresponding directions. Experimental results show that (1) the framework we proposed is promising for EEG construction and (2) methods we proposed are effective for both sequential relation and direction recognition.

Keywords: Event Evolutionary Graph · Sequential relation between events · Social media · Knowledge base

1 Introduction

The evolution and development of events have its underlying principles, making events happen sequentially. This kind of evolutionary patterns of events is of great value. For example, the sentence "After having lunch, Tom paid the bill and left the restaurant" shows a sequence of events evolution ① "have lunch", ② "pay the bill" and ③ "leave the restaurant". This event series is a very ordinary pattern for the scenario of having lunch in a restaurant. This kind of pattern is very common in our daily life, which usually indicate the basic patterns of events evolution and human behaviors. Hence, these patterns of evolutionary events are of great value and important for many tasks, such as event prediction, decision-making and scenario design of dialog system.

Numerous efforts have been dedicated to extracting temporal and causal relations from texts. As the most commonly used corpus, the TimeBank corpus [15] has been adopted in a lot of temporal relation extraction studies. Mani et al. [9]

© Springer Nature Singapore Pte Ltd. 2017
X. Cheng et al. (Eds.): SMP 2017, CCIS 774, pp. 40–52, 2017.
https://doi.org/10.1007/978-981-10-6805-8_4

applied the temporal transitivity rule to greatly expand the corpus. Chambers et al. [4] used previously learned event attributes to classify the temporal relationship. For causality relation extraction, Zhao et al. [19] extracted multiple kinds of features to recognize causal relations between two events in the same sentence. Radinsky et al. [16] automatically extracted cause-effect pairs from large quantities of news titles by predefined causal templates, and then they used them to predict news events. However, these studies have some limitations. First, this line of work can only extract relations from single sentences. Second, these studies extract relations based on the semantic of specific context rather than discover the underlying patterns of event evolution from large-scale user generated documents.

In order to discover patterns of evolutionary events, we propose **Event Evolutionary Graph** (EEG) and the framework to construct EEG. Specifically, our definition of EEG involves evolutionary patterns and development logics of events. The EEG is composed of events and relations between them. For the sake of generality, we consider sequential and causal relations in EEG. The value of each relation denotes the transition probability between events. Therefore, the construction of EEG can be simplified as two key problems. The first is to recognize relations between each two events. The second is to distinguish the direction of each relation between events. Both problems can be solved based on the classification framework.

The main contributions of this paper are as follows. First, we propose EEG and give its detailed definitions. Second, we propose a promising construction framework to build EEG from large-scale unstructured web corpus. Third, extensive experiments are conducted to solve the central problems of sequential relation and direction recognition. Experimental results show that the methods we developed are effective for both sequential relation and direction recognition.

2   Related Work

2.1   Statistical Script Learning

The use of scripts in Artificial Intelligence dates back to the 1970s [10]. In this conception, scripts are composed of complex events without probabilistic semantics. In recent years, a growing body of research has investigated learning probabilistic co-occurrence-based models with simpler events. Chambers et al. [3] proposed a co-occurrence-based model of (verb, dependency) pairs, which can be used to infer such pairs from documents. Pichotta et al. [12] described a method of learning a co-occurrence-based model of verbs with multiple arguments.

There have been a number of recent published neural models for script learning. Pichotta et al. [13] showed that an LSTM event sequence model outperformed previous methods for predicting verbs with arguments. Pichotta et al. [14] used a Seq2Seq model directly operating on raw tokens to predict sentences. Mark and Clark [7] described a feed-forward neural network which composed verbs and arguments into low-dimensional vectors.

Script learning is very similar to EEG in concepts. However, script learning usually extracts event chains without considering their temporal orders. Event definition and representation are also different. EEG aims to organize event evolutionary patterns into a commonsense knowledge base, which is the biggest difference between them.

2.2   Temporal Relation Extraction

A lot of annotation efforts have been devoted to construct corpora for building event ordering models. However, most of existing corpora focus on English. As the commonly used corpus in temporal relation extraction, the TimeBank corpus [15] has been adopted in a series of TempEval competitions [17,18], facilitating the development and evaluation of temporal relation extraction systems. But TimeBank corpus only annotates a small subset of easily-identified event mention pairs, which much limit its applications. To overcome this problem, Do et al. [6] produced an event-driven corpus on the ACE 2005 English corpus. Cassidy et al. [1] enriched the TimeBank-Dense corpus on top of TimeBank. In comparison, there are few corpora for Chinese temporal relation extraction.

Due to the corpus limitation, previous studies on temporal relation extraction focus on inferring temporal relations between event mentions in the same sentence or neighboring sentences from English text, dominated by feature-based approaches [4,8,9]. Chambers et al. [2] proposed a sieve-based architecture to joint those different tasks of temporal relation extraction. Mirza and Tonelli [11] used multiple cascaded classifiers to simultaneously solve the temporal and causal relation classification, and achieved the best experimental results.

All these studies solve temporal relation extraction from specific context. However, we solve this problem by frequency-based inference from multiple sentences, which is a commonsense reasoning process.

3   Event Evolutionary Graph

In this section, we give the detailed definition of EEG, which consists of event and relations between them. We consider two types of relation here, i.e., sequential and causal relations.

In EEG, events are represented by **abstract, generalized and semantic complete verb phrases**. Each event must contain a trigger word, which mainly indicates the occurrence of the event, and some other necessary components, such as the subject, object or modifier, to ensure the semantic completeness. **Abstract and generalized** means that we don't focus on the accurate happening location and time, and the exact subject of the event. **Semantic complete** means that human beings can understand the meaning of the event without vague and ambiguity. For example, "have hot pot", "watch movies", "go to the airport", are reasonable verb phrases to represent events. While "go somewhere", "do the things", "eat" are unreasonable or incomplete event representations, as they are too vague to understand.

Fig. 1. Tree structured event evolutionary graph under the scenario of "plan a wedding".

Fig. 2. Chain structured event evolutionary graph under the scenario of "watch movies".

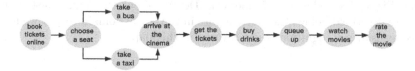

Fig. 3. Cyclic structured event evolutionary graph under the scenario of "fight".

The sequential relation between two events refers to their partial temporal orderings. For example, "After having lunch, Tom paid the bill and left the restaurant." "have lunch", "pay the bill" and "leave the restaurant" compose a sequential relation event chain.

Causality is the relation between one event (the cause) and a second event (the effect), where the second event is understood as a consequence of the first. It is obvious that the causal relation between events must be sequential. For example, "The nuclear leak in Japan leads to serious ocean pollution". "Nuclear leak in Japan" is the cause event, and "ocean pollution" is the effect event. Besides, the cause event happens before the effect event in temporal ordering. Hence, causal relation is a subset of sequential relation between events, and sequential relation is more general than causal relation.

EEG is a Directed Cyclic Graph, whose nodes are events, and edges stand for the sequential and causal relations between events. Essentially, EEG is a commonsense knowledge graph, which describes the event evolutionary patterns. Figures 1, 2 and 3 demonstrate three different event evolutionary subgraphs of three different scenarios. Concretely, Fig. 1 describe some event evolutionary patterns under the scenario of "plan a wedding", which happen repeatedly in people's daily life, and have evolved into some fixed human behavior patterns. For example, "plan a wedding" usually follows by "buy a house", "buy a car" and "plan a travel".

4   Methods for Constructing EEG

In this section, we propose a construction framework to construct EEG from large-scale unstructured text, including data cleaning, natural language processing, event extraction, event pair candidates extraction, sequential relation and direction recognition, causality recognition and transition probability computation. Figure 4 sketches this framework. Details about the main construction steps are described below. Causality is rare in the travel domain corpus we used. Hence, causality recognition is not covered in this paper.

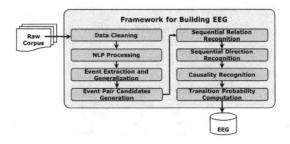

Fig. 4. Out proposed framework for building EEG from large-scale unstructured text.

4.1   Event Extraction

After cleaning the data, a series of natural language processing steps including segmentation, part-of-speech tagging, and dependency parsing are conducted for event extraction. Tools provided by Language Technology Platform [5] are used for this preprocessing.

Event extraction strategy is different from task to task, and it is mainly determined by the underlying task. As defined in Sect. 3, events in EEG are represented by abstract, generalized and semantic complete verb phrases, considering the corpus used in this paper. We extract verb-object phrases from the dependency-parsed tree. Though this is a simple strategy, we do find that a lot of high quality verb phrases are extracted.

We filter the low-frequency verb-object phrases by a proper threshold, to exclude the verb phrases extracted due to segmentation and dependency parsing errors. Some too general events such as "go somewhere" and "do something" are removed by regular expressions with a dictionary. Note that event extraction is not the key problem in this paper.

4.2 Event Pair Candidates Generation

Based on event extraction results, candidate event pairs are generated using two heuristic rules. First, every two events from single sentences are considered as an event pair candidate. Second, every two events from two consecutive sentences are considered as an event pair candidate as well. The event from the first sentence is taken as the first element of the pair, and the event from the second sentence as the second element.

For example, events A and B are extracted form the former sentence, and events C, D, E are extracted from the latter adjacent sentence. Ten event pairs are constructed as (A, B), (A, C), (A, D), (A, E), (B, C), (B, D), (B, E), (C, D), (C, E), (D, E).

Table 1. The features used for sequential relation and direction classification.

Count-based features	Ratio-based features
T1: count of (A, B)	R1: T2/T1, R2: T1/T4
T2: count of (A, B) where A occurs before B	R3: T1/T5, R4: T1/T6
T3: count of (A, B) where B occurs before A	R5: T1/T7, R6: T1/T8
T4: count of A	R7: T1/T9, R8: T6/T4
T5: count of B	R9: T7/T4, R10: T8/T5
T6: count of verb-A	R11: T9/T5
T7: count of object-A	
T8: count of verb-B	
T9: count of object-B	
Context-based features	**Association-based features**
C1: count of all unique contexts	A1: PMI of A and B
C2: average length of all contexts	A2: PMI of verb-A and verb-B
C3: count of one-sentence contexts	A3: PMI of verb-A and object-B
C4: count of two-sentence contexts	A4: PMI of object-A and verb-B
C5: C3/C1	A5: PMI of object-A and object-B
C6: contain relation of verb-A and verb-B	
C7: contain relation of object-A and object-B	
C8: postag of object-A	
C9: postag of object-B	

4.3 Sequential Relation and Direction Recognition

Given an event pair candidate (A, B), sequential relation recognition is to judge whether they have a sequential relation or not. For the ones having a sequential relation, direction recognition should be conducted to distinguish the direction. For example, we need to recognize there is a directed edge from "buy tickets" to "watches movies". We regard the sequential relation and direction recognition as two separate binary classification tasks. Previous temporal relation classification studies judge the relation and direction from single sentences. Alternatively, we solve this problem by frequency-based inference from multiple sentences. Specifically, we achieve this by forcing event A and B to co-occur more than k times. It is reasonable because event A and B are strongly associated with a high-frequency co-occurrence.

Multiple kinds of features are extracted for these two supervised classification tasks. All the features used are listed in Table 1. Details about the intuition why we choose these features are described below.

Count-Based Features: For a strongly associated event pair (A, B), there are many different statistics for them. They are directly counted from contexts that they co-occur or the whole corpus. We believe these statistics are effective to measure how strong and in which way event A and B are associated with each other. Therefore, they can be useful features for sequential relation and direction classification. These statistics include co-occur counts (T1 to T3), counts for events in the whole corpus (T4 and T5), and counts for verb and object in the events (T6 to T9).

Ratio-Based Features: The count-based features are numbers that directly counted from the context and whole corpus. Some meaningful combinations between these count numbers may provide extra information that is useful for sequential relation and direction classification. For example, if T2/T1 is close to 1, we can believe that A almost always occurs before B in the context they co-occur. It is a significant signal that the sequential direction is from A to B. These combination features are listed in Table 1 as ratio-based features (R1 to R11).

Context-Based Features: We believe that contexts exist where sequential candidates are more likely to appear. We developed features that capture the characteristics of likely contexts for sequential relations. In a nutshell, they include the count of unique context that A and B co-occur (C1), average length of all contexts (C2), the count of contexts that A and B are in the same sentence or not (C3 and C4), the ratio of one-sentence context (C5), the contain relation between verbs and objects (C6 and C7), and postag of objects (C8 and C9).

Association-Based Features: These features measure the association strength between event A and event B, including PMI scores from two different aspects. First, PMI score of (A, B) is computed as the macro measure of how strong event A and event B are associated with each other (A1). Second, we further consider the fine-grained PMI scores of (verb-A, verb-B), (verb-A, object-B), (object-A, verb-B) and (object-A, object-B), which measure the partial association strength of event components (A2 to A5).

4.4 Transition Probability Computation

Given an event pair (A, B), we use the following equation to approximate the transition probability from event A to event B:

$$P(B|A) = \frac{count(A, B)}{count(A)},$$ (1)

where $count(A, B)$ is the co-occurrence count of event pair (A, B), and $count(A)$ is the occurrence count of event A in the whole corpus.

5 Experiments

In this section, we conducted two kinds of experiments. The first is to judge whether two events has sequential relation. And the second is to judge the partial ordering between two sequential events. These two steps are crucial and central for constructing clean and accurate EEG.

5.1 Dataset Description

We crawled 320,702 question-answer pairs from travel topic on Zhihu[1] as our experimental dataset. Travel is a relatively high level topic, which covers a wide range of things about traveling. Therefore, a lot of commonsense event evolutionary knowledge can be discovered from this data source.

Table 2. The detailed data statistics.

	Total	Positive	Negative
Sequential relation	2,173	1,563	610
Sequential direction	1,563	1,349	214

We annotate 2,173 event pairs with high co-occurrence frequency ($>= 5$) as our experiment corpus. Each event pair (A, B) is ordered that A occurs before B with a higher frequency than B occurs before A. In the annotation process, the annotators are provided with the event pairs and their corresponding contexts. They need to judge whether there is a sequential relation between two events from a commonsense perspective. If true, they also need to give the sequential direction. For example, "watch movies" and "listen to music" are tagged as no sequential relation (negative), while "go to the railway station" and "by tickets" are tagged as having a sequential relation (positive), and the sequential direction is from the former to the latter (positive). The detailed corpus statistics are listed in Table 2. The positive and negative examples are very imbalanced. So we over sample the negative examples in training set to ensure the number of positive and negative training examples are equal.

---

[1] https://www.zhihu.com/.

5.2   Compared Methods and Evaluation Metrics

For sequential relation recognition, PMI score of an event pair is used as the baseline method. For sequential direction recognition, if event A occurs before B with a higher frequency than B occurs before A, we think the sequential direction is from event A to event B. This is called the **Preceding Assumption**, which is used as the baseline method for sequential direction recognition.

For two experiments, four classifiers are used for these classification tasks, which are naive bayes classifier (NB), logistic regression (LR), multiple layer perceptron (MLP) and support vector machines (SVM). We explored different feature combinations to find the best feature set for both classification tasks. All experiments are conducted using five-fold cross validation. The final experiment result is the average performance of ten times of implementations.

Two kinds of evaluation metrics are used to evaluate the performance of our proposed methods. They are accuracy, and the precision, recall and F1 value.

5.3   Results and Analysis

Table 3 shows the experimental results for sequential relation classification, and we find that the pure PMI baseline achieves very good performance. Indeed, due to the imbalance of positive and negative test examples, PMI baseline chooses a threshold to classify all test examples as positive, and get a recall of 1. Four different classifiers with all features in Table 1 achieve poor results, and only the NB achieves higher performance than the baseline method. We explored all combinations of four kinds of features to find the best feature set for different classifiers. Still, the NB classifier achieves the best performance with a 0.776 accuracy and a 0.857 F1 score.

Table 3. Sequential relation classification results. Baseline result is given at the top row. Results of four classifiers with all features in Table 1 are in the middle. Results of four classifiers with the best feature combinations are given at the bottom.

Features	Methods	Accuracy	Precision	Recall	F1
	Baseline	0.719	0.719	1.000	0.837
All features	NB	**0.763**	0.784	**0.924**	**0.848**
	LR	0.690	0.795	0.765	0.779
	MLP	0.683	0.841	0.692	0.756
	SVM	0.523	**0.849**	0.409	0.551
Ratio+Association	NB	**0.776**	0.789	**0.939**	**0.857**
Ratio	LR	0.770	0.800	0.907	0.850
Association	MLP	0.747	**0.808**	0.852	0.829
Ratio	SVM	0.765	0.789	0.919	0.849

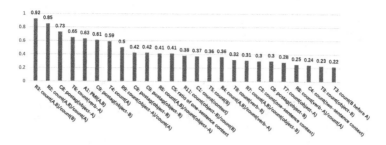

Fig. 5. The 25 most important features for relation classification and their relative importance scores.

Besides, we compute the relative importance scores for all features in Table 1, and the top 25 most important features for relation classification and their relative importance scores are illustrated in Fig. 5. These importance scores are computed by six different methods, including chi-square test, ANOVA F-value, maximal information coefficient, random forest, recursive feature elimination and stability selection. Their individual scores are normalized into the range of 0 and 1, and the average of six scores is computed as the last importance score. We find that ratio-based features are the most important features for sequential relation classification. Experimental results in Table 3 verify this conclusion, with three classifiers achieve the best performance using ratio-based features. And the association-based features are the second most important features.

Table 4. Sequential direction classification results. Baseline result is given at the top row. Results of four classifiers with all features in Table 1 are in the middle. Results of four classifiers with the best feature combinations are given at the bottom.

Features	Methods	Accuracy	Precision	Recall	F1
	Baseline	0.861	0.866	0.993	0.925
All Features	NB	0.803	0.891	0.880	0.885
	LR	0.642	0.894	0.663	0.761
	MLP	0.787	**0.903**	0.844	0.872
	SVM	**0.864**	0.866	**0.997**	**0.927**
Association	NB	0.862	0.863	**0.999**	0.926
Ratio+Association	LR	0.713	0.861	0.796	0.826
All Features	MLP	0.787	**0.903**	0.844	0.872
Association+Context	SVM	**0.870**	0.877	0.988	**0.929**

Table 4 shows the experimental results for sequential direction classification, from which we find that the **Preceding Assumption** is a very strong baseline for direction classification, and achieves a accuracy of 0.861 and F1 of 0.925.

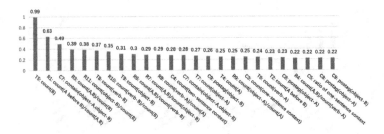

Fig. 6. The 25 most important features for direction classification and their relative effectiveness scores

Four classifiers with all features in Table 1 achieve poor results, and only the SVM achieves higher performance than the baseline method. We explored all combinations of four kinds of features, to find the best feature set for different classifiers. Still, the SVM classifier achieves the best performance with a 0.870 accuracy and a 0.929 F1 score, using the association and context based features.

We also compute the relative feature importance scores for all features in Table 1, and the top 25 most important features for direction classification and their relative importance scores are showed in Fig. 6. We find that the most important two features are T5: count of event B, and R1: count(A before B)/count(A, B). But three of four classifiers achieve their best performance without these two features, which is very interesting. We further experiment with adding these two features, finding that it doesn't help. This is mainly because the feature importance scores examine each feature individually to determine the strength of the relationship between the feature and the response variable. Therefore, they may degrade the performance when combined together due to their opposite correlation with each other.

Based on the experimental results listed above, some useful conclusions can be reached as follows:

- The more features the better performance is not true, and different classifiers capture different kinds of features.
- The two simple baseline methods used in our experiments achieve very good experimental results. However, our proposed feature-based supervised methods achieve the best performance.
- Though certain features are important individually, the performance can be degraded when they are combined, due to their opposite correlation.

6   Case Study

Based on the construction framework proposed, we construct a Chinese travel domain EEG[2] from large-scale unstructured web corpus. A subgraph in this EEG

---

[2] http://202.118.250.16:60810.

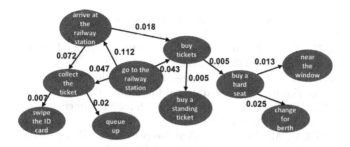

Fig. 7. Subgraph in our constructed travel domain EEG under the scenario of "buy train tickets".

under the scenario of "buy train tickets" is illustrated in Fig. 7. We can find that it covers a lot of important events, such as "go to the railway station", "buy tickets" and "buy a hard seat". They are organized into a directed sequential relation graph, whose edges are labeled with transition probability.

7   Conclusion and Future Work

In this paper, we propose Event Evolutionary Graph (EEG), which reveals evolutionary patterns and development logics between events. We also propose a framework to construct EEG from large-scale unstructured corpus. Extensive experiments are conducted to solve the central problems of sequential relation and direction recognition. Experimental results show that the approaches we proposed are effective for both sequential relation and direction recognition.

To the best of our knowledge, EEG is first proposed by this paper. It is a knowledge base about event evolutionary patterns. Our final goal is to automatically mine this kind of knowledge from open domain large-scale unstructured documents. In future work, we will explore more robust event extraction technique and integrate causality recognition into our construction process. Applying EEG to real world applications is also an interesting research direction.

Acknowledgments. This work was supported by the National Key Basic Research Program of China via grant 2014CB340503 and the National Natural Science Foundation of China (NSFC) via grants 61472107 and 61632011. The authors would like to thank the anonymous reviewers for their insightful comments and suggestions.

References

1. Cassidy, T., McDowell, B., Chambers, N., Bethard, S.: An annotation framework for dense event ordering. Technical report, Carnegie-Mellon University, Pittsburgh, PA (2014)
2. Chambers, N., Cassidy, T., McDowell, B., Bethard, S.: Dense event ordering with a multi-pass architecture. TACL **2**, 273–284 (2014)

3. Chambers, N., Jurafsky, D.: Unsupervised learning of narrative event chains. In: ACL, vol. 94305, pp. 789–797 (2008)
4. Chambers, N., Wang, S., Jurafsky, D.: Classifying temporal relations between events. In: ACL, pp. 173–176. ACL (2007)
5. Che, W., Li, Z., Liu, T.: LTP: a Chinese language technology platform. In: ICCL, pp. 13–16. Association for Computational Linguistics (2010)
6. Do, Q.X., Lu, W., Roth, D.: Joint inference for event timeline construction. In: EMNLP, pp. 677–687. ACL (2012)
7. Granroth-Wilding, M., Clark, S.: What happens next? Event prediction using a compositional neural network model. In: AAAI (2016)
8. Laokulrat, N., Miwa, M., Tsuruoka, Y., Chikayama, T.: UTTime: temporal relation classification using deep syntactic features. In: SemEval-2013, pp. 88–92 (2013)
9. Mani, I., Verhagen, M., Wellner, B., Lee, C.M., Pustejovsky, J.: Machine learning of temporal relations. In: ICCL and ACL, pp. 753–760. ACL (2006)
10. Minksy, M.: A framework for representing knowledge. Psychol. Comput. Vis. **73**, 211–277 (1975)
11. Mirza, P., Tonelli, S.: CATENA: CAusal and TEmporal relation extraction from NAtural language texts. In: ICCL, pp. 64–75 (2016)
12. Pichotta, K., Mooney, R.J.: Statistical script learning with multi-argument events. In: EACL, vol. 14, pp. 220–229 (2014)
13. Pichotta, K., Mooney, R.J.: Statistical script learning with recurrent neural networks. In: EMNLP, p. 11 (2016)
14. Pichotta, K., Mooney, R.J.: Using sentence-level LSTM language models for script inference. In: ACL (2016)
15. Pustejovsky, J., Hanks, P., Sauri, R., See, A., Gaizauskas, R., Setzer, A., Radev, D., Sundheim, B., Day, D., Ferro, L., et al.: The timebank corpus. In: Corpus Linguistics, Lancaster, UK, vol. 2003, p. 40 (2003)
16. Radinsky, K., Davidovich, S., Markovitch, S.: Learning causality for news events prediction. In: WWW, pp. 909–918. ACM (2012)
17. Verhagen, M., Gaizauskas, R., Schilder, F., Hepple, M., Katz, G., Pustejovsky, J.: SemEval-2007 task 15: TempEval temporal relation identification. In: SemEval-2007, pp. 75–80. ACL (2007)
18. Verhagen, M., Sauri, R., Caselli, T., Pustejovsky, J.: SemEval-2010 task 13: TempEval-2. In: SemEval-2010, pp. 57–62. ACL (2010)
19. Zhao, S., Liu, T., Zhao, S., Chen, Y., Nie, J.Y.: Event causality extraction based on connectives analysis. Neurocomputing **173**, 1943–1950 (2016)

Prediction of Cascade Structure
and Outbreaks Recurrence in Microblogs

Zhenhua Huang[1], Zhenyu Wang[1(✉)], Yingbo Zhu[2], Chengqi Yi[3],
and Tingxuan Su[1]

[1] School of Software Engineering, South China University of Technology,
Guangzhou, China
wangzy@scut.edu.cn
[2] Tianyi Music Culture Technology Co. Ltd., Guangzhou, China
[3] State Information Center, Beijing, China

Abstract. Cascades are formed as messages diffuse among users. Diffusion process of viral cascades is analogous to spread of virus infection. However, until recently structure properties of viral cascades are quantified and characterized due to available diffusion datasets and increasing knowledge towards it. The virality of structure is a notion for characterizing structural diversity of cascades, but relationship between structural virality and shape of cascades is not highly revealed. We address this problem in a more intuitively way under the help of visualization methods and define a new problem to predict future structure of cascades from the perspective of time. Whether structure of cascades can be predicted, how early it is perceived and which features play key roles in the future propagation structure are discussed in details. Results obtained have a precision rate ranging from 86% in an hour to 97% in a day, indicating future cascade structure can be predicted when proper features are chosen. And the hierarchical tree features which related with structural virality are proven to play an important role in cascade size prediction. Viral cascades often lead to phenomenon of outbreaks recurrence, which has never been discovered formally before. Prediction in outbreak recurrence can also achieve non-trivial performance under the same prediction framework. Moreover, outbreak recurrence is shown to play significant impacts on structure virality of cascades. Our research is especially useful in understanding how viral cascades and events are formed, as well as exploring intrinsic factors in cascade prediction.

Keywords: Information diffusion tree · Cascade prediction · Outbreaks recurrence · Tree-based features · Structural virality

1 Introduction

Online social networks facilitate the spread of information. Users post and share messages about life, moods and thoughts in these social networks. Weibo or Microblog is a Twitter-like platform, in which fans and friends can share and forward microblogs, therefore many Cascades are formed. Cascade is also referred as information dissemination tree or diffusion tree and is common occurred in main social networks. While some scholars ever believed that cascades are inherently unpredictable. Recent works

© Springer Nature Singapore Pte Ltd. 2017
X. Cheng et al. (Eds.): SMP 2017, CCIS 774, pp. 53–64, 2017.
https://doi.org/10.1007/978-981-10-6805-8_5

has shown that some properties of cascades, such as growth, shape and size can be predicted by training models on a mixture features [6]. Most research about cascades mainly focus on predicting size [12] or popularity [3], and retweet behavior [13] of cascades while ignoring the structure of diffusion. In recent years, structure of cascades has piqued interest of scholars. According to Cheng et al. [6], predicting cascade structure is a much more challenge task compared with cascade growth prediction. To understand structural property, data that can reconstruct process of information must be collected, which is possible in some online social networks. The data of blogs is first available to discovery how the information travels among users [9]. Researches about other kinds of cascades have emerged in recent years such as Facebook [1, 6], LinkedIn [2], Twitter [14], Microblog [7, 8], and WeChat.

Then how to describe and measure structure diversity of cascades? The notion of structural virality in online diffusion trees has been introduced [10]. Structural virality or virality of structure, can be regarded as a special property of cascades. Goel et al. [10] distinguished diffusion mechanism in two kinds: broadcast (Fig. 1a) and viral spreading (Fig. 1b). Some cascades between two extremes (Fig. 1c). Viral spreading is similar to a virus propagates from person to person, while broadcast makes information only available to people directly connected with original person. The value of structural virality increases as more branches or deep propagation paths are formed as shown in Fig. 1b and c. Wiener Index is applied to calculate structural virality and alternative possible measurements are also allowed which will be discussed latter. Wiener Index is a good measure most of time. However, the approach still has some drawbacks: it ignores direction of propagation and does not show how viral the propagation structure is intuitively. For example in Fig. 2a, b and c, the Wiener Index is 2.02, 2.52 and 3.77, without telling to what extend virality is. Measurements with values between 0 and 1 are expected for a more intuitive understanding. When depth of a cascade is infinite, the value is 1.0, while the value is 0.0 when a cascade is purely broadcast, no matter how big the size is.

Some microblogs in "hot events" have deep propagation paths or many big branches. These messages spread like a virus from person to person and form many ripple-like branches and deep propagation paths (see Fig. 2c). This kind of cascade is referred to as a viral cascade with viral structure in this paper. Another kind of cascade is the broadcast cascade or non-viral cascade with mostly or purely star-like structure, in which most paths are connected directly with the root node (Fig. 2a). This kind of structure is called a broadcast structure. Then the question is how to distinguish cascades with viral structures from those with broadcast structures? For example, is a cascade with Wiener Index value of 2.5 a viral cascade? And can we predict future cascade structure in early time? Which factors are related with viral cascades? These are questions remain open but is meaningful. In social network viral marketing, we tend to select people with a high capacity for creating messages that can go viral. The viral structure gives brands or products a wider range of exposure to potential users and improve their publicity. When structural virality reaches a threshold, the structure of cascades has apparent viral spreading and these cascades become intuitively viral, which makes those cascades viral cascades.

Observing the viral cascades, we can see that in addition to a big ripple-like outbreak surrounding the root node, many new "ripples" occurred in many forwarding or retweet nodes. The new ripple-like structure is named ***Outbreak Recurrence*** in this paper. It means another outbreak comes into being after some users forward a message. Outbreak Recur, somewhat similar to Cascades Recur in Cheng et al. [7]. However, we focus on analyzing the forwarding nodes that will form a new outbreak within a cascade, while Cheng [7] concerned when new cascades about the same topic will be formed. Outbreaks recurrence promote attendance of audiences and influence cascade structure. Can these outbreaks recurrence be predicted, and what kind of impacts they have on structure of cascades? The answers are very worthy of discovery. For example, we can prevent further spread of information by suppressing the occurrence of secondary outbreaks. In viral marketing, and we can take advantage of this phenomenon, making brands and products for known to more social groups, with the under help of some users that have a relatively small number of fans.

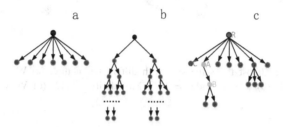

Fig. 1. (a) Broadcast. (b) Viral Spreading. (c) A cascade between two extremes.

2 Related Works

Some literatures have discussed virality of cascades or memes based on popularity or size, without discovering structure of cascades that is significant in propagation process [6, 14]. Most paper using audience size or retweet amount to define virality of cascades, while diffusion structure does not get enough concern. Cheng et al. [6] have proven that cascades can be predicted to some extent in Facebook. Their work mainly focused on predicting whether the final size of cascades will arrive $f(k)$ in the future when k nodes are observed. The structure of cascades is discussed briefly without discussing how a viral cascade is formed under description of structural virality in details. Moreover, the method does not tell how early time (e.g. in an hour) the prediction is made. Thus a new prediction task is defined in the perspective of time.

Most works about predicting outbreak in cascades are also based on size of them [8]. Cui [8] regarded outbreak prediction as binary classification problem and predict outbreaks based on the cascading behaviors of sensors. Yi [5] took structure of cascades into account when defining an outbreak. And some features such as entropy of eigenvector centrality and PapgeRank centrality are found to be key predictors in prediction without explain the how these features works. Although many works have discussed the outbreak problem, the outbreaks recurrence has not been explored with before.

Goel et al. [10] put forward the concept of the virality of diffusion tree structure and have applied Wiener Index to calculate the structural virality of cascades. The work has drawn attraction of many scholars. However, measurements to calculate virality of structure are not discussed in detail in their works. We point out the shortage of Wiener Index and improve it with normalized method. After comparing it with Modularity and under the help of a visualization tool, the relationship between shape and structural virality of cascade are intuitively revealed.

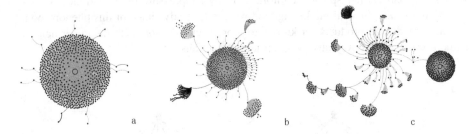

Fig. 2. Cascades of almost the same size with different structure. (a) Wiener Index = 2.02, Modularity = 0.024. (b) Wiener Index = 2.52, Modularity = 0.42. (c) Wiener Index = 3.77, Modularity = 0.74.

3 Preliminaries

3.1 Data Collection and Description

We randomly select 134279 microblogs between 2015.10 and 2016.12 and construct diffusion cascades of them. The propagation process is completely restored by tree-based structure with direction in each cascade. Retweet time, comment and user profile in information diffusion are collected. Figure 6a is size distribution of cascades under log-log axes. Figure 6b is distribution of structural virality with y-axe is value under logged. The size of distribution follows power-law distribution. Surprisingly, structural virality also almost follow power-law distribution, but there is an inexplicit turn at almost 2.5. After Wiener Index arrive 2.5, the downward trend becomes slower and proportion of cascades is relatively small (Fig. 3).

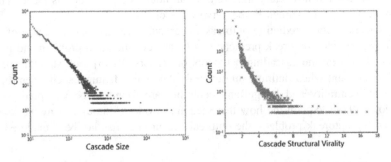

Fig. 3. (a) Distribution of cascade size. (b) Distribution of structural virality.

3.2   Structure Virality of Cascades

Under the visualization proper layout, structure of cascades is easily revealed. We also calculate Modularity under the community detection algorithm [4]. The approach can also be used to assess structural virality of cascades. Although it is not easy to prove arithmetically why the modularity and virality structural are closely related. However, in the actual experiments, after detecting communities, the modularity of entire graph can be a good description of structural virality. The correlation between Wiener Index and Modularity is 0.9. The Q-Q Figure of Wiener Index and Modularity show the same increasing tendency. An advantage of Modularity is the values fall into 0.0 and 1.0, which can be regard as degree of structural virality intuitively. If we normalized the Wiener Index directly, say, $(w - min)/(max - w)$, will cause misunderstanding. Minority large value will make Wiener Index very small. For example when a cascade with Wiener Index value 3.0, the largest value is 18.0, so the normalized value of the cascade is $(3.0 - 2.0)/(18.0 - 3.0) = 0.06$. While in fact a cascade with Wiener value 3.0 (as shown in Fig. 4c) is quite rare and virality of structure is quite large, a proportion of around 5%. The normalized value should be reconsidered. Thus, we compare it with Modularity and calculate functions in Q-Q Figure, and summarize a function. The proportion over 4.0 is 0.017, means it is a very large value. By this method, Wiener value of cascades in Fig. 4 are mapped into 0.15, 0.34 and 0.6, which are more reasonable. For cascades with Wiener value over 3.67 or more, they can be interpreted as extremely viral cascades.

$$f(w) = \begin{cases} 0.6 * w - 1.2, & w \leq 3.67 \\ 1.0, & w > 3.67 \end{cases}$$

From large observation, cascades with Modularity or Wiener Index large than a threshold, deep paths and outbreaks recurrence will appear. The threshold is around 0.25 for Modularity and 2.5 for Wiener Index. And larger value Modularity or Wiener Index usually means more deep path and outbreaks recurrence.

It should be noted that the threshold is not fixed and more like a transition zone, but for categories and prediction, we have to make such division. Based on the observation, we can category cascade structure or shape into several categories and predict future structure in early time, like viral or un-viral structure, or virality degrees, which indicate popularity of information. Firstly, we try to predict whether a message will be a Viral Cascade or not and explore how early and precise the prediction can be made.

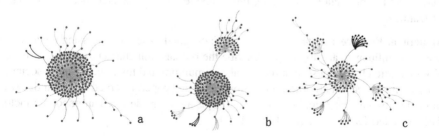

Fig. 4. (a) Wiener Index = 2.25, Modularity = 0.18. (b) Wiener Index = 2.57, Modularity = 0.45. (c) Wiener Index = 3.0, Modularity = 0.66.

4 Predicting Cascade Structure

4.1 Problem Statement and Definition

We try to predict future cascade structure dynamically. $T = (V, E)$ represents a diffusion tree. V represents nodes releasing or retweeting messages on the tree, E are edges indicating directions information flow. Diffusion tree grows with time. In time t, the tree is noted as T_t, and in the last time, the tree is noted as T_{Final}. If $v(T_{Final}) \geq \delta$, it is regarded as a viral cascade with viral structure, where v is structure virality function and δ is the threshold. Otherwise the diffusion tree is labelled as non-viral cascade with broadcast structure. So the problem is:

$$f : (T^t, t = 1 \ldots N) \to v(T_{Final})$$

The time interval of predicting points is set to 3600 s (1 h). Since 70% retweet behavior happens in a day, we only consider prediction within a day. Cascades with over 30 nodes are selected, δ is set to 0.25, and 35 features are extracted to trained classification models and make prediction.

4.2 Feature Description

We investigate the factors that contribute to future structure of cascades. We group these features into four classes: Content features, user features, temporal features and structural features.

Content Features: The content itself is first natural factor comes into mind. Messages with innovative ideas and original microblogs are easier to spread deep and become viral. Mentioned in the literature [11], emotion impacts propagation. SO emoji is extracted as a symbol of emotion. The length of content may also have an impact on propagation. Special symbols such as "@" and "\#" are also considered.

User Features: The influence of root user and their friends or fans will play an important role in message diffusion. And several other root user features are selected, such as verification, VIP member and so on.

Temporal Features: Temporal characteristics are found to have a significant impact on predicting future size [6]. We also want to test whether they are good features for predicting future structure. The time intervals of retweeting reflect the spread speed that may affect the propagation structure, thus some temporal characteristics are calculated as features.

Structural Features: Index-10 and h-index, originally used for assessing the author's academic influence, are applied to describe the propagation characteristics of cascades. The literature [14] suggests that the initial communities and final virality are associated. Community detection algorithm [4] is applied in this paper, reserved communities with more than two nodes. Initial structure type [6] is also considered as an important factor. The extracted features are listed in Table 1.

Table 1. Features used in prediction.

Feature description			
User Features		F18	Num. of intervals <10 min
F1	Fans num. of root user	*Structural Features*	
F2	Whether root user is verified	F19	h-index of resharers' degree
F3	Member type of root user	F20	Index-10 of resharers' degree
F4	Member rank of root user	F21	Avg. top five resharers' degree
F5	Varied type of user	F22	Avg. top ten resharers' degree
Content Features		F23	Avg. of resharers' degree
F6	Length of content	F24	Entropy of resharers' degree
F7	Whether an emoji is contained	F25	Community num
F8	Num. of emoji	F26	Rate of 2rd-level mediate nodes
F9	Whether a url is contained	F27	Rate of nodes with 2rd level
F10, 11	Whether "@/#" is contained	F28	Ratio of mediate nodes
F12, 13	Whether "!/?" is contained	F29	Num of mediate nodes
Temporal Features		F30	Avg of top-5 depths
F14	Min of time interval	F31, 32	Max/avg. of depth
F15	Avg reshare time	F33	Diameter
F16	Avg. time between reshares	F34	Initial type
F17	Entropy of time interval	F35	Reshare number

4.3 Results and Analysis

A variety of learning methods including the random forests, logistic regression, SMO, decision trees, and AdaboostM1 are employed. After 10-fold cross-training, the results of random forests are as Fig. 5a. It can be seen that in the first hour, the prediction accuracy rate and recall rate reached 88% under the random forests. After that, accuracy and recall rapidly increased at first, then held steady. In the second hour, the accuracy rate and recall rate both exceeded 90%. The precision and recall, F1 values ranged from 88% to 97%. ROC curve in the first hour went up to 93% and in the third hour to 96%, and rose steadily to 98.6% after that. Similar results can be also obtained using different classifiers. The F-1 value of different classifiers are shown in Fig. 5b. It can be seen within the first hour, F1 value of random forests algorithm is slightly better than other algorithms. After that, the results are almost the same among several classifiers and the curve show the same upward trend, indicating that features extracted for classification have a high stability.

To discovery key features we applied CfsSubset (CFS), a feature-selection method, and the BestFirst evaluator to the choose subsets of features. After using CFS to select the 35 features, there are 15 features occur at least once in prediction time points. The important features chosen at each moment are different. Nine features with no less than nine times, which are shown in Fig. 5d. The proportion of intermediate nodes with depth one (or level two) promotes the spreading of the propagation more deeply. Whether it has a "#" symbol has a significant impact because the "#" symbol implies the microblog is related with a topic and can be searched by "Weibo Search", and is

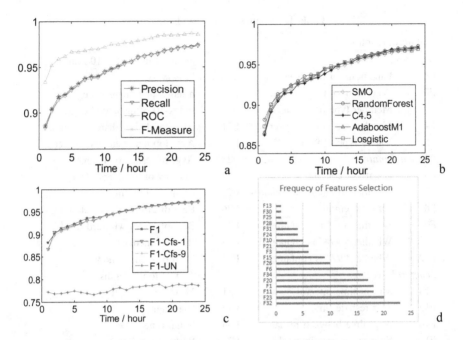

Fig. 5. (a) Results of RandomForest. (b) F1 value of different classifiers. (c) Results when only part of features are used. F1-Cfs-1 and F1-Cfs-9 represent results using features chosen at least once and nice times by CFS respectively. (d) Frequency of features chosen by CFS.

more likely to get attention and be retweeted. The average forwarding time reflects the forwarding speed. The faster speed may make it easier to form a deeper propagation path. The initial structure, average depth, average degree, index-5 of degree have great impacts on future virality of structure. Text that is too long is a barrier to depth of spreading. Viral cascades are relatively more concentrated in the short text area, while non-viral cascades are significantly higher than viral cascades in long text area. When the nine features with the number of occurrences greater than nine were selected and 15 features occur at least one time were used to make comparison. As can be seen from Fig. 5c, the F-Measure values are just slightly affected after feature selection. It shows that the nine features are key factors in predicting future structure.

It should be noted that the threshold is not limited to the values we mentioned, they should be set properly according to experience and observation of data.

Suppose the cascades are not with hierarchical tree-based structure, which is the kind of data happens mostly in previous research. After removing the features of hierarchical tree-based features, the F1 value sharply decreased, as shown in Fig. 5c.

Now we turn to a problem about observing an early set of nodes to predict final structure [6]. And the method benefits cascades with smaller size. Initial graph of k shares were extracted. The same parameters with an early work [6], k from 5 to 100 were set. The same threshold δ was used. The results are as following. Results varied from 78.4% when k is 5 to 92% when k is 100. The accuracy has achieved non-trivial performance compared with results in work [6], which may due to full using of hierarchical tree-based features. When minimum cascades size R is set to 100, k varying from 5 to 100, accuracy increased almost linearly from 77.8% to 92.9%. It also shows that the features extracted can be a good predictor of future structure. The cascade structure is easier to predict when more nodes are extracted.

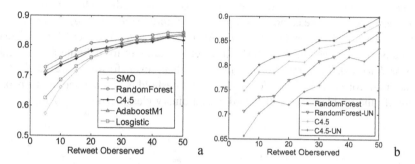

Fig. 6. (a) Predicting cascade structural virality when R is set to 100. (b) Predicting whether a message will be popular, RandomForest-UN and C4.5-UN represents results without structural virality related features, the μ is set to 70%.

Then we discovery whether hierarchical tree-based structure works in prediction future size or popularity of cascade. It is hard for us to make precision prediction due to size or popularity of cascade follow power-law distribution and complexity mechanism works in cascade growth. Most literature transform the problem into classification problem. One problem is called "virality prediction" related with size prediction. We choose the way in paper [14] and cascade of larger than μ percentage is viral in retweet amount. Also cascades with more than 50 nodes are chosen. After removing features of 'virality'. The accuracy has a different degree of decline in classifiers. It indicates that the initial propagation structure has certain influence on the future size. This may because the information with different topic has aroused interests of various users that organized in diverse structure. When the cascade structure goes viral, the exposure of the message is growing large, and it is easier for messages to form a larger cascade. However, the feature of size or retweet amount does not lead to a viral cascade, which is more related with content and friendship organization.

To sum up, these experiments prove that future structure of cascades can be predicted at early time, and the method to diving different types of structure is effective.

5 Predicting Outbreaks Recurrence

5.1 Problem Description

Outbreak Recurrence refers to an outbreak occurring around forwarding nodes but not root node, shown in red circle in Fig. 7c. Different from cascade recur [7], outbreaks recur happens within the same cascade. The outbreak recurrence can be also referred to as Secondary Outbreak. Past research concerned about when new cascades are formed about the same topic again. Here, we are mainly focus on predicting which one will lead to new outbreaks beyond the root node. If a node lead to outbreaks recurrence, it is referred as *secondary outbreak nodes*, such as nodes A, B, C and D in Fig. 7c. Given a threshold θ, the task is forecasting whether a new outbreak will come into being after a user retweeting a microblog in cascades.

Suppose at time t, user u forwards the message, predicting whether user u will become secondary outbreak node or not in the future, noted as $S(u)$. Suppose forwarding nodes directly connected with u is $R(u)$. If $R(u) \geq \theta$, u is a secondary outbreak node and $S(u) = 1$, wherein θ is threshold of retweet amount for defining an outbreak, otherwise $S(u) = 0$. Forward Weibo with more than seven forwards are labelled as Popular or hot forwards in Sina Weibo, the same threshold is also applied in this paper. To simplify the problem, we still regard it as a binary classification problem.

Fig. 7. (a) A sketch of outbreak recurrence. (b) Example of many outbreaks recur in a cascade. (c) Cascade failures after removing secondary outbreak points. (Color figure online)

5.2 Experiments and Results

Intuitively, the influence of root users and topic of content have impacts on the outbreak recurrence. In addition, the temporal factor is very important. Users with a small number of fans can also lead a very large outbreak, even beyond person with much more fans. In some case, number of retweets even beyond number of fans. Early reviewing and forwarding may benefit new outbreaks. Such as a user forwarded Mr. Hawking's weibo that blessing students who are going to attend the Chinese College Entrance Test. The number of retweets from the user is 4798, far more than his fans number which is 2746. He had forwarded the original weibo within less than two minutes after it was posted in the morning, and a brief comment was also added to attract the attention of others. Features selected are as following:

User features: gender, fans number, member rank, member type, whether verified, verified type, microblog number, length of self-introduction.

Temporal features: time interval with parent node, time interval with root node.

Content features: same with those in predicting structure.

Obviously, the proportion of the popular forwarding microblogs is far less than the non-popular. To solve this imbalance classification problem, we down-sample negative samples randomly to make sure classification models could be trained properly. The ratio of the popular/non-popular was maintained around 1:2. Forecasting results are in the Table 2. Under other sampling rates such as 1:1 or 1:3, similar results were also obtained to show its stability. The performance of different classification algorithms vary a little, with the highest accuracy rate of 91.5%. The forecast is finished immediately after a user forwards a microblog. It shows that the outbreaks recurrence can be accurately predicted early.

Table 2. Results of predicting outbreak recurrence.

	AdabsoostM1	C4.5	Logistic	RandomForest
Precision	0.898	0.887	0.843	0.903
Recall	0.896	0.886	0.836	0.902
F-Measure	0.895f	0.886	0.836	0.902
ROC	0.94	0.9	0.915	0.959

5.3 Influence to Structural Virality

The distribution of retweets from forwarding nodes directly follows a long-tail distribution (root node removed). Only a small portion of retweets can form secondary outbreaks. However, the small part plays a significant role on the overall structural virality. Suppose secondary outbreak nodes in Fig. 1b and c are eliminated, the structural virality of diffusion tree is sharply reduced. After the removal of the nodes, structural virality of the cascade in Fig. 2b falls from 0.26 to 0.09, and in Fig. 2c from 0.88 to 0.12, which indicates that secondary nodes are essential for structural virality in most cases. When many "slim" and deep propagation paths occur in a cascade, virality can still maintain a relatively high value after removing secondary outbreak nodes. However, this case does not appear often.

6 Conclusion

Structural virality as well as size of cascades shows how popular a message is. Structure of cascades have recorded how messages propagate among users and can be predicted at an early time in a relatively high accuracy rate when proper feature are selected. We propose a dynamic prediction framework for cascade structure from perspective of time and try to find out key reasonable factors in virality prediction. The structural virality is shown to play important role in improving performance in cascade popularity or size prediction. Outbreak recurrence often happens in viral cascades and is a main reason to facilitate messages diffusion like virus. However, an outbreak usually caused by an influence person or a message relates with hot event or contains valuable information. And hierarchical tree-based structure is especially important in cascade structure prediction. Overall, our research is meaningful in discovering the structure of information propagation and can be applied in viral market, event prediction and propagation control.

Acknowledgments. We are thankful for the support of Guangdong Province Major Projects (No. 2015B010131003), Guangzhou Major Projects (No. 201604010017), MOE Humanities and Social Sciences Research Planning Fund (No. 15YJA710035).

References

1. Dow, P.A., Adamic, L.A., Friggeri, A.: The anatomy of large facebook cascades. In: Proceedings of ICWSM (2013)
2. Anderson, A., Huttenlocher, D., Kleinberg, J., Leskovec, J., Tiwari, M.: Global diffusion via cascading invitations: Structure, growth, and homophily. In: Proceedings of World Wide Web, pp. 66–76, May 2015
3. Bao, P., Shen, H.-W., Jin, X., Cheng, X.-Q.: Modeling and predicting popularity dynamics of microblogs using self-excited hawkes processes. In: Proceedings of the 24th International Conference on World Wide Web, pp. 9–10 (2015)
4. Blondel, V.D., Guillaume, J.-L., Lambiotte, R., Lefebvre, R.: Fast unfolding of communities in large networks. J. Stat. Mech: Theory Exp. **2008**(10), P10008 (2008)
5. Yi, C., Bao, Y., Xue, Y.: Mining the key predictors for event outbreaks in social networks. Phys. A: Stat. Mech. Appl. **447**, 247–260 (2016)
6. Cheng, J., Adamic, L.A., Dow, P.A., Kleinberg, J., Leskovec, J.: Can cascades be predicted? In: Proceedings of WWW, pp. 925–936. ACM (2014)
7. Cheng, J., Adamic, L.A., Kleinberg, J., Leskovec, J.: Do cascades recur? In: Proceedings of WWW, pp. 671–681, April 2015
8. Cui, P., Jin, S., Yu, L., Wang, F., Zhu, W., Yang, S.: Cascading outbreak prediction in networks: a data-driven approach. In: Proceedings of SIGKDD, pp. 901–909. ACM (2013)
9. Adar, E., Adamic, L.A.: Tracking information epidemics in blogspace. In: Proceedings of IEEE/WIC/ACM International Conference on Web Intelligence, pp. 207–214 (2005)
10. Goel, S., Anderson, A., Hofman, J., Watts, D.J.: The structural virality of online diffusion. Manage. Sci. **62**(1), 180–196 (2015)
11. Jenders, M., Kasneci, G., Naumann, F.: Analyzing and predicting viral tweets. In: Proceedings of WWW, pp. 657–664 (2013)
12. Kupavskii, A., Ostroumova, L., Umnov, A., Usachev, S., Serdyukov, P., Gusev, G., Kustarev, A.: Prediction of retweet cascade size over time. In: Proceedings of CIKM (2012)
13. Zhang, Q., Gong, Y., Guo, Y., Huang, X.: Retweet behavior prediction using hierarchical dirichlet process. In: Proceedings of AAAI. AAAI Press (2015)
14. Weng, L., Menczer, F., Ahn, Y.-Y.: Virality prediction and community structure in social networks. Scientific report (2013)

Exploring Effective Methods for On-line Societal Risk Classification and Feature Mining

Nuo Xu[1,2] and Xijin Tang[1,2(✉)]

[1] Academy of Mathematics and Systems Science, Chinese Academy of Sciences, Beijing 100190, China
xunuo1991@amss.ac.cn, xjtang@iss.ac.cn
[2] University of Chinese Academy of Sciences, Beijing 100049, China

Abstract. China has to face lots of societal conflicts during periods of social and economic transformation. It is crucial to exactly detect societal risk for the mission to a harmonious society. On-line community concerns have been mapped into respective societal risks and support vector machine model has been used for risk multi-classification on Baidu hot news search words (HNSW). Different from traditional text classification, societal risk classification is a more complicated issue which relates to socio-psychology. Conditional random fields (CRFs) model is applied to access to societal risk perception more accurately. We regard the risks of all the terms throughout a hot search word as a sequential flow of risks. The experimental results show that CRFs model has superior performance with capturing the contextual constraints on HNSW. Besides, state features can be extracted based on CRFs model to study distributions of terms in each risk category. The distribution rules of geographical terms are found and summarized.

Keywords: Societal risk classification · HNSW · Paragraph Vector · Conditional random fields · Feature mining

1 Introduction

In the Web 2.0 era, Internet users are both content viewers and content producers. Search engines have been the most common tool to access to information. The contents of high searching volume of search engine reflect the netizens' attention. Baidu is now the biggest Chinese search engine. Baidu hot news search words (HNSW) are based on real-time search behaviors of hundreds of millions of Internet users and released at Baidu News Portal, reflecting the Chinese current concerns and ongoing societal topics. In such way, we utilize HNSW as a perspective to analyzing societal risk which refers to the risk problems raising the concern of the whole society. Traditional research on societal risk was studied from the angle of cognitive psychology based on the psychometric paradigm and questionnaires [1], which is generally expensive and time-consuming to be conducted. Zheng et al. constructed a framework of societal risk indicators including 7 categories which are national security, economy/finance, public morals, daily life,

© Springer Nature Singapore Pte Ltd. 2017
X. Cheng et al. (Eds.): SMP 2017, CCIS 774, pp. 65–76, 2017.
https://doi.org/10.1007/978-981-10-6805-8_6

social stability, government management, and resources/environment [2]. Tang tried to map HNSW into either risk-free event or one event with risk label from the 7 risk categories and aggregate all risky events over the whole concerns as the on-line societal risk perception [3]. By labeling those HNSW with relevant societal risk categories, we may get a general perception of online societal risks. An automated way to carry out societal risk classification by machine learning is necessary. Moreover, the results directly affect the accuracy of evaluating the level of societal risk. It is of great significance to monitor societal risk timely and efficiently.

This paper focuses on two points of the societal risk classification problem. Firstly, societal risk classification is a more complicated issue which relates to socio-psychology compared with traditional text classification. Different individuals may have different subjective perception of risks. Meanwhile, more challenges are confronted including the emerging words with risks, the transfer of the word's risk and widely usage of argots and proverbs [3]. Besides, the data set of societal risks is seriously unbalanced. More than 50% of the hot words are labeled as "risk-free". Therefore, improve the performance of automatic risk identification by traditional machine learning methods is with a big challenge. Secondly, HNSW are short texts with no punctuations and spaces, which makes it more difficult to deal with. Relevant news texts are crawled and extracted simultaneously to provide corpus for machine learning. Experiments were conducted which carried out societal risk multiple classifications on news contents, while the accuracy was barely needed to be improved [4]. As a result of these two points, conditional random fields (CRFs) model is firstly applied to societal risk classification directly dealing with short texts without news texts compared with previous studies. We regard the risk classification as a sequence labeling problem and use CRFs model to capture the relations among terms in hot words. In this paper, support vector machine (SVM) based on Paragraph Vector is also introduced in order to get better results of risk classification. SVM based on bag-of-words (BOW) used in previous study is chosen as baseline [4].

This paper is organized as follows: Sect. 2 introduces different models for societal risk multi-classification of Baidu hot news search words. Section 3 presents the risk multi-classification experiments and carries out the results analysis. Section 4 illustrates feature terms analysis in each risk category according to state features of CRFs model. Conclusions and future work are given in Sect. 5.

2   Societal Risk Classification Methods

Baidu hot news search words are provided in forms of 10 to 20 hot query news words updated every 5 min automatically which refer to bring the most search traffic. Each of HNSW corresponds to 1–20 news whose URLs are at the first page of hot words search results, as shown in Fig. 1. "HotWord Vision 2.0" was developed to hourly download HNSW and their corresponding news texts since November of 2011. HNSW serve as an instantaneous corpus to maintain a view of netizens' empathic feedback for social hotspots, etc. Therefore, we utilize HNSW as a perspective to analyze societal risk. The task for societal risk classification is conducted from two perspectives. On one hand, we map these Baidu hot news into eight categories. One hot search word belonging to one risk category is determined by the votes of risk categories for hot news. On the other hand, we directly map one hot search word into one risk category. Two different approaches to the societal risk classification will be discussed in the following subsections.

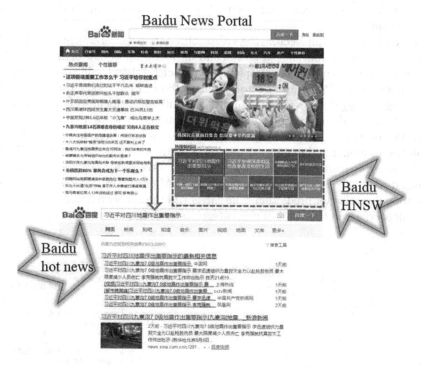

Fig. 1. HNSW released at Baidu News Portal and the corresponding news texts

2.1   Societal Risk Classification Based on Baidu Hot News

We try to investigate multi-classification problem of societal risk through mapping Baidu hot news texts into eight categories. Generally the most common fixed-length vector representation for texts is the BOW. Hu and Tang carried out multiple classifications on hot news utilizing SVM algorithm based on BOW [4]. What kinds of risk categories the HNSW belong to are determined by the largest number of risk categories of Baidu hot news. However, with the volume of news accumulated, BOW method is prone to dimension disaster. Besides, BOW method does not take semantic of the sentence and word order into consideration. Neutral networks approaches have overcome these problems by implementing unsupervised word embedding for feature representations [5]. Paragraph Vector model was proposed as an unsupervised framework that learned continuous distributed vector

representations for pieces of texts [6]. The texts can be of variable-length, ranging from sentences to documents. Chen and Tang had applied Paragraph Vector
model to societal risk classification on the corpus of posts crawled from Tianya
Forum and the performance was better than basic machine learning methods [7,8].
Paragraph Vector model has demonstrated obvious superiority in the issue of text
classification with its merits in capturing the semantics of paragraphs. Therefore,
we adopt the learning algorithm Paragraph Vector for societal multi-classification
on news contents.

As to the Paragraph Vector, the vector of a paragraph is concatenated with
several word vectors from the paragraph and the following word is predicted in the
given context [6]. The process of implementing societal risk classification based on
Baidu hot news by Paragraph Vector is illustrated in Fig. 2. Take the hot search
word "安徽多县遭遇虫灾" (Many counties had suffered pests in Anhui) for example. First, the Baidu hot news ID and the corresponding news text are fed to Paragraph Vector model. After the vector representations have been learned by Paragraph Vector model, n-dimensional vectors of Baidu hot news are acquired. Next,
the risk categories are concatenated with the vectors of Baidu hot news which are
extended to $n + 1$ dimensions. Finally, train SVM classifiers based on $(n + 1)$-
dimensional vectors for prediction. The categories the hot search words belonging
to are dependent on the votes of risk categories of Baidu hot news.

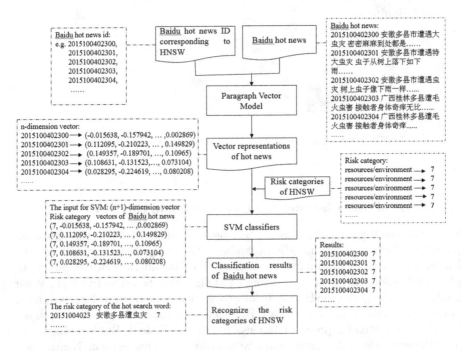

Fig. 2. Process for risk classification of HNSW by Paragraph Vector

2.2   Societal Risk Classification Based on Hot Words

Most of researchers focus on how to extract useful textual features for text classification using traditional machine learning algorithm as well as deep learning. Since HNSW consist of fewer words, traditional classification methods face the challenges of feature sparseness. Thus, CRFs model is adopted to deal with this problem.

CRFs model is an undirected graphical model used to calculate the conditional probability of a set of labels given a set of input variables [9], which has better performance in most natural language processing (NLP) applications, such as sequence labeling, part-of-speech tagging, syntactic parsing, and so on. Both maximum entropy and hidden Markov model, which are regarded as the theoretical foundations of CRFs model, have been successfully applied to text classification and achieved good performance [10, 11]. CRFs model was previously used for short text classification and sentiment classification. The results proved that CRFs outperformed the traditional methods like SVM and MaxEnt [12–15]. In this paper, we utilize CRFs model for societal risk classification. For capturing the contextual influence, we treat original societal risk classification as a sequence labeling problem.

Linear chain conditional random field (LCCRF) is the most simply and commonly used form of CRFs model. We choose LCCRF to carry out societal risk classification. HNSW and their corresponding risk categories are respectively represented as the observed sequences and state sequences. In view of the risk classification process, $X = (x_1, x_2, \cdots, x_n)$ is a set of input random variables. $Y = (y_1, y_2, \cdots, y_n)$ is a set of random labels. We have a collection of hot search words sequences D where each hot search word sequence $d \in D$ is a sequence of tuples $[(x_1, y_1), (x_2, y_2), \cdots, (x_T, y_T)]$. Each tuple$(x_T, y_T)$ is respectively presented as segmented word x_T and risk label y_T. The sequence length T varies for each sequence. For example, the hot search word "河北连日强降雨" (There are heavy rainfalls in Hebei for days.) can be expressed as the observation sequence and state sequence. The observation sequence is $X = $ (河北,连日,强,降雨). The state sequence is $Y = $ (resources/environment, resources/environment, resources/environment, resources/environment). Hot search words and their corresponding risk categories can be turned into the risk tagging sequence. In a given observation sequence X, the probability distribution of generating the output sequence can be described as follows:

$$P_w(Y|X) = \frac{exp(w \cdot F(Y,X))}{\sum_Y exp(w \cdot F(Y,X))}. \tag{1}$$

Here, $F(Y,X) = (f_1(Y,X), f_2(Y,X), \cdots, f_K(Y,X))^T$ is the feature vector, where $f_i(Y,X)$ is a binary indicator feature function with $f_i(Y,X) = 1$ when both the feature and label are presented in a hot word and 0 otherwise; w is a learned weight for each feature function as well as the main parameter to be optimized. Figure 3 shows the framework for CRFs applied on risk multi-classification.

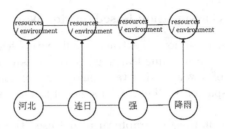

Fig. 3. Label sequences in CRFs model training

It is necessary to define the template for feature exaction to train LCCRFs model. We use the example above to illustrate the process of feature generation. Assume the current token is "强", the feature templates and corresponding features are defined as Table 1.

Table 1. Feature template and corresponding features

Template	Implication	Feature
U00: %x[-2,0]	the second term before current token	河北
U01: %x[-1,0]	the previous term	连日
U02: %x[0,0]	current token	强
U03: %x[1,0]	the previous term	降雨
U04: %x[2,0]	the previous term	/

We define two variables, namely L and N. L represents the number of categories including 7 risk categories and risk-free category, N represents the number of features generated by the template. There are $L * N$ feature functions, that is to say, there are 80 feature functions in the above example. The training of CRFs is based on maximum likelihood principle. The log likelihood function is

$$L(w) = \sum_{Y,X}[\widetilde{P}(Y,X)w \cdot F(Y,X) - \widetilde{P}(Y,X)log \sum_{Y} exp(w \cdot F(Y,X))]. \quad (2)$$

Limited-memory BFGS (L-BFGS) algorithm is used to estimate this nonlinear optimization parameters.

3   Societal Risk Classification of HNSW

3.1   Data Description and Data Processing

In this paper, we perform risk multi-classification respectively on Baidu HNSW collected from November 1, 2011 to December 31, 2016 and corresponding news corpus collected from April 1, 2013 to December 31, 2016 based on "HotWord

Table 2. Descriptive statistics of hot words and hot news

Risk category	Train dataset		Test dataset	
	#hot words	#hot news	#hot words	#hot news
National security	2258	18472	178	1568
Economy/finance	1222	8403	119	1205
Public morals	3368	25004	399	3440
Daily life	4920	32037	656	5870
Social stability	5342	58890	364	3087
Government management	5552	52748	339	3428
Resources/environment	1716	14653	358	3156
Risk-free	24587	274669	4855	42978

Vision 2.0". Table 2 shows the quantity distribution of each risk category respectively on hot search words and hot news.

We choose Ansj[1] as the segmentation tool to deal with hot words and corresponding news texts. We then remove stopwords and only reserve verbs, nouns, adjectives and adverbs. In the experiment, CRFs model, SVM based on Paragraph Vector and SVM based on BOW are compared as follows:

(1) **SVM-BOW:** We use SVM model based on vector representation BOW for text. The feature extraction method Chi-square is chosen, and the top 20% features are selected. We use LinearSVC in sklearn package for SVM model, whose parameters are set as default values. We then choose the news texts from April 1, 2013 to December 31, 2015 as the training set while all the news in 2016 as the testing set. The votes of risk categories of hot news identify which categories the hot search words belong to.

(2) **SVM-PV:** We perform news texts from April 1, 2013 to December 31, 2016 to learn vector representations. We also choose LinearSVC in sklearn package for SVM, whose parameters are set as default values. Once the vector representations have been learned, we feed them to the SVM to predict the risk label. The process is as shown in Fig. 2. The parameters are set as follows: the learned vector representations are set 100 dimensions, the optimal window size is 8, CBOW is chosen for vector representations. The votes of risk categories of hot news identify which categories the hot search words belong to.

(3) **CRFs:** Each hot word is represented as a label sequence. The template defined in Sect. 2.2 is chosen for feature extraction. We then choose the hot words from November 1, 2011 to December 31, 2015 as training set, while all the hot words in 2016 as testing set. L-BFGS algorithm is introduced to optimize the objective function. We use *sklearn_crfsuite*[2] package for CRFs model. We set the iteration number to 100 in the training process of the method based on CRFs.

---

[1] http://www.demo.ansj.com/.

[2] https://pypi.python.org/pypi/sklearn-crfsuite/0.3.3/.

3.2 Results

We utilize accuracy, macro-average and micro-average as the evaluation metrics to evaluate the overall performance of each model. Precision, recall and F-measure are used for performance measurement of each societal risk category. The accuracy of CRFs model, SVM based on Paragraph Vector and SVM based on BOW are 0.78, 0.68 and 0.74 respectively. CRFs have achieved the best performance. Table 3 shows the results of three models.

Table 3. Comparison results with different models

Risk category	Precision			Recall			F-measure		
	BOW	PV	CRFs	BOW	PV	CRFs	BOW	PV	CRFs
National security	0.56	0.00	0.66	0.29	0.00	0.43	0.38	0.00	0.52
Economy/ finance	0.58	0.00	0.68	0.16	0.00	0.34	0.25	0.00	0.46
Public morals	0.43	0.00	0.54	0.14	0.00	0.23	0.21	0.00	0.32
Daily life	0.63	0.89	0.70	0.25	0.01	0.51	0.36	0.03	0.59
Social stability	0.47	0.40	0.56	0.49	0.37	0.51	0.48	0.39	0.54
Government management	0.49	0.52	0.58	0.35	0.20	0.56	0.41	0.29	0.57
Resources/ environment	0.82	0.87	0.93	0.54	0.12	0.56	0.65	0.22	0.70
Risk-free	0.78	0.70	0.81	0.94	0.98	0.93	0.85	0.82	0.87
Macro-average	0.60	0.42	0.68	0.40	0.21	0.51	0.47	0.28	0.58
Micro-average	0.74	0.68	0.78	0.74	0.68	0.78	0.74	0.68	0.78

As is shown, SVM-PV has got rather poor performance on both precision and recall especially for the risk category "national security", "economy/finance" and "public morals". The phenomena are found in Table 2 that the corpus generated by the netizen's online search behavior is severely unbalanced, the "risk-free" category takes the absolute majority in the corpus. Besides, there is little difference between the semantic information of two corpora from different kinds of societal risk categories, such as "public morals" and "risk-free", which leads to a high probability classifying a hot search word to the majority category. For the risk category "national security", "economy/finance" and "public morals", although the precision and recall of SVM-PV on hot news are not zero, the votes of risk categories of hot news causes no sample to be correctly labeled on hot search words. As far as the task of societal risk classification is concerned, it is essential to find out risky words as many as possible. In other words, we pay more attention to recall for evaluation. The recall of SVM based on Paragraph Vector on "risk-free" category is 0.98, tending to find hot words whose categories are risk-free. In contrast, the recall of

CRFs model on risk category "national security", "economy/finance" and "public morals" are respectively 0.43, 0.34 and 0.23. Meanwhile, the three values are in turn increased by 0.48, 1.10 and 0.64 compared with the SVM-BOW. In other words, CRFs model tends to capture risky hot words. As can be seen from the overall scores of the whole data for three methods, CRFs method achieves better performances in each risk category than the other two methods apparently. Overall, CRFs model shows the discriminatory power of predictive models in societal risk multi-classification. Moreover, it has obvious superiority in data processing which is relatively easy and captures comprehensive text semantics.

4   Analysis of Feature Terms on Societal Risk

CRFs model has demonstrated its superiority for risk multi-classification. Besides, we obtain state features after CRFs model completing parameters learning on the training set. The state features can be expressed as the distribution of terms' weight values in each risk category. The magnitudes of the weight values represent their contribution to predicting which risk categories the hot words belonging to. As a result, we could select terms with greater weight values in each risk category as the factors or characteristics on behalf of each risk.

4.1   Analysis of Feature Weight

We now perform feature terms analysis in each risk category according to state features and their corresponding weight values. The corpus is chosen from November 1, 2011 to December 31, 2016 including 56,233 hot news search words for training. When the training process is completed, the state features and weight values will be expressed as the distribution of terms' weight values in each risk category. The occurrence frequency of term "雾霾" (haze) under "daily life", "government management", "resources/environment" and "risk-free" are respectively 1, 2, 105, 7. And the corresponding weight values are -0.35, 0.71, 6.65, -0.05. The significance of these weight values can be explained from an aspect of CRFs formula. Since $\sum_Y exp(w \cdot F(Y, X))$ is the normalization factor, values of $P_w(Y|X)$ depend on values of $exp(w \cdot F(Y, X))$. Take "haze" for example, we assume that the sequence only has one word "haze" for simplicity.

$$P_{w_3}(resources/environment|haze) > P_{w_2}(governmentmanagement|haze)$$

$$> P_{w_4}(risk - free|haze) > P_{w_1}(dailylife|haze).$$

As is seen, weight values represent the contribution to the prediction of risk category. The higher the weight values of terms in one risk category, the greater the contributions of terms to conditional probability. For instance, the weight values of "房价" (house prices) in "finance/economy" and "daily life" are respectively -0.82 and 4.88. The larger weight value of "房价" (house prices) contributes greater to conditional probability on "daily life". Then we try to investigate the distribution pattern of place names in feature terms in each risk category.

4.2   Distribution of Feature Terms

We first use Ansj to do the Chinese hot news search words segmentation and part-of-speech tagging. Terms that are tagged "ns" (geographical name) and "nt"(institutional name) are selected. Then we build the dictionary of Chinese regional areas which has a total of 34 provinces, including provinces, municipalities and autonomous regions. At last, the geographical terms with their weight values in each category are picked out according to the dictionary. The distribution pattern of geographical terms in each category is as shown in Fig. 4. The horizontal axis is the geographical terms, while the vertical axis is the eight risk categories. Each small colored cell in the figure represents weight values of the geographical terms in each category. The deeper the color is, the higher the weight value is. Here we list and analyze geographical terms results for illustration. By the visualized results, we summary the distribution patterns of geographical terms in each category as follows:

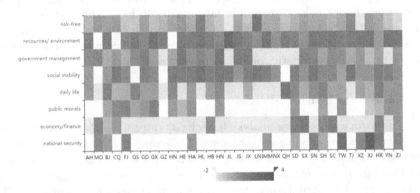

Fig. 4. Distribution pattern of regional terms in each category

(1) As to the risk of national security, the highest weight value of feature terms is Xinjiang, with Taiwan, Hong Kong, Fujian and Tibet decreasing in turn. This is because there have been hundreds of terrorist attacks happened in Xinjiang in recent years, including hijacking plane and attacking the police station so on. Terrorist attacks pose a great threat to social stability of Xinjiang and national security. In addition, there are patriotic movements like protecting the Diaoyu Islands and the South China Sea happened in Taiwan. Major political events such as sanctions against the Philippines and impeaching Chun-ying Leung have occurred in Hong Kong;

(2) As to the risk of finance/economy, weight values of Shanxi, Sichuan, Hong Kong and Shanghai decrease accordingly. Since anti-corruption movement in Shanxi, the economy of Lvliang city had crashed. Sichuan Province launched four trillion investment plans. Hong Kong and Shanghai are often mentioned in the risk of finance/economy owing to the Shanghai Composite Index and the Hong Kong Hang Seng Index, both of which may reflect the situation of stock market volatility to some extent;

(3) As to the risk of public morals, the issues such as integrity and social mode in Guangxi, Fujian, Chongqing and Henan are more salient and could not be neglected;

(4) As to the risk of daily life, Beijing and Shanghai mainly focus on housing issues such as property restriction and the rising price;

(5) As to the risk of social stability, weight values of Heilongjiang, Tianjin and Liaoning are higher relative to other areas in China. That is because the events such as coal mine explosion, the explosion of Tianjin harbor and the school bus rollover accident occurred respectively in Heilongjiang, Tianjin and Liaoning;

(6) As to the risk of government management, a number of top officials from provinces including Hunan, Hebei, Jiangxi, Guangdong and Shanxi were investigated by the commission for discipline inspection of the central committee due to the tough anti-corruption policy;

(7) As to the risk of resources/environment, there are earthquakes frequently occurred in Jilin, Yunnan and Tibet. And the snowstorm occurred in Inner Mongolia in November, 2012. As is known, haze pollution in Beijing is also prominent.

5   Conclusions

Societal risk refers to the risk problems raising the concerns of the whole society. The subjective perception of societal risk reflects the public attitudes to social issues as well as government decision-making. It is of great significance for government management and decision-making to monitor either the potential or the ongoing societal risk events. In this paper, CRFs and SVM-PV model are applied to obtain the subjective societal risk perception automatically and timely.

We conduct the research on societal risk perception based on HNSW. According to the current research, CRFs model is more effective in response to "subjective perception of societal risks" and "short texts". The main contributions are summarized as follows.

(1) CRFs model is first applied to societal risk classification directly dealing with short text, which tackles the challenge of feature sparseness and improves the performance.

(2) CRFs model is used to capture the contextual constraints on HNSW with obvious superiority in text processing.

(3) The geographical distribution rules of societal risks are found and summarized by studying distributions of place terms in each risk category by state features.

Lots of works need to be done. In the future, the combination of feature representation and CRFs will be developed to improve the performance. Besides, terms with greater weight values in each risk category may also be picked out either as the factors on behalf of risk events or as feature words to construct the risk lexicon.

Acknowledgments. This research is supported by National Key Research and Development Program of China (2016YFB1000902) and National Natural Science Foundation of China (61473284 & 71371107).

References

1. Xie, X.F., Xu, L.C.: The study of public risk perception. Psychol. Sci. **6**, 723–724 (2002). (in Chinese)
2. Zheng, R., Shi, K., Li, S.: The influence factors and mechanism of societal risk perception. In: Zhou, J. (ed.) Complex 2009. LNICSSITE, vol. 5, pp. 2266–2275. Springer, Heidelberg (2009). doi:10.1007/978-3-642-02469-6_104
3. Tang, X.J.: Exploring on-line societal risk perception for harmonious society measurement. J. Syst. Sci. Syst. Eng. **22**, 469–486 (2013)
4. Hu, Y., Tang, X.: Using support vector machine for classification of baidu hot word. In: Wang, M. (ed.) KSEM 2013. LNCS, vol. 8041, pp. 580–590. Springer, Heidelberg (2013). doi:10.1007/978-3-642-39787-5_49
5. Mikolov, T., Sutskever, l., Chen, K., Corrado, G., Dean, J.: Distributed representations of words and phrases and their compostitionality. In: Advances in Neural Information Processing Systems, pp. 3111–3119 (2013)
6. Le, Q.V., Mikolov, T.: Distributed representations of sentences and documents. Comput. Sci. **4**, 1188–1196 (2014)
7. Chen, J.D., Tang, X.J.: The challenges and feasibility of societal risk classification based on deep learning of representations. In: 2015 IEEE International Conference on Systems, Man, and Cybernetics, pp. 569–574 (2015)
8. Chen, J.D., Tang, X.J.: Societal risk classification of post based on paragraph vector and KNN method. In: Proceedings of the 15th International Symposium on Knowledge and Systems Sciences, pp. 117–123. JAIST Press (2014)
9. Lafferty, J., McCallum, A., Pereira, F.: Conditional random fields: probabilistic models for segmenting and labeling sequence data. In: Proceedings of The Eighteenth International Conference on Machine Learning, ICML, pp. 282–289 (2001)
10. Nigam, K., Lafferty, J., Mccallum, A.: Using maximum entropy for text classification. In: Proceedings of the IJCAI-99 Workshop on Information Filtering, pp. 61–67. San Fransisco (1999)
11. Yi, K., Beheshti, J.: A hidden Markov model based text classification of medical documents. J. Inform. Sci. **35**, 67–81 (2009)
12. Zhao, J., Liu, K., Wang, G.: Adding redundant features for CRFs-based sentence sentiment classification. In: Conference on Empirical Methods in Natural Language Processing, pp. 117–126 (2008)
13. Sudhof, M., Goméz Emilsson, A., Maas, A. L., Potts, C.: Sentiment expression conditioned by affective transitions and social forces. In: Proceedings of the 20th ACM SIGKDD International Conference on Knowledge Discovery and Data Mining, pp. 1136–1145 (2014)
14. Li, T.T., Ji, D.H.: Sentiment analysis of micro-blog based on SVM and CRF using various combinations of features. Appl. Res. Comput. **32**, 978–981 (2015). (in Chinese)
15. Zhang, C.Y.: Text categorization model based on conditional random fields. Comput. Technol. Dev. **21**, 77–80 (2011). (in Chinese)

A Markov Chain Monte Carlo Approach for Source Detection in Networks

Le Zhang, Tianyuan Jin, Tong Xu$^{(\boxtimes)}$, Biao Chang, Zhefeng Wang, and Enhong Chen

Anhui Province Key Laboratory of Big Data Analysis and Application,
School of Computer Science and Technology,
University of Science and Technology of China, Hefei, China
{laughing,jty123,chbiao,zhefwang}@mail.ustc.edu.cn,
{tongxu,cheneh}@ustc.edu.cn

Abstract. The source detection task, which targets at finding the most likely source given a snapshot of the information diffusion, has attracted wide attention in theory and practice. However, due to the hardness of this task, traditional techniques may suffer biased solution and extraordinary time complexity. Specially, source detection task based on the widely used Linear Threshold (LT) model has been largely ignored. To that end, in this paper, we formulate the source detection task as a Maximum Likelihood Estimation (MLE) problem, and then proposed a Markov Chain Monte Carlo (MCMC) algorithm, whose convergence is demonstrated. Along this line, to further improve efficiency of proposed algorithm, the sampling method is simplified only on the observed graph rather than the entire one. Extensive experiments on public data sets show that our MCMC algorithm significantly outperforms several state-of-the-art baselines, which validates the potential of our algorithm in source detection task.

Keywords: Social network · Source detection · Markov Chain Monte Carlo

1 Introduction

Recent years have witnessed a massive increase in the popularity of social networks, which emerged as important platforms for information propagation. Some studies focus on modeling the diffusion process of information [10,21]. Some studies focus on the application of information diffusion [13,15,22]. *Information source detection* [2,15,20] is one of the popular application, which aims to find the source of given information based on a snapshot of the propagation network, and further support many propagation-related scenarios, such as rumor source tracing and virus source identification.

Since firstly proposed in [15] as a preliminary study, the information source detection task has been discussed by many prior arts, which are based on centrality [15–17], sampling [7,11,13,18,20], spectral [6], or belief propagation [1].

© Springer Nature Singapore Pte Ltd. 2017
X. Cheng et al. (Eds.): SMP 2017, CCIS 774, pp. 77–88, 2017.
https://doi.org/10.1007/978-981-10-6805-8_7

Usually, most efforts have been made only on a simplified network topology, such as a tree or a grid. But it is hard for these approaches to guarantee high accuracy in general graphs. At the same time, some other related works assumed that information diffusion process followed the basic epidemic models, like the Susceptible Infected Recovered (SIR) model [10] or the Susceptible Infected (SI) model [15]. However, finding source in the liner threshold (LT) model which is one of the most popular models for social network analysis is rarely studied.

Recent studies [11,13] have noticed these mentioned facts, they attempted to solve the problem via maximizing the similarity between estimated and observed diffusion graph. However, these studies may be misled by the network structure. As an example, in Fig. 1(a), suppose that the observed nodes set $A = \{1, 2, 3, 5\}$ is given. v_2 will probably be estimated as the source by the K-effectors [11] and SISI [13]. In fact, only v_1 can activate all the nodes successfully. Therefore, the problem [11,13] studied is substantially different from the source detection. Clearly, more comprehensive method is still required.

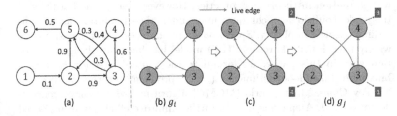

Fig. 1. (a): A social network instance in LT model. (b), (c) and (d) describe three possible diffusion process for given observed node set $A = \{2, 3, 4, 5\}$.

In light of the above, we will explore how to efficiently estimate the source in large social networks without bias. To be specific, the linear threshold (LT) model, one of the most popular models for social network analysis, will be used to describe the diffusion process. Then, the task of information source detection will be formalized as a maximum likelihood estimation (MLE) problem, where we try to pick up a node, i.e., the source, to maximize the likelihood of diffusion graph. To deal with that, considering that the size of the sample space is exponentially large compared with the number of edges in the network, we designed a Markov Chain Monte Carlo (MCMC) to efficiently detect the source. The contributions of our paper could be summarized as follows:

1. We formalize the problem of source detection under the LT model by maximum likelihood estimation (MLE). And to the best of our knowledge, we are the first to adapt the MCMC approach to solve the problem.
2. To further improve efficiency of the proposed algorithm, we reduce the sample space by sampling on the observed subgraph rather than the entire graph.
3. We conduct comprehensive experiments on a real network to validate the performance of the proposed approach. The results demonstrate the effectiveness of the proposed approach.

2   Related Work

In general, our related work could be divided into two categories, i.e., Source Detection and Markov Chain Monte Carlo Approach.

Source Detection. Due to the widespread applications in many real-world scenarios, a series of techniques for source detection have been studied. In the seminal works, Shah et al. [15,17] is the first to formulate the source detection and established the concept of rumor centrality under the SI model. Then they further improved the effectiveness of rumor centrality for source detection in generic trees [16]. Dong et al. [4] inferred a source with the maximum posteriori estimator on regular tree-type networks given a priori information. Wang et al. [19] found that multiple independent observations can significantly increase the detection probability for trees. Lappas et al. [11] formulated the source detection as the K-effectors which selects k active nodes that can best explain the observed nodes set in social networks. Fanti et al. [5] introduced a messaging protocol, which guarantees obfuscation of the source under the assumption that the network administrator utilizes the maximum likelihood (ML) estimator to identify the source. Chang et al. [3] derived a maximum posteriori estimator to find the source under the SI model. Nguyen et al. [13] used a similar formulation of K-effectors for source detection and provided an efficient guaranteed method.

Markov Chain Monte Carlo. Markov Chain Monte Carlo is a randomized sampling method which could sample from a probability distribution based on constructing a Markov chain that has the desired distribution as its equilibrium distribution. Due to its widespread applications in many important problems like rare event sampling [8] and Bayesian Inference [14], this approach has been richly studied in recent years. Along this line, a series of twist methods have been proposed like Gibbs sampling and Slice sampling.

In conclusion, on the one hand, most existing works assume that information diffusion process followed the basic epidemic models, such as the SI model or SIR model. Whereas that solving the source detection problem based on the LT model is rarely studied. On the other hand, although MCMC has been wildly used in rare event sampling, it is rarely used in source detection. Hence, we explore the MCMC approach to efficiently estimate the source in general graphs based on the LT model.

3   Problem Formulation

In this section, we first give the formal definition of the information source detection problem. Then, we introduce the widely used diffusion model called the LT model. At last, the basic Monte Carlo method is proposed to solve it.

3.1   Problem Definition

Given a social network $G = (V, E)$, where V is the node set, $E \subseteq V \times V$ is the set of edges and a snapshot of the information diffusion when diffusion process ends. Denote by A the set of infected nodes observed in V. Assuming that node v is the source, denote by $P(A|G, v)$ the probability that the infected node set we observe in G is A. According to the principle of Maximum Likelihood Estimation (MLE), the estimated source node is $arg\ max_v\ P(A|G, v)$. Formally, we define the source detection problem as follows.

Problem 1. **Information Source Detection Problem.** Given a graph $G = (V, E)$ and the infected node set A when diffusion process ends, the information source detection problem asks for finding a single source node $\hat{s}$ that most likely start the diffusion, that is $\hat{s} = arg\ max_v\ P(A|G, v)$.

3.2 Diffusion Model

To case the modeling, the widely-used Linear-Threshold (LT) model will be introduced to describe the information diffusion. In this model, we have an extra influence weight matrix B to evaluate the importance of edges. In particular, for a given edge (u, v), it corresponds to a weight $B_{u,v}$ such that $\sum_{u \in N^{in}(v)} B_{u,v} \leq 1$ where the $N^{in}(v)$ is the set of in-neighbors of v. The dynamic of information propagation in the LT model unfolds as follows.

A source node $s \in V$ is assumed to be the initial disseminator of the information. Denote by S_t the nodes activated in step t $(t = 0, 1, 2, \ldots)$ and $S_0 = s$. Initially, each node u selects a threshold θ_u in range $[0, 1]$ uniformly at random. At step $t > 0$, an inactive node u would be activated if $\sum_{w \in N^{in}(u) \cap (\bigcup_{i < t} S_i)} B_{w,u} \geq \theta_u$.

The process terminates when $S_t = \emptyset$.

Kempe [9] proved that the LT model is equal to the live-edge graph processes. In the LT model, a live-edge graph instance can be obtained by following rules. Each node v picks one incoming edge with the probability of $\sum_{u \in N^{in}(v)} B_{u,v}$. The selected edges are called live and the others are called dead. For a given live-edge graph denoted by g, each node i reachable from source s is active.

3.3 Basic Solution

We use $R(g, v)$ to denote a set of nodes that are reachable from v in g. Thus,

$$P(A|G, v) = \sum_{g \subseteq G} [\Upsilon_G(g) I(A = R(g, v))], \qquad (1)$$
$$= \mathbb{E}_{g \sim G}[I(A = R(g, v))]$$

where I is an indicator function defined as:

$$I(c) = \begin{cases} 1 \text{ if } c \text{ is true;} \\ 0 \text{ otherwise;} \end{cases}$$

and $\Upsilon_G(g)$ denotes the probability distribution of G, i.e.,

$$\Upsilon_G(g) = \prod_{v \in g \wedge (u,v) \in g} B_{u,v} \prod_{v \in g \wedge N_g^{in}(v) = \emptyset} \vartheta_v, \qquad (2)$$

where $N_g^{in}(v)$ denotes the set of in-neighbors of v in g and $\vartheta_v = \sum_{u \in N_G^{in}(v)} B_{u,v}$. Nevertheless, the evaluation of $P(A|G, v)$ is computationally prohibitive since it is related to counting the number of linear extensions of a ordered node set. A trivial method-Monte Carlo Method(MC) could be used to estimate $P(A|G, v)$ by drawing graph $g \sim G$. However, due to the often exponential small number $P(A|G, v)$, the running time of MC is not acceptable. To show this, consider a linear graph $G(V, E)$, where $V = \{v_1, v_2, \cdots, v_n\}$ and $E = \{(v_i, v_{i+1})|1 \leq i < n\}$. Suppose that $V = V'$ and $B_{v_i, v_{i+1}} = 1/2$, then we have $P(A|G, v_1) = 1/2^{n-1}$. We call a sampled graph positive sample, if there exist a node $v \in A$ could activate A having $R(g, v) = A$. To overcome the above problem, an ideal method is to make that all the output samples positive samples. In the next section, we will achieve this with a Markov Chain Monte Carlo (MCMC) method.

4 Solving Source Detection with MCMC

In this section, we explore the MCMC method for source detection.

We define a set of positive samples as

$$G' = \{g | \exists v, \ s.t. \ I(R(g, v) = A)\}. \tag{3}$$

and the distribution of G' as

$$\Upsilon_{G'}(g) = \begin{cases} \Upsilon_G(g)/Z & g \in G'; \\ 0 & \text{otherwise}; \end{cases}$$

where $Z = \sum_{g \in G'} \Upsilon_G(g)$

In MCMC, we sample in G' rather than G. The key property that makes MCMC approach work well is that $\hat{s} = arg \ max_v \mathbb{E}_{g \sim G'}[I(R(g, v) = A)]$. Now, we sample in G' with the MCMC approach.

First, we need to generate an instance of live-edge graph $g_1 \in G'$. We randomly pick a node in A and do the BFS in $G(A)$, where $G(A)$ is the infected subgraph induced by node set A and their inter edges. Till the observed nodes set in BFS process is A, we get a live-edge graph g. Then, by adding the dead-edges we expand the corresponding live-edge graph g in $G(A)$ to the graph g_1 in G. At last, we get a graph $g_1 \in G'$.

Second, we use the Algorithm 1 to generate a Markov chain. Algorithm 1 gives a local move in G'. Each local move will change a node's in-live-edge from the previous subgraph g_i.

At last, we measure $I(R(g_i, v) = A)$ on each graph g_i and figure out the maximum one as the source. Formally, we describe it in Algorithm 2. The following theorem demonstrates the correctness of the MCMC approach.

Theorem 1. *The Markov chain created by Algorithm 1 has a only stationary distribution and has* $\hat{s} = arg \ max_v \mathbb{E}_{g_i \sim G'}[I(R(g_i, v) = A)]$.

Proof. To prove Theorem 1, we give the following two lemmas first.

Algorithm 1. Local move
___
Require: G, A, g_i.
1: Choose a node $v \in V$ uniformly randomly; $g_{i+1} = g_i$;
2: Let $\sum_{w \in N^{in}(v)} B_{w,v} = \vartheta(v)$;
3: Delete the in-live-edge of node v in g_{i+1};
4: In g_{i+1}, pick v's in-live-edge (u,v) with probability $B_{u,v}$;
5: **if** $g_{i+1} \notin G'$ **then**
6:     $g_{i+1} = g_i$;
7: **end if**
Ensure: g_{i+1}.

Algorithm 2. MCMC for source detection
___
Require: G, A, g_1, parameter K;
1: Create new array *count* with size $|A|$;
2: $k = 0$, $g = g_1$;
3: **while** $k < K$ **do**
4:     $T = \{v | R(v,g) = A\}$;
5:     **for** $v \in T$ **do**
6:         $count[v] = count[v] + 1$;
7:     **end for**
8:     Local $move(G, A, g)$;
9: **end while**
10: $s = arg\ max_v count[v]$;
Ensure: s.

Lemma 1. *The Markov chain created by Algorithm 1 is irreducible.*

Proof. Given two instance of state $g_i, g_j \in G'$, we set the g_i as the input in Algorithm 1 and we want to prove that there exist a sequence that contains both g_i and g_j. By the local move method, we establish the sequence as follows.

First, for each edge $e \in g_i \wedge e \notin G(A)$, we can change the edge into the dead edge by a local move. Also, we do the same operation to g_j. Thus, without generality, suppose that all the edges in g_i and g_j are also the edges in $G(A)$.

Second, we give the order of changing the node's edges in g_i by following method. Since $g_i \in G'$, we can find out a node v which could activated A in g_i and mark it as 1. Then, we mark the node u that can reach v via a live edge in g_j as 2. We do above process repeatedly, till the present node is marked or the present node cannot be reached from other nodes in g_j. Next, we mark the left nodes by the pre-order traversal in g_j.

At last, we do the local move in g_i by the marked order. Each local move will make a node's in-live-edge in g_i be the same as g_j. Figure 1 shows an example of above process. It is obvious that each intermediate graph belongs to G'. Lemma follows.

Lemma 2. *The Markov chain is aperiodic.*

Proof. Given state g_i, for any possible next state g_{i+1}, the probability of $P(g_{i+1} = g_i) \neq 0$. Thus the Markov chain is aperiodic.

Since the Markov Chain is irreducible and aperiodic, it only has a stationary distribution. Then, we have Lemma 3.

Lemma 3. *In Algorithm 1, the transition probability of graph g_i and g_{i+1} satisfy with*

$$\frac{P(g_{i+1}|g_i)}{P(g_i|g_{i+1})} = \frac{\Upsilon_{G'}(g_{i+1})}{\Upsilon_{G'}(g_i)}, \tag{4}$$

where $P(g_{i+1}|g_i)$ is the probability of transiting from state g_i to next possible state g_{i+1}.

Proof. Suppose that the different live-edges in g_i and g_{i+1} are (u_1, v) and (u_2, v). There are three different cases.

Case 1: $(u_1, v) \in g_i$ and $(u_2, v) \in g_{i+1}$. In this case, we have $\frac{\Upsilon_{G'}(g_{i+1})}{\Upsilon_{G'}(g_i)} = \frac{B_{u_2,v}}{B_{u_1,v}}$.
From Algorithm 1, $\frac{P(g_{i+1}|g_i)}{P(g_i|g_{i+1})} = \frac{B_{u_2,v}/n}{B_{u_1,v}/n} = \frac{B_{u_2,v}}{B_{u_1,v}}$.

Case 2: $(u_1, v) \in g_i$ and $(u_2, v) \notin g_{i+1}$. In this case, $\frac{\Upsilon_{G'}(g_{i+1})}{\Upsilon_{G'}(g_i)} = \frac{1-\vartheta(v)}{B_{u_1,v}}$. From Algorithm 1, $\frac{P(g_{i+1}|g_i)}{P(g_i|g_{i+1})} = \frac{(1-\vartheta(v))/n}{B_{u_1,v}/n} = \frac{\Upsilon_{G'}(g_{i+1})}{\Upsilon_{G'}(g_i)}$.

Case 3: $(u_1, v) \notin g_i$ and $(u_2, v) \in g_{i+1}$. Similar to case 2, $\frac{P(g_{i+1}|g_i)}{P(g_i|g_{i+1})} = \frac{\Upsilon_{G'}(g_{i+1})}{\Upsilon_{G'}(g_i)}$. Thus, $\frac{P(g_{i+1}|g_i)}{P(g_i|g_{i+1})} = \frac{\Upsilon_{G'}(g_{i+1})}{\Upsilon_{G'}(g_i)}$.

From Lemma 3, π is the stationary distribution. From the definition of G', obviously, $\hat{s} = arg\ max_v \mathbb{E}_{g \sim G'}[I(R(v, g) = A]$. Theorem follows.

5 Sampling in Infected Subgraph

In this section, to improve the efficiency of MCMC, we simplify the sampling method that scales according to observed graph size rather than the size of the entire graph.

When an infected subgraph is obtained at the end of the diffusion process, the snapshot of the process is observed only at $G(A)$. We constrain information diffusion to $G(A)$. In this case, let $P(A|G(A), v)$ be the probability that the infected node set we observe in $G(A)$ is A. We arrange the elements in set $\{P(A|G(A), v)|v \in A\}$ by descending order and let $Rank[P(A|G(A), v)]$ be the rank of $P(A|G(A), v)$ in $\{P(A|G(A), v)|v \in A\}$. We obtain the following theorem:

Theorem 2. $Rank[P(A|G, v)] = Rank[P(A|G(A), v)]$.

Proof. By setting the nodes in A that are initially activated, let β be the probability that no node in $V \setminus A$ has been activated. Since the snapshot is observed when the information diffusion process terminated, each node in A try to activate the node in $V \setminus A$. Thus, $P(A|G,v) = \beta P(A|G(A),v)$. We normalize $P(A|G,v)$ to $N(A|G,v) = \frac{P(A|G,v)}{\sum_{u \in A} P(A|G,v)}$. Then,

$$N(A|G,v) = \frac{P(A|G,v)}{\sum_v P(A|G,v)} = \frac{\beta P(A|G(A),v)}{\sum_v \beta P(A|G(A),v)} = N(A|G(A),v).$$

Since $Rank[P(A|G,v)] = Rank[N(A|G,v)]$, $Rank[P(A|G(A),v)] = Rank[N(A|G(A),v)]$ and $N(A|G,v) = N(A|G(A),v)$. Thus $Rank[P(A|G,v)] = Rank[P(A|G(A),v)]$.

This means we can simplify the sampling strategy only in the infected subgraph.

6   Experiment

In this section, we conduct experiments of our source detection algorithm on a real network dataset. Our experiments are conducted on a machine with an Intel i5 CPU and 16 GB of memory. All experiments are implemented in JAVA.

6.1   Experimental Setting

Data Sets. We conduct our experiments on a real network dataset: **Wiki-Vote** [12]. This dataset is composed of all Wikipedia voting data from the inception of Wikipedia till January 2008. And it contains 7115 nodes and 103,689 directed edges.

In our experiments, the probability of the propagation of an edge (u, v) is set to $\frac{\alpha}{indegree(v)}$, where the $indegree(v)$ is the set of in-neighbors of v. To make our experiments more compelling, we set the value of α to 0.5, 0.75 and 1 respectively. We randomly choose a node as source and run the LT model until no more nodes are infected, then we obtain the set of these infected nodes. We repeat the process 1,000 times to get 1,000 random datasets. In order to challenge the effectiveness of our algorithm, we guarantee that each dataset contains more than 5 infected nodes and contains at least two candidate sources whose likelihood to infect others is larger than 0.

Algorithms and Their Explanations. To compare the proposed algorithm with existing algorithms, we also implement some other algorithms. The algorithms used in the experiments are as follows:

- *JC*: Jordan center [23]. This selects an activated node with the maximal distance to the others as source.
- *Ef*: A heuristic algorithm proposed in [11]. In [11], it proposes the "DP", "Sort", "OutDegree" algorithms to find the k sources. However, to the single source, "DP=Sort". In the experiments, we use this algorithm and call it *Ef*.
- *MCMC*: the method proposed in Sect. 4.

Evaluation Index. We apply the following widely used measures to evaluate the effectiveness of our methods.

– *Detection Rate.* Detection rate is the probability that the node identified by the algorithm is the actual source.
– *γ − accuracy.* γ-accuracy is the probability that the actual source ranked among top γ.
– *Error Distance.* Error Distance is defined as the distance between detected source node and true source node assuming edges are undirected.

6.2 Effectiveness Validation

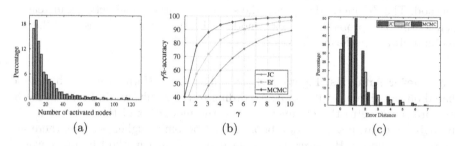

(a)   (b)   (c)

Fig. 2. When α is equal to 1.0. Statistics histogram of activated subgraph used in experiments (a) The γ − *accuracy* result (b) and the error distance (c) of different algorithms.

Comparison of Results. Figure 2(b) shows the γ−*accuracy* computed by each method when α is set to 1. Note that the intersection of the polyline and the y-axis indicates the detection rate. To some degree, detection rate is on behalf of the precision of a method. The higher the value is, the better performance the algorithm achieves. Obviously, the *MCMC* method yields the best performance. In more than 40% of the total experiments, the *MCMC* method can find the true source. The detection rate of the *MCMC* method is 20% higher than that of the *Ef* method and even two times more than that of the *JC* method. For the γ − *accuracy*, we make our algorithm output a list of candidate source nodes sorted in descending order of likelihood, the value of γ represents the index of actual node in the list which is no more than γ. According to Fig. 2(b), it is clear that when the value of γ is fixed, the accuracy of the *MCMC* method is higher than that of the other two methods. In more than 90% of the total experiments, the rank of the actual source is among top 3 in the list of candidates which is produced by the *MCMC* method. In about 75% of the total experiments, the *Ef* method can make sure the rank of the actual source is among top 3. In addition, the *JC* method can satisfy the same condition in no more than 50% of the total experiments.

Figure 2(c) shows the distribution of error distances when α is set to 1. Error distance reflects how far the detected nodes is away from the actual source. Note that the larger of the error distance is, the worse performance the corresponding method achieves. It is clear that all source nodes identified by the proposed algorithm are four hops around the source node. In more than 90% of the total experiments, the detected nodes detected by the $MCMC$ method are two hops around the source node. At the same time, the Ef method and JC method not only have fewer results with 0 or 1 error distance, but have heavier tail as well.

In comparison, the $MCMC$ method always yields the best performance, and the Ef method is slightly worse than the $MCMC$ method. Because the idea of the Ef method is to maximize the similarity between estimated and observed diffusion graph which is different from the fundamental meaning of source detection. Contrarily, the JC method is obviously worse than the $MCMC$ method. The JC method just uses the structure of the social network instead of considering the probability of each node to be the source. This is why the JC method always performs the worst.

Parameter Analysis. Figures 2, 3 and 4 show the performance of each method based on the different value of α. According to the result, we find that all methods are sensitive to the parameter. With the increment of the value of α, the probability of an inactive node be activated becomes higher, so that there are more active nodes. However, the performance of each method becomes worse. Because the difficulty of the source detection problem is related to the number of activated nodes. Even so, the performance of the $MCMC$ method just drops a little, it always holds on to the number-one spot among the three methods. At the same time, the gap between the $MCMC$ method and the other two methods gets larger and larger. To some extent, the $MCMC$ method is more stable than the other in large network.

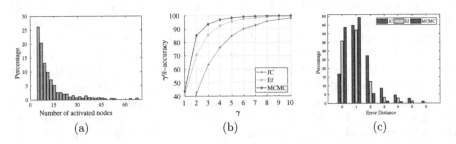

(a)   (b)   (c)

Fig. 3. When α is equal to 0.75. Statistics histogram of activated subgraph used in experiments (a) The $\gamma - accuracy$ result (b) and the error distance (c) of different algorithms.

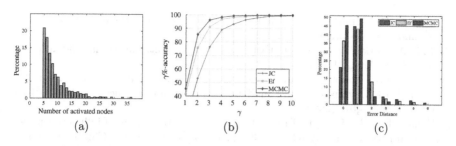

(a)   (b)   (c)

Fig. 4. When α is equal to 0.5. Statistics histogram of activated subgraph used in experiments (a) The $\gamma - accuracy$ result (b) and the error distance (c) of different algorithms.

7   Conclusion

In this paper, we address the source detection problem in the LT model, and then derive a MCMC approach to improve the accuracy for general graphs. To further improve the efficiency, we reduce the sample space by sampling on the observed subgraph rather than the entire graph. Experiments on real social network demonstrate the performance of our approach.

In future, parallelizing our method in source detection problem based on other diffusion models is an interesting direction of our work. And the efficiency of our method can be further improved. At the same time, our work just concentrates on the single source detection problem, expanding our method for multiple source detection problem is a meaningful work.

Acknowledgments. This research was partially supported by grants from the National Natural Science Foundation of China (Grant No. U1605251) and the National Science Foundation for Distinguished Young Scholars of China (Grant No. 61325010). Also, this research was supported by the Anhui Provincial Natural Science Foundation (Grant No. 1708085QF140), and the Fundamental Research Funds for the Central Universities (Grant No. WK2150110006).

References

1. Altarelli, F., Braunstein, A., DallAsta, L., Ingrosso, A., Zecchina, R.: The patient-zero problem with noisy observations. J. Stat. Mech. **2014**, P10016 (2014)
2. Altarelli, F., Braunstein, A., DallAsta, L., Lage-Castellanos, A., Zecchina, R.: Bayesian inference of epidemics on networks via belief propagation. Phys. Rev. Lett. **112**, 118701 (2014)
3. Chang, B., Zhu, F., Chen, E., Liu, Q.: Information source detection via maximum a posteriori estimation. In: ICDM 2015, pp. 21–30. IEEE Press, Atlantic City (2015)
4. Dong, W., Zhang, W., Tan, C.W.: Rooting out the rumor culprit from suspects. In: ISIT 2013, pp. 2671–2675. IEEE Press, Turkey (2013)
5. Fanti, G., Kairouz, P., Oh, S., Viswanath, P.: Spy vs. spy: rumor source obfuscation. In: SIGMETRICS 2015, pp. 271–284. ACM Press, Portland (2015)

6. Fioriti, V., Chinnici, M.: Predicting the sources of an outbreak with a spectral technique. arXiv preprint arXiv:1211.2333 (2012)

7. Jain, A., Borkar, V., Garg, D.: Fast rumor source identification via random walks. Soc. Netw. Anal. Min. **6**, 62 (2016)

8. Kelner, J.A., Madry, A.: Faster generation of random spanning trees. In: 50th Annual IEEE Symposium on Foundations of Computer Science, pp. 13–21. IEEE Press, Washington (2009)

9. Kempe, D., Kleinberg, J., Tardos, É.: Maximizing the spread of influence through a social network. In: SIGKDD 2013, pp. 137–146. ACM Press, Chicago (2003)

10. Kermack, W.O., McKendrick, A.G.: Contributions to the mathematical theory of epidemics. II. The problem of endemicity. Proc. Roy. Soc. London A Math. Phys. Eng. Sci. **138**, 55–83 (1932)

11. Lappas, T., Terzi, E., Gunopulos, D., Mannila, H.: Finding effectors in social networks. In: SIGKDD 2010, pp. 1059–1068. ACM Press, Washington (2010)

12. Leskovec, J., Huttenlocher, D., Kleinberg, J.: Predicting positive and negative links in online social networks. In: WWW 2010, pp. 641–650. Raleigh (2010)

13. Nguyen, H.T., Ghosh, P., Mayo, M.L., Dinh, T.N.: Multiple infection sources identification with provable guarantees. In: CIKM 2016, pp. 1663–1672. ACM Press. Indianapolis (2016)

14. Schoot, R., Kaplan, D., Denissen, J., Asendorpf, J.B., Neyer, F.J., Aken, M.A.: A gentle introduction to Bayesian analysis: applications to developmental research. Child Dev. **85**, 842–860 (2014)

15. Shah, D., Zaman, T.: Detecting sources of computer viruses in networks: theory and experiment. In: SIGMETRICS 2010, vol. 38, pp. 203–214 (2010)

16. Shah, D., Zaman, T.: Rumor centrality: a universal source detector. In: SIGMETRICS 2012, pp. 199–210. ACM Press, London (2012)

17. Shah, D., Zaman, T.: Rumors in a network: who's the culprit? TIT **2011**(57), 5163–5181 (2011)

18. Tong, G., Wu, W., Guo, L., Li, D., Liu, C., Liu, B., Du, D.-Z.: An Efficient Randomized Algorithm for Rumor Blocking in Online Social Networks. arXiv preprint arXiv:1701.02368 (2017)

19. Wang, Z., Dong, W., Zhang, W., Tan, C.W.: Rumor source detection with multiple observations: fundamental limits and algorithms. In: SIGMETRICS 2014, pp. 1–13. ACM Press, Austin (2014)

20. Zhai, X., Wu, W., Xu, W.: Cascade source inference in networks: a Markov chain Monte Carlo approach. Comput. Soc. Netw. **2**, 17 (2015)

21. Yang, Y., Chen, E., Liu, Q., Xiang, B., Xu, T., Shad, S.A.: On approximation of real-world influence spread. In: ECML-PKDD 2012, pp. 548–564. UK (2012)

22. Xu, T., Zhong, H., Zhu, H., Xiong, H., Chen, E., Liu, G.: Exploring the impact of dynamic mutual influence on social event participation. In: SDM 2015, pp. 262–270. Canada (2015)

23. Jordan, C.: Sur les assemblages de lignes. J. Reine Angew. Math. **70**, 81 (1869)

Natural Language Processing

Neural Chinese Word Segmentation as Sequence to Sequence Translation

Xuewen Shi, Heyan Huang, Ping Jian[⊠], Yuhang Guo, Xiaochi Wei, and Yi-Kun Tang

School of Computer Science and Technology, Beijing Engineering Research Center of High Volume Language Information Processing and Cloud Computing Applications, Beijing Institute of Technology, Beijing 100081, China
{xwshi,hhy63,pjian,guoyuhang,wxchi,tangyk}@bit.edu.cn

Abstract. Recently, Chinese word segmentation (CWS) methods using neural networks have made impressive progress. Most of them regard the CWS as a sequence labeling problem which construct models based on local features rather than considering global information of input sequence. In this paper, we cast the CWS as a sequence translation problem and propose a novel sequence-to-sequence CWS model with an attention-based encoder-decoder framework. The model captures the global information from the input and directly outputs the segmented sequence. It can also tackle other NLP tasks with CWS jointly in an end-to-end mode. Experiments on Weibo, PKU and MSRA benchmark datasets show that our approach has achieved competitive performances compared with state-of-the-art methods. Meanwhile, we successfully applied our proposed model to jointly learning CWS and Chinese spelling correction, which demonstrates its applicability of multi-task fusion.

Keywords: Chinese word segmentation · Sequence-to-sequence · Chinese spelling correction · Natural language processing

1 Introduction

Chinese word segmentation (CWS) is an important step for most Chinese natural language processing (NLP) tasks, since Chinese is usually written without explicit word delimiters. The most popular approaches treat CWS as a sequence labelling problem [14,21] which can be handled with supervised learning algorithms, e.g. Conditional Random Fields [11,14,18,24]. However the performance of these methods heavily depends on the design of handcrafted features.

Recently, neural networks for CWS have gained much attention as they are capable of learning features automatically. Zheng et al. [25] adapted word embedding and the neural sequence labelling architecture [7] for CWS. Chen et al. [4] proposed gated recursive neural networks to model the combinations of context characters. Chen et al. [5] introduced Long Short-Term Memory (LSTM)

© Springer Nature Singapore Pte Ltd. 2017
X. Cheng et al. (Eds.): SMP 2017, CCIS 774, pp. 91–103, 2017.
https://doi.org/10.1007/978-981-10-6805-8_8

into neural CWS models to capture the potential long-distance dependencies. The aforementioned methods predict labels of each character in the order of the sequence by considering context features within a fixed-sized window and limited tagging history [2]. In order to eliminate the restrictions of previous approaches, we cast the CWS as a sequence-to-sequence translation task.

The sequence-to-sequence framework has successful applications in machine translation [1,9,19,20], which mainly benefits from (i) distributed representations of global input context information, (ii) the memory of outputs dependencies among continuous timesteps and (iii) the flexibilities of model fusion and transfer.

In this paper, we conduct sequence-to-sequence CWS under an attention-based recurrent neural network (RNN) encoder-decoder framework. The encoder captures the whole bidirectional input information without context window limitations. The attention based decoder directly outputs the segmented sequence by simultaneously considering the global input context information and the dependencies of previous outputs. Formally, given an input characters sequence $\mathbf{x}$ with T words i.e. $\mathbf{x} = (x_1, x_2, ..., x_{T_x})$, our model directly generates an output sequence $\mathbf{y} = (y_1, y_2, ..., y_T)$ with segmentation tags inside. For example, given a Chinese sentence " 我爱夏天" (I love summer), the input (我, 爱, 夏, 天) and the output (我, </s>, 爱, </s>, 夏, 天) where the symbol '</s>' denotes the segmentation tag. In the post-processing step, we replace '</s>' with word delimiters and join the characters sequence into a sentence as $\mathbf{s} =$ "我爱夏天". In addition, considering that the sequence-to-sequence CWS is an end-to-end process of natural language generation, it has the capacity of jointly learning with other NLP tasks. In this paper, we have successfully applied our proposed method to jointly learning CWS and Chinese spelling correction (CSC) in an end-to-end mode, which demonstrates the applicability of the sequence-to-sequence CWS framework.

We evaluate our model on three benchmark datasets, Weibo, PKU and MSRA. The experimental results show that the model achieves competitive performances compared with state-of-the-art methods.

The main contributions of this paper can be summarized as follows:

- We first treat CWS as a sequence-to-sequence translation task and introduce the attention-based encoder-decoder framework into CWS. The encoder-decoder captures the whole bidirectional input information without context window limitations and directly outputs the segmented sequence by simultaneously considering the dependencies of previous outputs and the input information.
- We let our sequence-to-sequence CWS model simultaneously tackle other NLP tasks, e.g., CSC, in an end-to-end mode, and we well validate its applicability in our experiments.
- We propose a post-editing method based on longest common subsequence (LCS) [12] to deal with the probable translation errors of our CWS system.

This method solves the problem of missing information in the translation process and improves the experiment results.

2   Method

2.1   Attention Based RNN Encoder-Decoder Framework for CWS

Our approach uses the attention based RNN encoder-decoder architecture called RNNsearch [1]. From a probabilistic perspective, our method is equivalent to finding a character sequence $\mathbf{y}$ with segmentation tags inside via maximizing the conditional probability of $\mathbf{y}$ given a input character sequence $\mathbf{x}$, i.e., $argmax_y p(\mathbf{y}|\mathbf{x})$.

The model contains (i) an bidirectional RNN encoder to maps the input $\mathbf{x} = (x_1, x_2, ..., x_{T_x})$ into a sequence of annotations $(h_1, h_2, ..., h_{T_x})$, and (ii) an attention based RNN decoder to generate the output sequence $\mathbf{y} = (y_1, y_2, ..., y_T)$. Figure 1 gives an illustration of the model architecture.

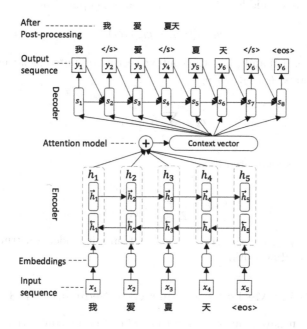

Fig. 1. Illustration of the presented model for CWS. The tag '<eos>' refers to the end of the sequence.

2.2   Bidirectional RNN Encoder

The bidirectional RNN encoder consists of forward and backward RNNs. The forward RNN $\overrightarrow{f}$ reads the input sequence in the order of (from x_1 to x_{T_x}) and calculates the sequence $(\overrightarrow{h}_1, \overrightarrow{h}_2, ..., \overrightarrow{h}_{T_x})$, while the backward RNN $\overleftarrow{f}$ reads the input sequence in the reverse order of (from x_{T_x} to x_1) and calculates the sequence $(\overleftarrow{h}_1, \overleftarrow{h}_2, ..., \overleftarrow{h}_{T_x})$. Finally, the annotation h_j for each x_j is obtained by $h_j = \left[\overrightarrow{h}_j^T ; \overleftarrow{h}_j^T\right]^T$.

2.3 Attention-Based RNN Decoder

The attention-based RNN decoder estimates the conditional probability $p(\mathbf{y}|\mathbf{x})$ as

$$p(\mathbf{y}|\mathbf{x}) = \prod_t^T p(y_t|y_1, ..., y_{t-1}, \mathbf{x}).$$ (1)

In Eq. (1), each conditional probability is defined as:

$$p(y_t|y_1, ..., y_{t-1}, \mathbf{x}) = g(y_{t-1}, s_t, c_t),$$ (2)

where s_t is the RNN hidden state for time t and computed by

$$s_t = f(s_{t-1}, y_{t-1}, c_t).$$ (3)

The c_t in Eqs. (2) and (3) is the context vector computed as a weighted sum of the annotations $(h_1, h_2, ..., h_{T_x})$:

$$c_t = \sum_{j=1}^{T_x} \alpha_{t,j} h_j.$$

The weight $\alpha_{t,j}$ is computed by:

$$\alpha_{t,j} = \frac{exp(e_{t,j})}{\sum_{k=1}^{T_x} exp(e_{t,k})},$$

where $e_{t,j} = a(s_{t-1}, h_j)$, therein, $a(\cdot)$ is an attention model constructed with a feedforward neural network.

2.4 Post-editing Method for Sequence-to-Sequence CWS

We found some negative outputs of our model caused by translation errors such as missing words and extra words. The cause of the errors is mostly due to out-of-vocabulary or rare Chinese characters of input sequence.

Table 1 shows an example with translation errors of our sequence-to-sequence CWS system. The original input comes from the Weibo dataset (seen in Sect. 3.1). The output missed three Japanese characters "極" (extreme), "の" (of) and "親" (parent), and introduced three extra characters "UNK" instead which means 'unknown word' in the vocabulary.

Table 1. An example of translation errors in our CWS system and post-editing results.

Original input	岛国一超精分的小品《極道の親子》，看完之后我想说，为什么我没有这么"通情达理"的老爸呢？
System output	岛国 一 超 精分 的 小品 《 UNK道 UNK UNK子 》 ， 看完 之后 我 想 说 ， 为什么 我 没有 这么 " 通情达理 " 的 老爸 呢 ？
After post-editing	岛国 一 超 精分 的 小品 《 極道 の 親子 》 ， 看完 之后 我 想 说 ， 为什么 我 没有 这么 " 通情达理 " 的 老爸 呢 ？
Gold standard	岛国 一 超 精分 的 小品 《 極道 の 親子 》 ， 看完 之后 我 想 说 ， 为什么 我 没有 这么 " 通情达理 " 的 老爸 呢 ？

Algorithm 1. Post-editing algorithm for our CWS system

Input:

The original input character sequence: s_{ori};

The segmented word sequence with translation errors from sequence-to-sequence CWS system: s_{seg};

Output:

Segmentation labels set: $L \leftarrow \{B, M, E, S\}$;

Labeling characters in s_{seg} with labels in L gets lab_{seg};

$length_{ori} \leftarrow getLength(s_{ori})$, $length_{seg} \leftarrow getLength(s_{seg})$

if $length_{ori} \neq length_{seg}$ **then**

    Labeling characters in s_{ori} with position labels;

    Extracting the longest common subsequences between s_{ori} and s_{seg} using longest common subsequence (LCS) algorithm: $s_{sub} = LCS(s_{seg}, s_{ori})$;

    Taking s_{ori} as a reference, filling the missing characters in s_{sub} and labeling them with label X;

    Replacing label X with labels in L according to manually prepared rules;

else

    Labeling s_{ori} according to lab_{seg};

end if

Merging the characters in s_{ori} into word sequence s_{pe} according to their segmentation labels;

return s_{pe};

Inspired by Lin and Och [12], we proposed an LCS based post-editing algorithm[1] (seen in Algorithm 1) to alleviate the negative impact to CWS. In the algorithm, we define an extended word segmentation labels set $\{B, M, E, S, X\}$. $\{B, M, E\}$ represent begin, middle, end of a multi-character segmentation respectively, and S represents a single character segmentation. The additional label X in L can be seen as any other labels according to its context. For example, given a CWS label sequence (S, S, B, E, B, X, E), the X should be transformed into label M and in the other case of (S, X, B, E, B, M, E), the X should be treated as label S. The above transformation strategy can be based on handcraft rules

---

[1] Executable source code is available at https://github.com/SourcecodeSharing/CWSpostediting.

or machine learning methods. In this paper, we use the transformation rules written manually. Table 1 also gives an example of post-editing results.

3   Experiments

3.1   Datasets

We use three benchmark datasets, Weibo, PKU and MSRA, to evaluate our CWS model. Statistics of all datasets are shown in Table 2.

Weibo[2]: This dataset is provided by NLPCC-ICCPOL 2016 shared task of Chinese word segmentation for Micro-blog Texts [16]. The data are collected from Sina Weibo[3]. Different with the popular used newswire dataset, the texts of the dataset are relatively informal and consists various topics. Experimental results on this dataset are evaluated by eval.py scoring program[1], which calculates standard precision (P), recall (R) and F1-score (F) and weighted precision (P), recall (R) and F1-score (F) [15, 16] simultaneously.

PKU and MSRA[4]: These two datasets are provided by the second International Chinese Word Segmentation Bakeoff [8]. We found that the PKU dataset contains many long paragraphs consisting of multiple sentences, which has negative impacts on the training of the sequence translation models. To solve this problem, we divide the long paragraphs in the PKU dataset into sentences. Experiment results on those two datasets are evaluated by the standard Bakeoff scoring program[3], which calculates P, R and F scores.

Table 2. Statistics of different datasets. The size of training/testing datasets are given in number of sentences (Sents), words (Words) and characters (Chars).

Datasets	Training			Testing		
	Sents	Words	Chars	Sents	Words	Chars
Weibo	20,135	421,166	688,743	8,592	187,877	315,865
PKU	43,475	1,109,947	1,826,448	4,261	104,372	172,733
MSRA	86,924	2,368,391	4,050,469	3,985	106,873	184,355

3.2   Model Setup and Pre-training

We use the RNNsearch[5] model [1] to achieve our sequence-to-sequence CWS system. The model is set with embedding size 620, 1000 hidden units and an

---

[2] All data and the program are available at https://github.com/FudanNLP/NLPCC-WordSeg-Weibo.

[3] http://www.weibo.com.

[4] All data and the program are available at http://sighan.cs.uchicago.edu/bakeoff2005/.

[5] Implementations are available at https://github.com/lisa-groundhog/GroundHog.

alphabet with the size of 7190. We also apply the Moses' phrase-based (Moses PB) statistical machine translation system [10] with 3-gram or 5-gram language model as sequence-to-sequence translation baseline systems.

Since our sequence-to-sequence CWS model contains large amount numbers (up to ten million) of free parameters, it is much more likely to be overfitting when training on small datasets [17]. In fact, we make an attempt to train the model on the benchmark datasets directly and get poor scores as shown in Table 3. To deal with this problem, a large scale pseudo data is utilized to pre-train our model. The Weibo, PKU and MSRA datasets are then used for fine-tuning. To construct the pseudo data, we label the UN1.0 [26] with LTP[6] [3] Chinese segmentor. The pseudo data contains 12,762,778 sentences in the training set and 4,000 sentences in the validation set and the testing set. The testing set of the pseudo data is used to evaluate the pre-training performance of the model, and the result P, R and F scores are **98.2**, **97.1** and **97.7** respectively w.r.t the LTP label as the ground truth.

Table 3. Experimental results on benchmark datasets w/o pre-training.

Datasets	P	R	F
Weibo	89.8	89.5	89.6
PKU	87.0	88.6	87.8
MSRA	95.1	93.2	94.1

3.3 CWS Experiment Results

Weibo: For Weibo dataset, we compare our models with two groups of previous works on CWS as shown in Table 4. The LTP [3] in group A is a general CWS tool which we use to label pseudo data. S1 to S8 in Group B are submitted systems results of NLPCC-ICCPOL 2016 shared task of Chinese word segmentation for Micro-blog Texts [16]. Our works are shown in Group M. Since the testing set of Weibo dataset has many out-of-vocabulary (OOV) words, our post-editing method shows its effective for enhancing our CWS results for its abilities to recall missing words and replace extra words.

PKU and MSRA: For the two popular benchmark datasets, PKU and MSRA, we compare our model with three groups of previous models on CWS task as shown in Table 5. The LTP [3] in group A is same as Table 4. Group B presents a series of published results of previous neural CWS models with pre-trained character embeddings. The work proposed by Zhang et al. [23] in group C is one of the state-of-the-art methods. Our post-editing method dose not significantly enhance the CWS results for PKU and MSRA datasets comparing with Weibo dataset. The reason is that the text style in the two datasets is formal and

---

[6] Available online at https://github.com/HIT-SCIR/ltp.

Table 4. Experimental results on the CWS dataset of Weibo. The contents in parentheses represent the results of comparison with other systems.

Groups	Models	Standard scores			Weighted scores		
		P	R	F	P	R	F
A	LTP [3]	83.98	90.46	87.09	69.69	80.43	74.68
B [16]	S1	94.13	94.69	94.41	79.29	81.62	80.44
	S2	94.21	95.31	94.76	78.18	81.81	79.96
	S3	94.36	95.15	94.75	78.34	81.34	79.81
	S4	93.98	94.78	94.38	78.43	81.20	79.79
	S5	93.93	94.80	94.37	76.24	79.32	77.75
	S6	93.90	94.42	94.16	75.95	78.20	77.06
	S7	93.82	94.60	94.21	75.08	77.91	76.47
	S8	93.74	94.31	94.03	74.90	77.14	76.00
	S9	92.89	93.65	93.27	71.25	73.92	72.56
M	Moses PB w/3-gram LM	92.42	92.26	92.34	76.74	77.23	76.98
	Moses PB w/5-gram LM	92.37	92.26	92.31	76.58	77.25	76.91
	RNNsearch w/o fine-tuning	86.10	88.82	87.44	68.88	75.20	71.90
	RNNsearch	92.09	92.79	92.44	75.00	78.27	76.60
	RNNsearch w/post-editing	93.48(>S9)	94.60(>S6)	94.04(>S8)	76.30(>S5)	79.99(>S5)	78.11(>S5)

the OOV words are less common than Weibo dataset. In addition, the sequence translation baselines of Moses PB also gained decent results without pre-training or any external data.

According to all experimental results, our approaches still have gaps with the state-of-the-art methods. Considering the good performance (F1-score 97.7) on the pseudo testing data, the sequence-to-sequence CWS model has shown its capacity on this task and the data scale may be one of main limitations for enhancing our model.

3.4   Learning CWS and Chinese Spelling Correction Jointly

As a sequence translation framework, the model can achieve any expected kinds of sequence-to-sequence transformation with the reasonable training. It hence leaves a lot of space to tackle other NLP tasks jointly.

In this paper, we apply the model to jointly learning CWS and Chinese spelling correction (CSC). To evaluate the performance of spelling correction, we use automatic method to build two datasets, modified PKU and MSRA, based on assumptions that (i) most spelling errors are common with fixed pattern and (ii) the appearance of spelling errors are randomly. The details are as follows:

Table 5. Experimental results on the CWS benchmark datasets of PKU and MSRA.

Groups	Models	PKU			MSRA		
		P	R	F	P	R	F
A	LTP [3]	95.9	94.7	95.3	86.8	89.9	88.3
B	Zheng et al. [25]	93.5	92.2	92.8	94.2	93.7	93
	Pei et al. [13]	94.4	93.6	94.0	95.2	94.6	94.9
	Chen et al. [4]	96.3	95.9	96.1	96.2	96.3	96.2
	Chen et al. [5]	96.3	95.6	96.0	96.7	96.5	96.6
	Cai and Zhao [2]	95.8	95.2	95.5	96.3	96.8	96.5
C	Zhang et al. [23]	-	-	96.1	-	-	97.4
M	Moses PB w/3-gram LM	92.9	93.0	93.0	96.0	96.2	96.1
	Moses PB w/5-gram LM	92.7	92.8	92.7	95.9	96.3	96.1
	Moses PB w/3-gram LM w/CSC	92.9	93.0	92.9	95.3	96.5	95.9
	Moses PB w/5-gram LM w/CSC	92.6	93.2	92.9	95.9	96.3	96.1
	RNNsearch w/o fine-tuning	93.1	92.7	92.9	84.1	87.9	86.0
	RNNsearch	94.7	95.3	95.0	96.2	96.0	96.1
	RNNsearch w/post-editing	94.9	95.4	95.1	96.3	96.1	96.2
	RNNsearch w/CSC	95.2	94.6	94.9	96.1	96.1	96.1
	RNNsearch w/CSC and post-editing	95.3	94.7	95.0	96.2	96.1	96.2

firstly, we construct a correct-to-wrong word pair dictionary counting from the Chinese spelling check training dataset of SIGHAN 2014 [22] as a fixed pattern of spelling errors; secondly, we randomly select 50% sentences from PKU and MSRA training set respectively and replace one of the correct words with the wrong one according to the dictionary for each selected sentence. The testing set is generated in the same way.

Table 6. An example of modified data. The character with double underline is wrong and the characters with single underlines are correct.

Original input	在这个基础上，公安机关还从原料采购等方面加以严格控制，统一发放"准购证"。
Modified input	在这个基础上，公安机关还从源料采购等方面加以严格控制，统一发放"准购证"。
Gold standard	在 这个 基础 上 ， 公安 机关 还 从 原料 采购 等 方面 加以 严格 控制 ， 统一 发放 " 准购证 " 。

We treat the modified sentences and the original segmented sentences as the input sequence and the golden standard respectively in the training procedure. Table 6 gives an example of the modified data. In the testing procedure, we send

Table 7. Experimental results on modified PKU data. The numbers in parentheses represent the changes compared with the normal CWS results shown in Table 5.

Models	P	R	F
Modified testing data	99.0	99.0	99.0 (−1.0)
LTP [3]	94.0	93.2	93.6 (−1.7)
Moses PB w/3-gram LM	90.8	91.5	91.2 (−1.8)
Moses PB w/3-gram LM w/CSC	92.7	92.9	92.8 (−0.1)
Moses PB w/5-gram LM	90.6	91.3	91.0 (−1.7)
Moses PB w/5-gram LM w/CSC	92.3	93.0	92.6 (−0.3)
RNNsearch	93.2	93.2	93.2 (−1.8)
RNNsearch w/CSC	**95.0**	**94.5**	**94.8** (−0.1)

Table 8. Experimental results on modified MSRA data. The numbers in parentheses represent the changes compared with the normal CWS results shown in Table 5.

Models	P	R	F
Modified testing data	98.5	98.5	98.5 (−1.5)
LTP [3]	84.8	88.4	86.6 (−1.7)
Moses PB w/3-gram LM	93.7	94.6	94.2 (−1.9)
Moses PB w/3-gram LM w/CSC	95.0	96.3	95.6 (−0.3)
Moses PB w/5-gram LM	93.7	94.7	94.2 (−1.9)
Moses PB w/5-gram LM w/CSC	94.6	95.9	95.3 (−0.7)
RNNsearch	93.8	94.7	94.2 (−1.9)
RNNsearch w/CSC	**96.0**	**96.0**	**96.0** (−0.1)

the sentence with wrong words into the model, and expect to get the segmented sentence with all correct words. The results are shown in Tables 7 and 8. Since the general CWS tool LTP does not have the ability to correct spelling mistakes, the performance decreases. Whereas, the impact of the wrong words is limited in our models trained to do CWS and CSC jointly.

4   Related Work

CWS using neural networks have gained much attention in recent years as they are capable of learning features automatically. Collobert et al. [7] developed a general neural architecture for sequence labelling tasks. Zheng et al. [25] adapted word embedding and the neural sequence labelling architecture [7] for CWS. Pei et al. [13] improved upon Zheng et al. [25] by modeling complicated interactions between tags and context characters. Chen et al. [4] proposed gated recursive neural networks to model the combinations of context characters. Chen et al. [5] introduced LSTM into neural CWS models to capture the potential long-distance dependencies. However, the methods above all regard CWS as sequence labelling with local input features. Cai and Zhao [2] re-formalize CWS as a direct segmentation learning task without the above constrains, but the maximum length of words is limited.

Sequence-to-Sequence Machine Translation Models. Neural sequence-to-sequence machine translation models have rapid developments since 2014. Cho et al. [6] proposed an RNN encoder-decoder framework with gated recurrent unit to learn phrase representations. Sutskever et al. [19] applied LSTM for RNN encoder-decoder framework to establish a sequence-to-sequence translation framework. Bahdanau et al. [1] improved upon Sutskever et al. [19] by introducing an attention mechanism. Wu et al. [20] presented Google's Neural Machine Translation system which is serving as an online machine translation system. Gehring et al. [9] introduce an architecture based entirely on convolutional neural networks to sequence-to-sequence learning tasks which improved translation accuracy at an order of magnitude faster speed. Other efficient sequence-to-sequence models will be introduced into this task and compared with existing works in our future work.

5   Conclusion

In this paper, we re-formalize the CWS as a sequence-to-sequence translation problem and apply an attention based encoder-decoder model. We also make an attempt to let the model jointly learn CWS and CSC. Furthermore, we propose an LCS based post-editing algorithm to deal with potential translating errors. Experimental results show that our approach achieves competitive performances compared with state-of-the-art methods both on normal CWS and CWS with CSC.

In the future, we plan to apply other efficient sequence-to-sequence models for CWS and study an end-to-end framework for multiple natural language preprocessing tasks.

Acknowledgments. This work was supported by the National Basic Research Program (973) of China (No. 2013CB329303) and the National Natural Science Foundation of China (No. 61132009).

References

1. Bahdanau, D., Cho, K., Bengio, Y.: Neural machine translation by jointly learning to align and translate. arXiv preprint arXiv:1409.0473 (2014)
2. Cai, D., Zhao, H.: Neural word segmentation learning for Chinese. arXiv preprint arXiv:1606.04300 (2016)
3. Che, W., Li, Z., Liu, T.: LTP: a Chinese language technology platform. In: Proceedings of the 23rd International Conference on Computational Linguistics: Demonstrations, pp. 13–16. Association for Computational Linguistics (2010)

4. Chen, X., Qiu, X., Zhu, C., Huang, X.: Gated recursive neural network for Chinese word segmentation. In: ACL, vol. 1, pp. 1744–1753 (2015)
5. Chen, X., Qiu, X., Zhu, C., Liu, P., Huang, X.: Long short-term memory neural networks for Chinese word segmentation. In: EMNLP, pp. 1197–1206 (2015)
6. Cho, K., Van Merriënboer, B., Gulcehre, C., Bahdanau, D., Bougares, F., Schwenk, H., Bengio, Y.: Learning phrase representations using RNN encoder-decoder for statistical machine translation. arXiv preprint arXiv:1406.1078 (2014)
7. Collobert, R., Weston, J., Bottou, L., Karlen, M., Kavukcuoglu, K., Kuksa, P.: Natural language processing (almost) from scratch. J. Mach. Learn. Res. **12**, 2493–2537 (2011)
8. Emerson, T.: The second international Chinese word segmentation bakeoff. In: Proceedings of the Fourth SIGHAN Workshop on Chinese Language Processing, vol. 133 (2005)
9. Gehring, J., Auli, M., Grangier, D., Yarats, D., Dauphin, Y.N.: Convolutional sequence to sequence learning. ArXiv e-prints, May 2017
10. Koehn, P., Hoang, H., Birch, A., Callison-Burch, C., Federico, M., Bertoldi, N., Cowan, B., Shen, W., Moran, C., Zens, R., et al.: Moses: open source toolkit for statistical machine translation. In: Proceedings of the 45th Annual Meeting of the ACL on Interactive Poster and Demonstration Sessions, pp. 177–180. Association for Computational Linguistics (2007)
11. Lafferty, J., McCallum, A., Pereira, F., et al.: Conditional random fields: probabilistic models for segmenting and labeling sequence data. In: Proceedings of the Eighteenth International Conference on Machine Learning, ICML, vol. 1, pp. 282–289 (2001)
12. Lin, C.Y., Och, F.J.: Automatic evaluation of machine translation quality using longest common subsequence and skip-bigram statistics. In: Proceedings of the 42nd Annual Meeting on Association for Computational Linguistics, p. 605. Association for Computational Linguistics (2004)
13. Pei, W., Ge, T., Chang, B.: Max-margin tensor neural network for Chinese word segmentation. In: ACL, vol. 1, pp. 293–303 (2014)
14. Peng, F., Feng, F., McCallum, A.: Chinese segmentation and new word detection using conditional random fields. In: Proceedings of the 20th International Conference on Computational Linguistics. p. 562. Association for Computational Linguistics (2004)
15. Qiu, P., Qiu, X., Huang, X.: A new psychometric-inspired evaluation metric for Chinese word segmentation. In: Proceedings of the 54th Annual Meeting of the Association for Computational Linguistics, pp. 2185–2194 (2016)
16. Qiu, X., Qian, P., Shi, Z.: Overview of the NLPCC-ICCPOL 2016 shared task: Chinese word segmentation for micro-blog texts. In: Lin, C.-Y., Xue, N., Zhao, D., Huang, X., Feng, Y. (eds.) ICCPOL/NLPCC-2016. LNCS, vol. 10102, pp. 901–906. Springer, Cham (2016). doi:10.1007/978-3-319-50496-4_84
17. Srivastava, N., Hinton, G.E., Krizhevsky, A., Sutskever, I., Salakhutdinov, R.: Dropout: a simple way to prevent neural networks from overfitting. J. Mach. Learn. Res. **15**(1), 1929–1958 (2014)
18. Sun, X., Li, W., Wang, H., Lu, Q.: Feature-frequency-adaptive on-line training for fast and accurate natural language processing. Comput. Linguist. **40**(3), 563–586 (2014)
19. Sutskever, I., Vinyals, O., Le, Q.V.: Sequence to sequence learning with neural networks. In: Advances in Neural Information Processing Systems, pp. 3104–3112 (2014)

20. Wu, Y., Schuster, M., Chen, Z., Le, Q.V., Norouzi, M., Macherey, W., Krikun, M., Cao, Y., Gao, Q., Macherey, K., et al.: Google's neural machine translation system: bridging the gap between human and machine translation. arXiv preprint arXiv:1609.08144 (2016)
21. Xue, N., Shen, L.: Chinese word segmentation as LMR tagging. In: Proceedings of the Second SIGHAN Workshop on Chinese Language Processing, vol. 17, pp. 176–179 (2003)
22. Yu, L.C., Lee, L.H., Tseng, Y.H., Chen, H.H., et al.: Overview of SIGHAN 2014 bake-off for Chinese spelling check. In: Proceedings of the 3rd CIPSSIGHAN Joint Conference on Chinese Language Processing (CLP 2014), pp. 126–132 (2014)
23. Zhang, L., Wang, H., Sun, X., Mansur, M.: Exploring representations from unlabeled data with co-training for Chinese word segmentation In: Proceedings of the 2013 Conference on Empirical Methods in Natural Language Processing, pp. 311–321 (2013)
24. Zhao, H., Huang, C.N., Li, M., Lu, B.L.: A unified character-based tagging framework for chinese word segmentation. ACM Trans. Asian Lang. Inf. Process. (TALIP) 9(2), 5 (2010)
25. Zheng, X., Chen, H., Xu, T.: Deep learning for Chinese word segmentation and pos tagging. In: EMNLP, pp. 647–657 (2013)
26. Ziemski, M., Junczys-Dowmunt, M., Pouliquen, B.: The united nations parallel corpus v1.0. In: Proceedings of the Tenth International Conference on Language Resources and Evaluation, LREC, pp. 23–28 (2016)

Attention-Based Memory Network
for Sentence-Level Question Answering

Pei Liu[1]([✉]), Chunhong Zhang[1], Weiming Zhang[1], Zhiqiang Zhan[2],
and Benhui Zhuang[1]

[1] Key Laboratory of Universal Wireless Communications,
Ministry of Education, Beijing University of Posts
and Telecommunications, Beijing, China
{peiliu,zhangch,zhangwm,zhuangbenhui}@bupt.edu.cn
[2] School of Information and Communication Engineering,
Engineering Research Center of Information Networks, Ministry of Education,
Beijing University of Posts and Telecommunications, Beijing, China
zqzhan@bupt.edu.cn

Abstract. Sentence-level question answering (QA) for news articles is a promising task for social media, whose task is to make machine understand a news article and answer a corresponding question with an answer sentence selected from the news article. Recently, several deep neural networks have been proposed for sentence-level QA. For the best of our knowledge, none of them explicitly use keywords that appear simultaneously in questions and documents. In this paper we introduce the Attention-based Memory Network (Att-MemNN), a new iterative bi-directional attention memory network that predicts answer sentences. It exploits the co-occurrence of keywords among questions and documents as augment inputs of deep neural network and embeds documents and corresponding questions in different way, processing questions with word-level and contextual-level embedding while processing documents only with word-level embedding. Experimental results on the test set of NewsQA show that our model yields great improvement. We also use quantitative and qualitative analysis to show the results intuitively.

Keywords: Sentence-level question answering for news articles · Attention mechanism · Memory network · Deep learning

1 Introduction

Question answering (QA) for social media is a complex research problem in natural language processing because of the rapid growth of news articles and the diversity of text expressions in news article.

Our task is sentence-level QA which extracts an answer sentence from a document which is news article in QA for social media to answer a question based on the document. Many proposed works have focused on answer-sentence selection problems to leverage deep neural networks [1–4]. All of the models use datasets from the annual TREC evaluations [5] and WikiQA [6]. This kind of dataset provides a question and a set of candidate sentences and we should choose the best sentence from a candidate

© Springer Nature Singapore Pte Ltd. 2017
X. Cheng et al. (Eds.): SMP 2017, CCIS 774, pp. 104–115, 2017.
https://doi.org/10.1007/978-981-10-6805-8_9

sentence set that can answer the question. Most recently, Trischler et al. (2016) present a challenging new large-scale dataset for machine comprehension: NewsQA [7], whose source material was chosen from CNN articles. Unlike TREC and WikiQA dataset, NewsQA provides a document and a question based on the document and we should answer the question with a sentence from the document. To explore NewsQA task, we propose a new iterative bi-directional attention neural network architecture. In our model, we explicitly use the keywords which appear simultaneously in the question and corresponding document. This idea is inspired from the fact that when human do the task of reading comprehension, some semantic words including verbs, nouns and all other words except prepositions and conjunctions in the question are critical clues to find the correct answer from the document, which we call them keywords. After identifying keywords in the question, human always find the same words in the document to answer the question. For example, given a question: *Who is Barcelona playing against?*, human always focus on *"Barcelona"* and *"play against"* and find the same words or words that represent the same meaning in the document. In a segment of corresponding document: *Barcelona has been in indifferent recent form and a 1-1 draw at Athletic Bilbao on Saturday. Barca will certainly want the key pair to be fit for next Sunday's El Clasico against Real.* The same word is the name of football club *"Barcelona"*, and the *"Barca"* is the same meaning of *"Barcelona"*, the *"against"* is the same meaning of *"playing against"* but have different spellings. Inspired by this way, we exploit the co-occurrence of words among questions and documents as augment input of our model, which has been reported to be one of the most important features for modeling question answering problem using a logistic regression model [8]. For the best of our knowledge, none of previous proposed deep neural networks takes the information of keywords as augment input of neural network. We index every sentence in the document with the keywords information into an index-vector and apply it to hidden representations of our neural attention model, which gets a noticeable improvement.

Our model can also be seen as a kind of Memory Network, generalizes the original Memory Network, MemN2N [9]. Both of the models have memory components to read from and write to, which can make iterative attention process. Our model offers following improvements to the benchmark model [9]. First, it explicitly uses keywords information and applies it to hidden representations in the memory network. Especially, due to the diversity of text expression in news articles, we use text normalization to transform documents and questions into a single canonical form, which helps to avoid missing the matching keywords in the document. Second, instead of uni-directional weight calculation in the baseline model, we use bi-directional attention mechanism in our model. We use a similarity matrix to calculate two different weights on both of document and question. The attention mechanism in our model is similar with it in MPCM model which only encodes a weighted document and an original question [10]. It is also similar with bi-directional attention flow in Bi-DAF network [11] whose target is to produce a set of question-aware feature vectors for each word in the document while our target is to produce a weighted document and a weighted question. Third, Sentence-level QA system always focus on every sentence in a document and temporal interactions between words in the document have less effect on our model. Therefore, we process questions with word-level and contextual-level embedding while process documents only with word-level embedding.

In this paper we introduce the Attention-based Memory Network (Att-MemNN), a new iterative bi-directional attention memory network architecture. It explicitly applies the keywords information to hidden representations in deep neural network and embeds documents and questions in different way. We perform experiments on the high-quality NewsQA dataset and our approach outperforms baseline methods by a significant amount. We also use quantitative and qualitative analysis to show the results intuitively.

2 Related Work

Recent years, many deep neural networks have been proposed for the QA task [12, 13], which have accelerated the progress of QA system. In this work we propose Att-MemNN model for sentence-level QA, and there are three main works which we are related to.

2.1 Question Answering System

Based on information retrieval, early QA systems were designed to return a segment of text from the corresponding reading document to answer a question which usually stuck in employing linguistic tools, feature engineering or other simple networks [8, 14]. However, without the use of deep natural networks all of the systems make a poor performance because of errors of many NLP tools and limitations of additional resources. Recently, many deep natural network models have been proposed for QA. From the way of identifying answers, most of the models can be roughly categorized into two classes: selecting the answer from a set of alternatives [15, 16] and extracting the answer from corresponding documents [17, 18]. In the former kind of method, we always extract candidate answers and train the model to rank the correct to the top of the list. The latter can be divided into sentence-level QA whose answer is a sentence from the corresponding document and span-level QA whose answer is a segment of text from the document. For span-level QA, Vinyals et al. (2015) use the Pointer Network to return a list of positions from the document as the final answer [19]. However, we cannot guarantee the selected positions to be consecutive. Xiong et al. (2016) introduce Dynamic Coattention Network (DCN) for question answering, which can recover from local maxima corresponding to incorrect answers [18].

2.2 Attention Based Models

Attention mechanisms are important in natural networks, which can significantly improve the performance of QA systems. There are many works have been done to show the effect of attention mechanisms [2, 20]. In attention based QA models, the representation of document is always built with attention from the representation of question which is uni-directional attention mechanism. Wang et al. (2016) use uni-directional attention mechanism in its model, adjusting each word-embedding vector in the document by multiplying a relevancy weight computed against the question [10]. Sukhbaatar et al. (2015) proposed a recurrent attention model with a large external memory [9], which is also a kind of uni-directional attention mechanism. There are also some models represent questions with attention from the representation of documents [21]. To get a better performance, many QA systems start to use bi-directional attention mechanism in their model, which provide complimentary information to both of documents and the questions. Seo et al. (2017) proposed the Bi-Directional Attention Flow (BIDAF) network with the use of bi-directional attention flow mechanism which obtains the attentions and the attended vectors in both directions of document-to-question and question-to-document [11]. We use similar bi-directional attention mechanism in our model. However, the target of attention mechanism in BIDAF network is to produces a set of question-aware feature vectors for each word in the document while our target is to produce a weighted document and a weighted question.

2.3 Memory Networks

There are two difficulties in reading comprehension models: making multiple computational steps and representing long-term dependencies sequential. Many ways have been explored to exploit long-distance sequential information using RNNs or LSTM-based models which use the state of models to be memory [2, 22, 23]. However, the memory represented in that way is not stable over long timescales. Some works try to use global memory components in their models. Graves et al. (2014) proposed a Neural Turing Machine (NTM) model using a continuous memory representation [24]. However, the memory size in that model is small and the operation of sorting and recalling in NTM requires more complex models. Weston et al. (2014) proposed a Memory Network with a long-term memory component which enables multiple computational steps [25]. There are two deficiencies that the model requires supervision at each layer and is not easy to train via backpropagation algorithm. Sukhbaatar et al. (2015) proposed a continuous form of Memory Network, MemN2N which is trained end-to-end and requires less supervision [9]. Our model generalizes MemN2N model and offers some improvements to this benchmark model.

3   Model

In this section, we propose an Attention-based Memory Network (Att-MemNN) to estimate probability distribution P upon all of the sentences in the document to predict the answer sentence. Figure 1 shows the architecture of our model. Here the input of our model is a document and a corresponding question which are successively passed through embedding layer, multi-hops attention layer and output layer to get an answer sentence for the question as the output of our model. The keywords information module uses the document and the question to obtain an augment input for multi-hops attention layer and output layer.

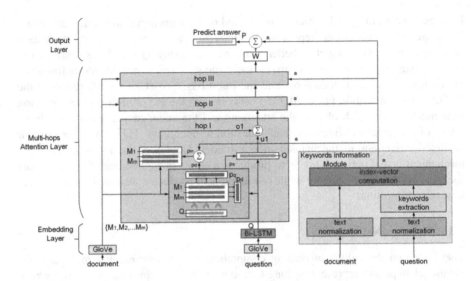

Fig. 1. Architecture of Attention Based Memory Network (Att-MemNN). Our model is stacked to multiple hops and is set to be 3 hops in this architecture.

3.1 Keywords Information Module

The target of this module is to represent keywords information. The "keywords" in this paper are semantic words that appear simultaneously in the document and corresponding question. To represent the keywords information, we propose an index-vector whose detail will be described in the following. The inputs of this module are raw texts of a document and corresponding question while the output is an index-vector which is sent to each hop of the attention layer and the output layer. The advantage of the module is to increase the weights of some sentences containing the same keywords with the question in the document. Though, for some examples, this way may results new noise, the adaptability of the model will reduce the impact of noise and experiments show that this way yields great improvement.

Firstly, we use text normalization to transform the document and question into a single canonical form [26], which makes the inputs of next step to be consistent texts to avoid missing the same keywords in document. Secondly, we extract keywords in the question. There are many ways to extract keywords such as simple statistic approach, linguistic approach, machine learning approach and hybrid approach [27]. In our model, the simple statistic approach is used to extract keywords in the question. To be simple, if words in the question are semantic words, they are chose to be keywords. Finally, determine which sentences in the document contain keywords. The event of word co-occurrence for each individual sentence is indicated by an (0, 1)-element of index-vector for the whole document. For example, given a question: *Where is Sonia Sotomayor?*, the keyword in this question is *Sonia Sotomayor*. Then, given a document: *Sonia Sotomayor goes to the bed room. Tom goes to the bathroom. Mary returns to the garden*, we index the first sentence containing *Sonia Sotomayor* with "1" and other sentences with "0" and obtain an index-vector [1, 0, 0]. In this module, index-vector is represented by $a \in R^m$ consisting of 0 and 1 for each document where m is the maximum number of sentences of all the documents. If the number of sentences in a document is less than m, the index-vector will be padded with 0.

3.2 Embedding Layer

The target of this layer is to embed the document and question in different ways. As is shown in Fig. 1, the inputs of embedding layer are a question $q \in R^w$ and a document represented by a discrete set $s_1, s_2, \ldots, s_m$ where $s_i \in R^w$ represents the i-th sentence in the document and w is the maximum number of words of the sentences in the documents and questions. Each of the s_i, q contains w symbols coming from a dictionary which indexes every word in NewsQA dataset with a unique number. If the number of words in a sentence is less than w, s_i and q will be padded with symbol 0. The outputs of this layer are a question vector $Q \in R^e$ obtained from question q and memory vectors $\{M_i\}(M_i \in R^e)$ to represent the document obtained from a discrete set $\{s_i\}$.

When processing the question, we use word-level and contextual-level embedding. In the word-level embedding, we use pre-trained word vectors, GloVe [28], to represent every symbol in q with an e-dimensional continuous vector and obtain an intermediate matrix $q' \in R^{w \times e}$ for the question. We take the intermediate matrix q' as input of contextual-level embedding. In contextual-level embedding, we place a Long Short-Term Memory Network (LSTM) in both directions to utilize contextual cues from surrounding words to refine the embedding of the words. We sum the outputs of the two LSTM together by which we get a matrix (of size $w \times e$). Then, in order to convert the matrix into a vector, elements of the matrix were summed in column. In this way, we convert the question q into a question vector $Q \in R^e$. On account that sentence-level QA system always focus on every sentence in a document and temporal interactions between words in the document have less effect on our model, we doesn't process the document with contextual-level embedding. With the same way using in word-level embedding of the question, we embed a discrete set $\{s_i\}$ into memory vectors $\{M_i\}$.

3.3 Multi-hops Attention Layer

This is the core layer in our model and there is a memory component with shared read and write functions. In typical memory model, there are many memory input/output operations in the same way using in MemN2N [9]. To be simple, we write representation of the document into memory in embedding layer and read the memory in multi-hops attention layer many times. In this layer, the continuous memory representation for document and continuous representation for question are processed via multiple hops. In Fig. 1, it shows a model stacked to 3 hops and we simplify the graphical representation of the second and the third hop.

In each hop, we use a bi-directional attention mechanism on both of the question and the document stored in the memory. To calculate the memory weight p_m on the document and question weight p_q on the question, we firstly calculate a similarity matrix $S \in R^{m \times e}$ by taking the inner product of Q and $\{M_i\}$. S_{ij} is a numerical value which indicates the similarity between i-th element in question and j-th element of i-th sentence in the document:

$$S_{ij} = Q_i M_{ij}. \tag{1}$$

By similarity matrix, we can easily get a document weight $p_d \in R^m$ which indicates which sentences in the document are more relevant to the question. Because the i-th line of the similarity matrix represents the similarity between each element of the question and i-th sentence in the document, we sum elements of the similarity matrix in row to get a document weight $p_d \in R^m$ for every sentence in the document:

$$p_d = Softmax(\sum\nolimits_j S_{ij}). \tag{2}$$

The index-vector $a \in R^m$ obtained by keywords information module is an augment input for this layer. It models the event of keywords co-occurrence and can also measure which sentences in the document are more relevant to the question. Although p_d is somewhat an indication of the similar relation between question and the sentences in document which is usually adopted by previous attention mechanism, the index-vector will enhance the similar relation as an explicit prior knowledge. Therefore, we sum index-vector to document weight p_d by which we increase the weights of sentences in document containing keywords of corresponding question. Then, the document weight p_d was updated to the memory weight $p_m \in R^m$:

$$p_m = Softmax(p_d + a). \tag{3}$$

By using attention mechanism on every sentence in the memory with the memory weight p_m, we obtain a response vector $o \in R^e$ from the memory vectors $\{M_i\}$:

$$o = \sum\nolimits_i p_{m_i} M_i. \tag{4}$$

In a similar way, we add elements of the similarity matrix in column to get a question weight $p_q \in R^e$ for the question vector Q and weight every element in the question to obtain an internal state $u \in R^e$ from question vector Q:

$$p_q = Softmax(\sum\nolimits_i S_{ij}). \tag{5}$$

$$u = p_q \circ Q. \tag{6}$$

where o in the formula represents Hadamard product.

The output of this hop is $(u \cdot H + o)$ where H is a trainable matrix of size $e \times e$. This output is inputted to the next hop as the question vector Q of the next hop. Every hop in our model has the same architecture and $\{M_i\}$ is obtained by memory output operations in each hop. The output of the modeling layer is the response vector o and the internal state u of the last hop.

3.4   Output Layer

The target of the output layer is to estimate probability distribution on all of the sentences in the document and predict the answer sentence. The sum of the output vector o and internal state u is then passed through a final weight matrix $W \in R^{e \times m}$ and a softmax to get an intermediate probability $P_i \in R^m$:

$$P_i = Softmax(W(o + u)). \tag{7}$$

The index-vector containing keywords information is utilized in the probability distribution, which can increase the weights of sentences containing keywords of corresponding question and eliminate interference from other sentences. The index-vector a is added to the intermediate probability to produce final predicted probability P:

$$P = Softmax(P_i + a). \tag{8}$$

The predicted probability P is used to predict the answer sentence.

During training, the training loss (to be minimized) is defined as the standard cross-entropy loss between predicted probability P and the true probability P'. The matrix W and H are jointly learned when the training is performed using stochastic gradient descent. During testing, the sentence with the maximum probability is chosen, computed by the predicted probability P.

4   Experiment

We conducted our experiments on the NewsQA dataset to evaluate the performance of our model.

4.1   Dataset

NewsQA is a crowd-sourced machine comprehension dataset on a large set of CNN articles. The number of average words per article is 616 from which we can see the article in NewsQA is large volumes of text. To evaluate our model, we use accuracy, can also be seen as "Exact match (EM)", which calculate the ratio of questions that are answered correctly. We also use F1-Measure to evaluate models which is calculated by precision and recall. In our experiment, searching the wide space of possible configurations is quite costly because of the size of the dataset. To alleviate this, we randomly select 3221 question-answer pairs to train the model and 546 question-answer pairs evaluated the performance of model. The NewsQA dataset is for span-level QA systems, we extract sentences containing answer spans to be the answer of corresponding questions. In batch tests, we randomly divide the test set into three parts. Each part contains 182 question-answer pairs. In addition similar results are obtained from all parts. The maximum number of sentences of all the documents (represent by m in our model) is 152 and the maximum number of words of the sentence in all the documents and questions (represent by w in our model) is 155.

4.2 Model Setup

In embedding layer, we use 50-dimensional vector to represent each word in documents and questions. We use the Adam optimizer, with an initial learning rate of 0.01 and an epsilon value of 1e−8. No momentum or weight decay was used. We use a batch size of 32 in all training, and the maximum gradient norm is 40. The gradient in training is clip to this norm. A dropout rate of 0.26 is used for the model. Since the number of sentences and the number of words are constrained into fixed size, we use some null symbol to pad them. The training process takes roughly 480 min on a single NVIDIA GPU.

Because of the random initialization, the result of every training process is different. To remedy this, we repeated the training for five times and picked the best result as the final result.

4.3 Results and Analysis

The results of our model and competing approaches are shown in Table 1. We evaluated our model with accuracy, can also be seen as "Exact match (EM)"and F1-Measure. In Table 1, the inverse sentence frequency (ISF) model is proposed by [11], which is a technique that resembles inverse document frequency (idf). The MemN2 N model is proposed by [13] used for bAbI task and we make minor modifications to apply it for sentence-level QA task. As we can see, the accuracy of our model is 61.3% exceeding MemN2N by 29.2% and ISF by 25.9%. Our model yields improved results.

Table 1. The performance of our model Att-MemNN and competing approach including ISF [7] and MemN2N [9]. Memory module of our model is set to 3 hops.

Model	Accuracy (EM)	F1
ISF [11]	35.4%	
MemN2N [9]	32.1%	38.3%
Att-MemNN (ours)	61.3%	69.1%

Model Ablation. We also propose an ablation subtask to evaluate the effectiveness of various improvements in our Att-MemNN model. In introduction of this paper, we have proposed three improvements of our model. We remove one improvement at a time to perform the experiment. When removing different embedding, we use a trainable embedding matrix to embed documents and questions which is used in benchmark model, MemN2N [9]. Table 2 shows the results of all ablation models and our full model on NewsQA. We can see that each of the components have effect on the model. Removing keywords information module reduces the accuracy by 22.7% and F1 by 22.5%. Changing bi-attention mechanism into uni-attention mechanism reduces the accuracy by 5.6% and F1 by 6.6%. Removing different embedding reduces the accuracy by 5.1% and F1 by 5.6%. Among all the components, removing the keywords information module decreases the performance significantly. It indicates that keywords information has the biggest promotion for our model among the three improvements.

Table 2. Test accuracy on ablation experiment. Memory module of our model is set to 3 hops.

Model	Accuracy	F1
Att-MemNN	61.3%	69.1%
No keywords information	38.6%	46.6%
No bi-attention	55.7%	62.5%
No different embeddings	56.2%	63.6%

Table 3. Effectiveness of multiply hops memory module on NewsQA.

Model	Accuracy	F1
1 hop	47.5%	55.7%
2 hops	55.6%	61.6%
3 hops	61.3%	69.1%

Analysis of Multiply Hops Memory Module. We put quantitative and qualitative analysis on multiply hops memory module which is an important part of our model. Table 3 shows the effectiveness of multiply hops memory module on NewsQA. Memory module of our model is set to 1 hop, 2 hops and 3 hops. We note that for NewsQA, multiply hops memory module which enable iterative attention are crucial to achieving high performance. Figure 2 shows the attention weights on every sentence in a document for 1-hop model and 3-hops model. This example demonstrates that the multi-hops memory module allows the model sharply focus on relevant sentences.

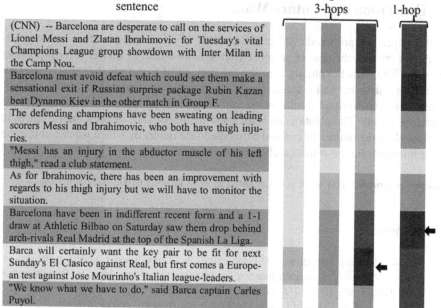

Fig. 2. Attention weights on every sentence in a document for 1-hop model and 3-hops model. In 3-hops model, it shows attention weights of the first hop, second hop and third hop from left to right. Color deepness in the picture means different weight. The sentence indicated by the arrow is the predicted answer sentence of the model. (Color figure online)

Table 4. Analysis on keywords information module.

Question	Answer
Q: What is going live on **Tuesday**?	A: Comcast rolled out a Web-based on-demand television and movie service on **Tuesday**
Q: Who is **Sonia Sotomayor**?	A: Judge **Sonia Sotomayor**, center, meets with staffers from the White House Counsel's Office
Q: What was the **space station crew forced** to take **shelter** from?	A: Last week, a piece of debris **forced** the **space station's** current **crew**
Q: What does **Gary go use** for **musical accompaniment**?	A: **Gary go uses** iPhone apps to help him compose new material – and provide instant **accompaniment**

Analysis of Keywords Information Module. To show the effect of keywords information, we output results of keywords extraction module in our model. Table 4 shows some examples on keywords information. We randomly select several question-answer pairs from test set of NewsQA and highlight keywords in questions extracted by keywords extraction module and the corresponding keywords in answer sentence. The result of Table 4 and statistics suggest that 70.5% correct sentences contain keywords in questions, which show that keywords information is useful for sentence-level QA system.

5 Conclusion and Future Work

In this paper, we proposed Att-MemNN, a new bi-directional attention memory network that predicts the answer sentence from a news article to answer a corresponding question. The model explicitly uses the information of keywords that appear simultaneously in questions and documents and represents documents and questions in different way. Experimental results on the test set of NewsQA show that our model yields improved results. The ablation analyses show the importance of each improvement in our model. In the future, we can add a module to the model by which we can obtain the exact answer from the sentence chosen from our model.

Acknowledgments. This work was supported by NSF Projects 61602048, 61302077.

References

1. Tymoshenko, K., Bonadiman, D., Moschitti, A.: Convolutional neural networks vs. convolution kernels: feature engineering for answer sentence reranking. In: Proceedings of NAACL-HLT, pp. 1268–1278 (2016)
2. Wang, B., Liu, K., Zhao, J.: Inner attention based recurrent neural networks for answer selection. In: The Annual Meeting of the Association for Computational Linguistics (2016)
3. Zheng, Z.: AnswerBus question answering system. In: Proceedings of the Second International Conference on Human Language Technology Research, pp. 399–404. Morgan Kaufmann Publishers Inc., March 2002

4. Tahri, A., Tibermacine, O.: DBPedia based factoid question answering system. Int. J. Web Semant. Technol. **4**(3), 23 (2013)
5. Wang, M., Smith, N.A., Mitamura, T.: What is the Jeopardy Model? A quasi-synchronous grammar for QA. In: EMNLP-CoNLL, vol. 7, pp. 22–32, June 2007
6. Yang, Y., Yih, W.T., Meek, C.: WikiQA: a challenge dataset for open-domain question answering. In: EMNLP, pp. 2013–2018, September 2015
7. Trischler, A., Wang, T., Yuan, X., Harris, J., Sordoni, A., Bachman, P., Suleman, K.: NewsQA: a machine comprehension dataset. arXiv preprint arXiv:1611.09830 (2016)
8. Rajpurkar, P., Zhang, J., Lopyrev, K., Liang, P.: Squad: 100,000+ questions for machine comprehension of text. arXiv preprint arXiv:1606.05250 (2016)
9. Sukhbaatar, S., Weston, J., Fergus, R.: End-to-end memory networks. In: Advances in Neural Information Processing Systems, pp. 2440–2448 (2015)
10. Wang, Z., Mi, H., Hamza, W., Florian, R.: Multi-perspective context matching for machine comprehension. arXiv preprint arXiv:1612.04211 (2016)
11. Seo, M., Kembhavi, A., Farhadi, A., Hajishirzi, H.: Bidirectional attention flow for machine comprehension. arXiv preprint arXiv:1611.01603 (2016)
12. Wang, W.H., Yang, N., Wei, F.R., et al.: Gated self-matching networks for reading comprehension and question answering. In: ACL (2017)
13. Hu, M., Peng, Y., Qiu, X.: Mnemonic reader for machine comprehension. arXiv preprint arXiv:1705.02798 (2017)
14. Richardson, M., Burges, C.J., Renshaw, E.: MCTest: a challenge dataset for the open-domain machine comprehension of text. In: EMNLP, vol. 3, p. 4, May 2013
15. Yu, Y., Zhang, W., Hasan, K., Yu, M., Xiang, B., Zhou, B.: End-to-end answer chunk extraction and ranking for reading comprehension. arXiv preprint arXiv:1610.09996 (2016)
16. Lee, K., Kwiatkowski, T., Parikh, A., Das, D.: Learning recurrent span representations for extractive question answering. arXiv preprint arXiv:1611.01436 (2016)
17. Wang, S., Jiang, J.: Machine comprehension using match-LSTM and answer pointer. arXiv preprint arXiv:1608.07905 (2016)
18. Xiong, C., Zhong, V., Socher, R.: Dynamic coattention networks for question answering. arXiv preprint arXiv:1611.01604 (2016)
19. Vinyals, O., Fortunato, M., Jaitly, N.: Pointer networks. In: Advances in Neural Information Processing Systems, pp. 2692–2700 (2015)
20. Zhang, W.M., Zhang, C.H., Liu, P., Zhan, Z.Q., Qiu, X.F.: Two-step joint attention network for visual question answering. In: BIGCOM (2017)
21. Sordoni, A., Bachman, P., Trischler, A., Bengio, Y.: Iterative alternating neural attention for machine reading. arXiv preprint arXiv:1606.02245 (2016)
22. Chung, J., Gulcehre, C., Cho, K., Bengio, Y.: Empirical evaluation of gated recurrent neural networks on sequence modeling. arXiv preprint arXiv:1412.3555 (2014)
23. Graves, A.: Generating sequences with recurrent neural networks. arXiv preprint arXiv: 1308.0850 (2013)
24. Graves, A., Wayne, G., Danihelka, I.: Neural turing machines. arXiv preprint arXiv:1410. 5401 (2014)
25. Weston, J., Chopra, S., Bordes, A.: Memory networks. Eprint Arxiv (2014)
26. Clark, E., Araki, K.: Text normalization in social media: progress, problems and applications for a pre-processing system of casual English. Procedia Soc. Behav. Sci. **27**, 2–11 (2011)
27. Bharti, S.K., Babu, K.S.: Automatic keyword extraction for text summarization: a survey. arXiv preprint arXiv:1704.03242 (2017)
28. Pennington, J., Socher, R., Manning, C.D.: Glove: global vectors for word representation. In: EMNLP 2014, pp. 1532–1543 (2014)

Entity Set Expansion on Social Media: A Study for Newly-Presented Entity Classes

He Zhao[1], Chong Feng[1(✉)], Zhunchen Luo[2], and Yuxia Pei[3]

[1] School of Computer Science and Technology, Beijing Institute of Technology,
Beijing 100081, China
zhaohe1995@outlook.com, fengchong@bit.edu.cn
[2] China Defense Science and Technology Information Center, Beijing 100142, China
zhunchenluo@gmail.com
[3] State Key Laboratory of Smart Manufacturing for Special Vehicles and
Transmission System, Beijing, China

Abstract. Online social media yields a large-scale corpora which is fairly informative and sometimes includes many up-to-date entities. The challenging task of expanding entity sets on social media text is to extract more unheard entities with several seeds already in hand. In this paper, we present a novel approach that is able to discover newly-presented objects by doing entity set expansion on social media. From an initial seed set, our method first explores the performance of embedding method to get semantic similarity feature when generating candidate lists, and detects features of connective patterns and prefix rules with specific social media nature. Then a rank model is learned by supervised algorithm to synthetically score each candidate terms on those features and finally give the final ranked set. The experimental results on Twitter text corpus show that our solution is able to achieve high precision on common class sets, and new class sets containing abundant informal and new entities that have not been mentioned in common articles.

Keywords: Social media mining · Information extraction · Entity set expansion

1 Introduction

Have you ever expected to find out a TV series as hot as the one that you are obsessed with? Or have you ever wanted to know the newly released product by the brand of which you have been a great fan? As is well known that information in social media, where all these topics are included, always keeps abreast of the times. In order to extract entity instances of a latent semantic class, a set expansion method takes an initial set of seed examples as inputs, expands the set and then outputs more entity instances of that class. For example, one expecting to find all fast-food restaurants may give two or three well-known names like "KFC" and "Taco Bell" as seeds, based on the seed set and the text corpus it

© Springer Nature Singapore Pte Ltd. 2017
X. Cheng et al. (Eds.): SMP 2017, CCIS 774, pp. 116–128, 2017.
https://doi.org/10.1007/978-981-10-6805-8_10

can discover other more restaurant names such as "Mcdonald's", "Pizza hut" and "Dairy Queen", which distinctly belong to that set. The task has been well-studied on news corpus and web-based text in the past decades. However, in recent years the booming user quantity of social media platforms such as Twitter and Facebook provides tremendous amount of noisy and informal short text, simultaneously presents a new challenge of expanding entity sets by usual method and technology. Furthermore, due to the low-barrier of posting and the proliferation with mobile devices, open-domain social media text is often more up-to-date and inclusive than news articles. Conceivably, in informally written text there are many newly created entities that are in abbreviated form or did not exist before. As an example, "BK" usually means "BurgerKing" which is a fast-food chain and "FB" usually means "FaceBook" which is a social media website or an Internet company name. Identifying these entities is profoundly significant to the downstream applications, for instance, constructing entity repositories for social media and delivering precise searching results in social media-oriented query suggestion systems.

It is worth-noting that most semantic classes studied in previous work can be explicitly defined by several entity instances, like given "LALakers" and "Chicago Bulls" as seeds it can wisely infer the intention is to get other entities which are also "NBA teams". Whereas plenty of newly-presented entities have more than one semantic layers, so that we cannot give an absolute definition to a set of entities. A more specific example is, for "Temple Run" and "Subway Surfers", it can be concluded both of them are mobile games, iOS Applications or parkour games. There is no denying that the two instances should be contained into one set in some sense but the semantic class description may be diverse from different point of view. Therefore, unlike previous work, our research purpose is to aggregate the entities into one set just considering their implicit semantic concepts, while not imposing restrictions on the possible class names.

Corpus-based method has been applied to solve entity set expansion problem on general-purpose web source data and promising results have been reported in the past years. Wang and Cohen [1–3] develop the SEAL system on semi-structured documents from web. He and Xin [4] propose SEISA system using lists from the HTML web pages and the search query logs. A more latest work by Dalvi et al. [5] relies on clustering terms found in HTML tables and mainly focused on tabular data in web pages. Other than the web corpus, limited in 140 characters in most cases the text from social media does not comprise distinct structured tables or lists. And that due to the short text length, there's often not enough maximal context to build reliable wrappers for each seed example mentions. Hence obviously those off-the-shelf web-based methods show ineffective when applied to solve the entity set expansion task on social media corpus. Another prominent line of work using bootstrapping approaches like that by Thelen and Riloff [6], the ASIA system [7] and KnowItAll [8]. They extract context patterns to capture information about the behavior of a term or utilize hyponym patterns (e.g.,"fruits such as apples, bananas and oranges") to find instances based on the "is-a" relation. But with informal writing style and unconventional grammar, social media text often lacks signal words like "such as" to make up hyponym patterns. And since the oral expression characteristics differ depending on individual users, context around an entity mention is rarely to be coincident to compose patterns.

In this paper, we present a novel entity set expansion method which especially orients to social media text and focus on some newfangled semantic classes that have not been sufficiently studied before. Our method uses a two-phrase architecture of extracting and ranking. In the first phase, to extract and generate candidate lists we adopt an embedding-based strategy to get scores on the feature of *semantic similarity*. Through the observation of various social media text, we find that many homogeneous entities are in appositive construction, connected by some conjunction words or symbols (e.g., ",", "and", "or", et al.). The intuition has been verified in the research of Widdows and Dorow [9] and been also depended on in the entity set expansion algorithm of Sarmento et al. [10]. Another observation is that in one single piece of message the entities, which synchronously have the prefixes "@" (to remind someone to notice) or "#" (to be a hashtag), are very likely in the same class. So here we propose the following hypotheses:

1. If a term is often connected with seed entities by the conjunction symbols that frequently occur between two seed entities in text corpus, the term is more likely to be a candidate instance of that class set.
2. If a term often occurs in the same piece of message with seed entities, both having "@" or "#" as prefix, the term is more likely to be a candidate instance of that class set.

Based on the two hypotheses, for each terms in candidate list, we respectively get scores on the feature of *connective patterns* and *prefix rules*. In the second phase, we present a rank model using a supervised learning method to combine the three features and provide the optimized coordination coefficient vector. We do experiments on outline Twitter text corpus and show that our approach can achieve high precision on some general entity classes and find multiple entities on some newfangled classes.

2   Related Work

Most previous work use two-phases method to expand sets, so as we do, which divides the task into two steps, extracting and ranking. In extracting, some common strategies like corpus-based, search-based, pattern-based and bootstrapping approaches are adopted extensively. And in ranking, most base on probability statistics, distributional similarity and some graph-based structures.

As has been mentioned in the introduction part, the SEAL system by Wang and Cohen [1–3] gets data source from web pages. In extracting phase they construct maximal context wrappers using given seeds and gather the candidate terms enclosed by these wrappers. Then in ranking phase they use a random walk process and rank by similarity metric on the basis of a graph containing seeds, wrappers and candidate terms.

The KnowItAll system by Etzioni et al. [8] also targets to web-scale data source. In extracting phase they use textual patterns like "colors such as pink, blue" as generic templates to generate candidate lists. And in ranking phase they adopt a bootstrapping method with the PMI measure computed by web search engines to estimate the likelihood of each candidate terms.

Similar to KnowItAll, the ASIA system by Wang and Cohen [4] is another pattern-based method. In extracting phase they also use hyponym patterns to get candidate terms and different from SEAL, they utilize a Bootstrapper to make the process iterative. Then in ranking phase supported by Random Walk with Restart on their graph it can determine the final ranking.

Talukdar et al. [11] propose a context pattern method that firstly find "trigger words" indicating the beginning of a pattern to extract candidate terms. And then they set evaluation mechanism to rank both the patterns and candidates.

Other existing work specially focuses on ranking the candidate sets. Bayesian Sets by Ghahramani and Heller [12], which is based on Bayesian inference, aims to estimate the probability that a candidate term belongs to a set by learning from a positive set P and an unlabeled candidate set U.

Sarmento et al. [10] propose a distributional similarity method according to a co-occurrence assumption that two elements consistently co-occur tent to be in similar semantic class. They find entity pairs by coordination structures and represent entities in vector space by encoding the co-occurrence frequency to compute the similarity with distance measure.

A PU-Learning method by Li et al. [13] transform set expansion into a two-class classification problem. A seed set is regarded as a set P of positive examples and candidate set is a set U containing hidden positive and negative cases. The task of filtering the candidate set turns to building a classifier to test if each candidate member is positive or not.

Beyond the work listed above, we want to mention another more recent work closed to set expansion on social media text. Qadir et al. [18] propose a novel semantic lexicon induction approach to learn new vocabulary from Twitter. Their method roots on the ability to extract and prioritize N-gram context patterns that are semantically related to the categories and is able to deal with multiword phrases of flexible term length. Starting with a few seed terms of a semantic category, they first explore the context around seeds a corpus, and identify context patterns relevant to that category, which then used to extract candidate terms. They experiment with three commonly discussed semantic categories, which are Food&Drinks, Games&Sports and Vehicles, and show good performance in these topics.

3   Entity Set Expansion

Our task can be defined as follows: given a text corpus T and a set S of seed entities of a hidden class C, we expand S by finding new terms t (we use the word "term" interchangeably with "entity" since textual terms are taken as candidate entities in expanded sets) from T, such that t is also semantically in the class C.

3.1 Entity Class

In this paper, we experimented with 24 semantic classes that are more concerned and widely discussed in social media platforms as is shown in Table 1. To demonstrate that our method can achieve high precision on some common classes and simultaneously find out entities of newfangled classes, we take both the *Common-class* sets containing more common entities with clearer semantics and *New-class* sets, in which most entities are proper and complex in semantic. In Table 1 we give 3 representative seed examples for each sets, but without any straightforward description about the class. And in the actual experiment, we construct initial candidate seed sets for each classes, which contains 9 seed entities on average.

Table 1. Semantic classes and some representative seed examples

Common-class	New-class
Sociology, Chemistry, Biology	Avatar, Transformers, The Dark Knight
lawyer, doctor, manager	iPhone, iPad, MacBook
hat, shirt, sweater	Microsoft, IBM, Apple
Canada, France, Morocco	keyboard, print, mouse
steak, pork, sausage	KFC, Taco Bell, Starbucks
fever, headache, migraine	Temple Run, Warfare, Dota
milk, juice, coffee	Amazon, eBay, Google
brother, sister, dad	Super Junior, BIGBANG, CNBlue
head, ear, mouth	mcflurry, frappe, smoothie
football, baseball, tennis	Twitter, Facebook, Instagram
piano, guitar, drum	Justin Bieber, Taylor Swift, Katy Perry
gym, store, library	Skype, FaceTime, kik

3.2 Corpus Description and Pre-processing

In this research, we use Twitter text to build our corpus dataset. From online Twitter platform, each tweet is a short message with a maximum length of 140 characters, in the nature of informal grammar, abbreviated expressions and misspellings. We collect 9,582,314 English tweets published in January, 2013, removing other redundant information and only keeping textual content left.

For pre-processing, we utilize Twitter NLP pipeline contributed by Ritter et al. [14] to chunk each tweet. For identified noun phrases, we delete the space between two words to integrate multiple words into a whole (e.g.,"Los Angeles

Lakers" to "LosAngelesLakers"). The purpose of this is in vectorization procedure the phrases can be represented with single word embedding. Another important cause is that we observe and discover many Twitter users do like the integrated expression instead of typing more spaces, so it can exactly make the two forms refer to the same semantic instance.

3.3   Generating Candidate Sets

To extract candidate terms for each sets of seeds, we first train a word embedding model on the corpus T using Word2Vec, which is a text deep distributional representation approach proposed by Mikolov et al. [19]. Then the semantic similarity between two terms turns to be computed by cosine distance in the vector space. For each seed term $s \in S$, we select the top n semantically similar terms add to the candidate set D and remove the duplicates. After that for each candidate term $item \in D$, we calculate mean similarity with all seed terms in S as its score on the feature of *Semantic Similarity*.

$$SC_{sim}(item) = \frac{\sum_{s \in S} Similar(s, item)}{|S|} \tag{1}$$

3.4   Connective Patterns and Prefix Rules

By observing tweets text we conclude a set CS of 24 conjunction symbols that are generally used to connect two congeneric terms as is showed in Table 2. For each conjunction symbol $c \in CS$, and a pair of seed terms $(s_1, s_2) \in S$, we respectively construct two reversed strings "$s_1 \ c \ s_2$" and "$s_2 \ c \ s_1$" that we called *searching patterns* (e.g., "KFC & Starbucks"). We count frequencies of these two *searching patterns* in corpus text and sum them to get the frequency value $f_{(c,pair)}$ of c in regard to pair (s_1, s_2). The sum of the values computed by all possible seed pairs in S indicates the weight of that symbol c.

Table 2. Conjunction symbols set CS

and	or	,	&	+	−
x	X	/	$\star$	>	<
|	vs	VS		:	//
·	\	\\	=	≪	≫

$$Weight(c) = \frac{\sum_{pair \in S} f_{(c,pair)}}{\sum_{c \in CS} \sum_{pair \in S} f_{(c,pair)}} \tag{2}$$

Then for each candidate term $item \in D$ and each seed term $s \in S$, again we use symbol c to construct two reversed strings "$s \ c \ item$" and "$item \ c \ s$" that we called *matching patterns* (e.g. "KFC & Mcdonald's", "Mcdonald's & KFC").

Counting and summing the frequencies of the *matching patterns*, and multiplying the weight of c, we get the frequency value $f_{(item,c,s)}$ of *item* weighted by c. So the sum of the values computed by all symbols in CS indicates the score of *item* in feature of *Connective Patterns*.

$$SC_{con}(item) = \sum_{c \in CS} \sum_{s \in S} Weight(c) \times f_{(item,c,s)} \qquad (3)$$

We use similar process to get the score in feature of *Prefix Rules*. First we conclude a set PS containing 4 groups of different prefix collocations by "@" and "#" as is showed in Table 3. For each prefix collocation $(p_1, p_2) \in PS$, and a pair of seed terms $(s_1, s_2) \in S$, we respectively construct a group of *searching patterns* which are "$p_1 s_1$" and "$p_2 s_2$" (e.g., "@KFC" and "@Starbucks"). We count frequencies of the two patterns in corpus text and sum them to get the frequency value $f_{(p,s)}$ of prefix collocation p in regard to seed s. The sum of the values computed by all seeds in S indicates the weight of that collocation p.

Table 3. Prefix collocations set PS

Prefix collocations	Seed terms	Candidate terms
(@, @)	@s	@$item$
(@, #)	@s	#$item$
(#, @)	#s	@$item$
(#, #)	#s	#$item$

$$Weight(p) = \frac{\sum_{s \in S} f_{(p,s)}}{\sum_{p \in PS} \sum_{s \in S} f_{(p,s)}} \qquad (4)$$

Then for each candidate term $item \in D$ and each seed term $s \in S$, we use prefix collocation p to construct two strings "$p_1 item$" and "$p_2 s$" that are also named as *matching patterns* (e.g., "@KFC" and "@Mcdonald's"). Counting and summing the frequencies of the *matching patterns*, and multiplying the weight of p, we get the frequency value $f_{(p,item,s)}$ of *item* weighted by p. So the sum of the values computed by all collocations in PS indicates the score of *item* in feature of *Prefix Rules*.

$$SC_{pre}(item) = \sum_{p \in PS} \sum_{s \in S} Weight(p) \times f_{(p,item,s)} \qquad (5)$$

3.5   Candidate Terms Ranking

With the scores of each candidate terms in feature of *Semantic Similarity, Connective Patterns* and *Prefix Rules*, now we can respectively rank the candidate set according to the three features, and get three ranking lists that are expressed as R_{sim}, R_{con} and R_{pre}. We present a synthetic ranking model R, learning a set of coordination coefficient vector $\boldsymbol{W} = (\alpha_1, \alpha_2, \alpha_3)$ to combine the each item's positions, which are $R_{sim}(item)$, $R_{con}(item)$ and $R_{pre}(item)$ in the three ranking lists.

$$R(\boldsymbol{W}) = \alpha_1 R_{sim}(item) + \alpha_2 R_{con}(item) + \alpha_3 R_{pre}(item) \qquad (6)$$

First we give some notations and explanations using those defined in AdaRank by Xu and Li [20] as reference, as is showed in Table 4.

Table 4. Notations and explainations

Notations	Explanations
$c_i \in C$	i^{th} hidden semantic class
$D_i = \{item_{i1}, \cdots, item_{i,n(c_i)}\}$	Candidate set of class c_i
$y_{ij} \in \{r_1, r_2, \cdots, r_l\}$	Rank of $item_{ij}$ w.r.t. c_i
$Y_i = \{y_{i1}, y_{i2}, \cdots, y_{i,n(c_i)}\}$	List of ranks for c_i
$S = \{(c_i, D_i, Y_i)\}_{i=1}^m$	Training set
$\boldsymbol{W} = (\alpha_1, \alpha_2, \alpha_3)$	The coordination coefficient vector
$f, R(\boldsymbol{W}), R_{sim}, R_{con}, R_{pre} \in \mathcal{R}$	Ranking models
$\pi(c_i, D_i, f)$	Permutation for c_i, D_i, and f
$E(\pi(c_i, D_i, f), Y_i) \in [-1, +1]$	Performance measure function

Here we use MAP (Mean Average Precision) as performance measure. For a given semantic class c_i, rank of candidate lists Y_i and a permutation π_i on candidate set D_i, average precision for c_i is defined as:

$$AvgP_i = \frac{\sum_{j=1}^{n(c_i)} P_i(j) \cdot y_{ij}}{\sum_{j=1}^{n(c_i)} y_{ij}}, \qquad (7)$$

where y_{ij} takes on 1 and 0 as values, which presents $item_{ij}$ is positive or negative instance and $P_i(j)$ is defined as precision at the position of d_{ij}:

$$P_i(j) = \frac{\sum_{k:\pi_i(k) \leq \pi_i(j)} y_{ij}}{\pi_i(j)}, \qquad (8)$$

where $\pi_i(j)$ presents the position of $item_{ij}$.

To learn coordinate coefficient for each rank model, to start with we assign the same value to α_1, α_2 and α_3. With a training set $S = \{(c_i, D_i, Y_i)\}_{i=1}^m$ as input, the algorithm takes the performance measure function E and runs at most T rounds. At each round, it creates a new rank model f by linearly combining the three feature rankers with $\boldsymbol{W}$. Each coefficient α_i will be updated to increase if the corresponding feature rank model outperforms model f, but decrease the opposite. Then we test if the rank model combined with updated $\boldsymbol{W}$ is better than f. If not, the algorithm reaches the maximum point at f, the loop ends and f is returned. The pseudocode of the whole process is shown in Algorithm 1, where Z_m is a standardization factor, to make $\sum \boldsymbol{W} = 1$:

$$Z_m = \sum_{k=1}^{3} \alpha_k \cdot exp\{\frac{\sum_{i=1}^{m} E(\pi(c_i, D_i, R_k), Y_i) - \sum_{i=1}^{m} E(\pi(c_i, D_i, f_t), Y_i)}{m}\} \quad (9)$$

Algorithm 1. The rank model algorithm

Input:
Training set $S = \{(c_i, D_i, Y_i)\}_{i=1}^{m}$
Performance measure function $E(\pi(c_i, D_i, f), Y_i)$
Max iterations T
Initialize:
for α_k in $\boldsymbol{W}$: $\alpha_k = \frac{1}{|\boldsymbol{W}|}$
$R_1 = R_{sim}, R_2 = R_{con}, R_3 = R_{pre}$
Algorithm:
1: **for** $t = 1$ to T **do**
2:     $f_t = \sum_{k=1}^{3} \alpha_k \cdot R_k$
3:     **for** α_k in $\boldsymbol{W}$ **do**
4:         $\alpha_k = \alpha_k \frac{exp\{\frac{\sum_{i=1}^{m} E(\pi(c_i, D_i, R_k), Y_i) - \sum_{i=1}^{m} E(\pi(c_i, D_i, f_t), Y_i)}{m}\}}{Z_m}$,
5:     **end for**
6:
7:     **if** $\sum_{i=1}^{m} E(\pi(c_i, D_i, \sum_{k=1}^{3} \alpha_k \cdot R_k), Y_i) < \sum_{i=1}^{m} E(\pi(c_i, D_i, f_t), Y_i)$ **then**
8:         **return** f_t
9:     **end if**
10: **end for**
Output:
$R(\boldsymbol{W}) = f_t$

---

4   Experiments

4.1   Baselines

To compare the performance of our set expansion algorithm with previous work, we use two pattern-based approaches. One is a Context Pattern Induction Method by Talukdar et al. [11], which we briefly name as **CPIM**, and another is a Twitter-oriented Semantic Lexicon Induction Method by Qadir et al. [18], briefly named as **TSLIM**. Due to that many previous works target to structured text data source from web and utilize tables, lists and markup tags to solve the problem, obviously they are not fit for our Twitter text. Hence we choose approaches based on context patterns that don't like that impose restrictions on specific corpus. Besides, to the best of our knowledge very few works study entity set expansion problem on social media, so we unavoidably take Qadir's lexicon induction method as baseline, which is more or less distinguishing compared with our task.

Furthermore, to demonstrate our synthetic ranking model is effective, we also compare the ranking results with that only using semantic similarity from Word2Vec, and show the increase in precision.

4.2   Evaluation

Since the output of all these approaches are ranked lists, we adopt *Rank Precision@N*, which is commonly used for evaluation of entity set expansion techniques [13] and defined as follow:

Precision@N: The percentage of correct entities among the top N entities in the ranked list.

For each entity class, we collected the top 100 terms in the ranked lists generated by each of the approaches, and provided the sets to three individual annotators. The annotators were given only the initial sets of seed examples as guidelines, without any redundant definitions or descriptions of semantic class, and were asked to determine if each term is a positive or negative instance of that set. At last we assign class membership to each term follow the majority voting rule, that is, a term will be judged as positive only if more than two annotators deem it is.

4.3   Results

In our experiment, we use a word embedding model of 200 dimensions, and set the window size to 5, minimum word count to 5. To train our synthetical rank model, we select 10 semantic class sets, being more generic and commonly-used in set expansion tasks, which are completely not same as the sets that are intentionally used to perform our entity extraction method. Finally we get the set of coordination coefficient vector as: $\boldsymbol{W} = (0.49578957, 0.26961497, 0.23459546)$.

First the number of initial seed entities is fixed to 3, and for four methods, we present the precisions at the top 5-, 10-, 20-, 50- and 100-ranked positions (i.e., precisions@5, 10, 20, 50 and 100). The detailed experimental results are shown in Table 5 for both the *Common-class* and the *New-class*.

Table 5. Precision @ top N for four methods (with 3 seeds).

Method	Common-class					New-class				
	5	10	20	50	100	5	10	20	50	100
CPIM	0.35	0.35	0.37	0.34	0.36	0.48	0.44	0.36	0.30	0.26
TSLIM	0.74	0.65	0.60	0.45	0.40	0.61	0.50	0.42	0.39	0.37
Word2Vec	1.00	0.99	0.96	0.86	0.73	1.00	0.99	0.92	0.80	0.67
Our method	**1.00**	**0.99**	**0.97**	**0.90**	**0.76**	**1.00**	**0.99**	**0.95**	**0.84**	**0.74**

From Table 5, we observe that in *Common-class*, on average our method outperforms CPIM by about 40–65%, TSLIM by about 26–45% and Word2Vec by about 1–4%. And in *new-class* our method outperforms more compared with all these three approaches, which is to 48–59%, 37–52% and 3–6%. It indicates that our method is more effective when dealing with newly-presented semantic class sets.

To test the sensitivity of initial seeds, we also vary the strategies of selecting seed entities from candidate seed sets. We adopt three different manners:

- *Random Manner:* Randomly select 3 entities from candidate seed set.
- *Maximum Similarity:* Select the most 3 semantically similar entities as seeds.
- *Maximum Frequency:* Select the most 3 frequent entities occurred in corpus as seeds.

Then we use different size of corpus to analyze the corpus effects on performance. Due to space constrains, we present only average results of *Common-class* sets and *New-class* sets. All that results are shown in Table 6. From Table 6 we can see that using MaxFreq to select seeds performs better. We speculate that the reason is, whether seed entities are common in corpus has a great influence on the precision result. A seed more frequent in corpus is able to bring more information in constructing patterns and rules, for example, "BurgerKing" and "KFC" as seeds are better than "WhataBurger". Hence uncommon and infrequent entities should be avoided when generating initial seed sets. In addition, we speculate that the relevancy in semantic among seed entities also affects much. It is probably because more similar seeds are often more representative for their class, and may prevent the semantic center drift. Such as the seed set { "ukulele", "violin", "guitar"} is better than { "piano", "violin", "guitar"}, where the former is more likely to have the center in "string instruments". Not surprisingly, enlarging corpus size have a significant impact on set expansion performance. One side the chunk parsing and word embedding model training process work better on larger corpus, and on the other side, most terms occur more times with the corpus growing, which greatly improves the entity sparseness problem.

Table 6. Precision @ top N when varying seed sets and corpus size (with 3 seeds).

Seed select	Corpus size	5	10	20	50	100
Random	1,398,511	0.91	0.85	0.77	0.64	0.51
MaxSim	1,398,511	0.99	0.95	0.86	0.73	0.58
MaxFreq	1,398,511	1.00	0.96	0.86	0.73	0.59
MaxFreq	4,080,031	1.00	0.97	0.93	0.85	0.73
MaxFreq	9,582,314	1.00	0.99	0.96	0.86	0.74

5   Conclusion

We present a novel entity set expansion method that is able to expand newfangled semantic class on social media text. We first adopt a word embedding model to generate candidate sets and get scores of semantic similarity. And the other two scores are acquired by using some connective patterns and prefix rules which are specific in social media text. Then we propose a synthetical rank model to integrate the feature scores and show evident improvement promotion in some newly-presented entity class.

The demarcation and definition of "newly-presented" entities need further investigation. And as further work direction, we will take advantage of more features from accessible social media platform, such as userinfos, geographical location information and published time attributes.

Acknowledgements. This work was supported by the National High-tech Research and Development Program (863 Program) (No. 2014AA015105) and National Natural Science Foundation of China (No. 61602490).

References

1. Wang, R.C., Cohen, W.W.: Language-independent set expansion of named entities using the web. In: IEEE International Conference on Data Mining, pp. 342–350. IEEE (2007)
2. Wang, R.C., Cohen, W.W.: Iterative set expansion of named entities using the web. In: Eighth IEEE International Conference on Data Mining, pp. 1091–1096. IEEE (2009)
3. Wang, R.C., Cohen, W.W.: SEAL. http://rcwang.com/seal
4. He, Y., Xin, D.: SEISA: Set Expansion by Iterative Similarity Aggregation. In: International Conference on World Wide Web, WWW 2011, Hyderabad, India, pp. 427–436 (2011)
5. Dalvi, B.B., Cohen, W.W., Callan, J.: WebSets: extracting sets of entities from the web using unsupervised information extraction. In: ACM International Conference on Web Search and Data Mining, pp. 243–252. ACM (2012)
6. Thelen, M., Riloff, E.: A bootstrapping method for learning semantic lexicons using extraction pattern contexts. In: Conference on Empirical Methods in Natural Language Processing, ACL 2002, pp. 212–221. ACL (2002)
7. Wang, R.C., Cohen, W.W.: Automatic set instance extraction using the web. In: ACL 2009, Proceedings of the Meeting of the Association for Computational Linguistics and the International Joint Conference on Natural Language Processing of the AFNLP, pp. 441–449. ACL, Singapore (2009)
8. Etzioni, O., Cafarella, M., Downey, D., Kok, S., Popescu, A.M., Shaked, T., Soderland, S., Weld, D.S., Yates, A.: Web-scale information extraction in KnowItAll. In: WWW, pp. 100–110 (2004)
9. Widdows, D., Dorow, B.: A graph model for unsupervised lexical acquisition. In: International Conference on Computational Linguistics, pp. 1093–1099 (2002)
10. Sarmento, L., Jijkuon, V., De Rijke, M., Oliveira, E.: More like these: growing entity classes from seeds. In: Sixteenth ACM Conference on Conference on Information and Knowledge Management, pp. 959–962. ACM (2007)
11. Talukdar, P.P., Brants, T., Liberman, M., Pereira, F.: A context pattern induction method for named entity extraction. In: Computational Natural Language Learning, CoNLL-X, pp. 141–148 (2006)
12. Ghahramani, Z., Heller, K.A.: Bayesian sets (2005)

13. Li, X.L., Zhang, L., Liu, B., Ng, S.K.: Distributional similarity vs. PU learning for entity set expansion. In: ACL 2010 Conference Short Papers, pp. 359–364. ACL (2010)
14. Ritter, A., Sam, C., Mausam, Etzioni, O.: Named entity recognition in tweets (2011)
15. Li, C., Weng, J., He, Q., Yao, Y., Datta, A., Sun, A., Bu-Sung, L.: TwiNER: named entity recognition in targeted twitter stream. In: Proceedings of the 35th International ACM SIGIR Conference on Research and Development in Information Retrieval, pp. 721–730 (2012)
16. Bontcheva, K., Derczynski, L., Funk, A., Greenwood, M.A., Maynard, D., Aswani, N.: TwitIE: an open-source information extraction pipeline for microblog text. In: Proceedings of the International Conference on Recent Advances in Natural Language Processing. Association for Computational Linguistics (2013)
17. GATE. https://gate.ac.uk/wiki/twitie.html
18. Qadir, A., Mendes, P.N., Gruhl, D., Lewis, N.: Semantic lexicon induction from twitter with pattern relatedness and flexible term length. In: Twenty-Ninth AAAI Conference on Artificial Intelligence, pp. 2432–2439 (2015)
19. Mikolov, T., Chen, K., Corrado, G., Dean, J.: Efficient estimation of word representations in vector space. Comput. Sci. (2013)
20. Xu, J., Li, H.: AdaRank: a boosting algorithm for information retrieval. In: International ACM SIGIR Conference on Research and Development in Information Retrieval, pp. 391–398 (2007)

An Effective Approach of Sentence Compression Based on "Re-read" Mechanism and Bayesian Combination Model

Zhonglei Lu[1], Wenfen Liu[2(✉)], Yanfang Zhou[1], Xuexian Hu[1], and Binyu Wang[1]

[1] State Key Laboratory of Mathematical Engineering and Advanced Computer, Zhengzhou 450001, Henan, China
lzl_xd6j@163.com, zyf_xd6j@163.com, hxx_xd6j@163.com, wby_xd6j@163.com
[2] Guangxi Key Laboratory of Cryptogpraphy and Information Security, School of Computer Science and Information Security, Guilin University of Electronic Technology, Guilin 541004, Guangxi, China
liuwenfen@guet.edu.cn

Abstract. As the mass social media texts and the increasingly popular "small screen" interaction mode are producing a huge text compression requirement, this paper presents an approach for English sentences compression based on a "Re-read" Mechanism and Bayesian combination model. Firstly, we build an encoder-decoder consisting of Long Short-Term Memory (LSTM) model. In the encoding stage, the original sentence's semantics is modeled twice. The result from the first encoder, as global information, and the original sentence are input into the second encoder together, obtaining a more comprehensive semantic vector. In the decoding stage, we adopt a simple attention mechanism, focusing on the most relevant semantic information to improve the decoding efficiency. Then, a Bayesian combination model combines the explicit prior information and "Re-read" models to enhance the use of explicit training data features. Experimental results on Google Newswire sentence compression dataset show that the method proposed in this paper can greatly improve the compression accuracy, and the F1 score reaches 0.80.

Keywords: Natural language processing · Sentence compression · Re-read · Combination model

1 Introduction

The popularity of smart phones, tablet PCs and other mobile terminals, gives birth to a large number of social media sources. Massive text messages are blowing. However, this open, free information sharing and propagation pattern has increased difficulty of obtaining effective information. The contradiction between sharp growth of social media texts and urgent desire to save time by streamlining information becomes increasingly acute. Recently, the development of natural language processing technology makes text compression possible to be a powerful tool which effectively

© Springer Nature Singapore Pte Ltd. 2017
X. Cheng et al. (Eds.): SMP 2017, CCIS 774, pp. 129–140, 2017.
https://doi.org/10.1007/978-981-10-6805-8_11

alleviates this contradiction. Sentence compression is one key technology among them. Sentence compression, also known as sentence pruning, aims to delete the redundant information in the original sentence via a variety of algorithms. In this way, we expect to get a simple sentence with proper grammar and core content of the original sentence automatically, which is also easy to read. It is widely used in automatic headline generation [1], automatic text summarization [2], automatic subtitle generation [3], small screen text display [4], etc. It accelerates the intellectualization, specialization and refinement of information technology.

Traditional sentence compression methods mainly include minimize the proportion of the grammatical mistakes [5] or pruning the syntax tree [6], etc., which heavily depends on the design and selection of features. The process needs much expert knowledge support, and consumes a lot of manpower or resources. Due to the powerful modeling and feature extraction capability, deep learning releases a new solution for sentence compression. Fillippova et al. [7] applied deep learning models to sentence compression task for the first time. They used a 3-layers LSTM model as component of an encoder-decoder, which achieved better results than traditional systems on a large-scale dataset. Tran et al. [8] improved Fillippova's model structure, and proposed a bidirectional LSTM compression model based on attention mechanism, performing better on a small dataset. In addition, Sigrid et al. [9] incorporated eye-tracking information into the sentence compression system, getting higher accuracy also with a 3-layers LSTM encoder-decoder architecture.

However, there are three shortcomings in the current research. Firstly, most of the models are too simple, which can't fully encode the original sentence semantics (especially for long sentences). This may bring about serious loss in semantic information followed with weak decoding. Secondly, single-model's generalization capability is insufficient, resulting in low compression accuracy without full capture of data characteristics. Thirdly, the above models haven't used the explicit prior information fully, and the description of training process is poor.

To address the above problems, this paper proposes a bionic mechanism called "Re-read" and a Bayesian combination model for sentence compression. Firstly, we model the original sentence semantics twice. The original sentence is encoded by a BiLSTM for the first time. The modeling result is straightly input into a LSTM together with the original sentence sequence. Outcomes from the previous semantic model, as the global information, enhances the local semantic modeling results, while we use simple attention mechanism for decoding. Then, based on the "Re-read" mechanism, we sample training data equally from a large number of training dataset to train models respectively. Further, we introduce a Bayesian method to combine the trained models and utilize the prior information plenarily.

The main contributions of this paper are as follows: (1) As far as we know, we are the first to use the "Re-read" mechanism in sentence compression task, which improves the semantic modeling effectively. (2) The method of sampling in training promotes the model's generalization capability to fully capture data recessive compression features. (3) The combined model, integrating the Bayesian ideas, strengthens the use of explicit prior information, and F1 score has been greatly improved.

The paper is organized as follows: In Sect. 2, we describe the preliminary knowledge. Section 3 presents the details of our model. The evaluation setup and results are discussed in Sect. 4 which is followed by the conclusions.

2 Preliminary Knowledge

2.1 The LSTM Model

Recurrent Neural Network (RNN) is an extension of Feedforward Neural Network (FNN) in time series, which can handle variable length input sequences [11]. It is widely used in machine translation [12], automatic text summarization [13], etc. Unfortunately, the traditional RNN is still hard to apply in practice due to the vanishing and exploding gradient problems [13] when modeling long sequences. Hochreiter et al. [10] improved the model with LSTM. The LSTM model consists of input gate, forget gate, output gate and memory cell. Its structure is shown in Fig. 1.

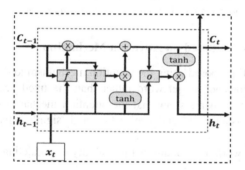

Fig. 1. The LSTM model structure

It updates as follows,

$$i_t = \sigma(W_i x_t + W_i h_{t-1} + b_i). \tag{1}$$

$$f_t = \sigma\left(W_f x_t + W_f h_{t-1} + b_f\right). \tag{2}$$

$$o_t = \sigma(W_o x_t + W_o h_{t-1} + b_o). \tag{3}$$

$$\tilde{C}_t = \tanh(W_c x_t + W_c h_{t-1} + b_c). \tag{4}$$

$$C_t = f_t \odot C_{t-1} + i_t \odot \tilde{C}_t. \tag{5}$$

$$h_t = o_t \odot \tanh(C_t). \tag{6}$$

Where i, f and o are input, forget and output gate, respectively. x, h and C represent the input layer, the hidden layer and the memory cell, respectively. W and b are the weight matrix and the bias, namely the network's parameters. σ is a sigmoid function, and $\odot$ is element-wise multiplication.

The LSTM model only encodes the forward semantic information. Furthermore, for the time t, the hidden layer output contains only the information before t, i.e. the above context information. However the following context information is equally important to characterize the whole semantics of sentences, such as "Smith will speak (German), because he comes from Germany". When we infer "German" in brackets, the following context information is integrant. In order to get a better representation of context information, the Bi-directional Long Short-Term Memory (BiLSTM) model is proposed. BiLSTM reads the input in both directions with 2 separate hidden layers: the forward and the backward. Two hidden states are $\vec{h}_t = LSTM\left(x_t, \vec{h}_{t-1}\right)$ and $\overleftarrow{h}_t = LSTM\left(x_t, \overleftarrow{h}_{t+1}\right)$. We summarize the information from the forward and the backward hidden states by concatenating them, i.e. $h_t = \left[\vec{h}_t, \overleftarrow{h}_t\right]$. By this way, the hidden state h_t contains the information of sentence not only in the reverse order but also in the original one. This improves the performance of the last words of the sentence which are too far away for the encoder to remember.

2.2 Encoder-Decoder Based on Attention Mechanism

Bahadanau et al. [14] proposed using the attention mechanism to dynamically generate the context vector for each target word rather than the fixed vector over the entire source sequence. Experiments show that the attention mechanism can deal with the problem of long distance dependence better and significantly improve NMT's performance.

Under this framework, the conditional probability of predicting each target word is

$$p(y_t|y_1,\ldots y_{t-1}, x) = g(y_{t-1}, s_t, c_t). \tag{7}$$

Where s_t represents the hidden state of the decoder-side at time t,

$$s_t = f(s_{t-1}, y_{t-1}, c_t). \tag{8}$$

The probability p is modeled in a way that is different from the conventional encoder-decoder, and the prediction of each target word y_t involves different context vectors. c_t depends on the hidden states $(h_1, h_2, \ldots, h_T)$,

$$c_t = \sum_{j=1}^{T_x} \alpha_{tj} h_j. \tag{9}$$

The weight α_{tj} of each hidden state h_j is,

$$\alpha_{tj} = \frac{\exp(e_{tj})}{\sum_{k=1}^{T_x} \exp(e_{tk})}. \tag{10}$$

This weight indicates the degree of alignment of the j-th word from the input sequence with the t-th word from the output sequence.

3 Our Model

The sentence compression task can be described briefly as follows: Given the original input sentence $x = (x_1, x_2, \ldots, x_n)$, where n is the sentence's length. After compression, we can get $y = (y_1, y_2, \ldots, y_m)$, where m is the length of the compression sentence, $m < n$ and $|y| \subseteq |x|$.

For simplicity, the task can be transformed into a problem labelling "0" or "1" for each word of the sentence. That is, the output sequence is $l = (l_1, l_2, \ldots, l_n), l_t \in \{0, 1\}$, $t = 1, 2, \ldots, n$, where "0" for deletion, "1" for reservation. Like this,

---

Original sentence: A plane got stuck in the snow at YLW Monday morning.
Composed sentence: A plane got stuck in the snow.
Output sequence: 1, 1, 1, 1, 1, 1, 1, 0, 0, 0, 0, 1

Based on the sequence-sequence paradigm proposed by Sutskever et al. [12], we adopt the end-to-end strategy to train the model, which is to maximize the probability of the correct output when given the input sentence. Furthermore, for each training sample (X, Y), the following optimization problem is solved by using the Stochastic Gradient Descent (SGD), to learn model parameters θ^*,

$$\theta^* = arg \max_\theta \sum\nolimits_{X,Y} logp(Y|X; \theta). \tag{11}$$

We use the chain rule to model the probability p,

$$p(Y|X; \theta) = \prod\nolimits_{t=1}^{T} p(Y_t|Y_1, \ldots, Y_{t-1}, X; \theta). \tag{12}$$

Once we find the optimal θ^*, the compression result Y can be predicted,

$$\hat{Y} = arg \max_Y logp(Y|X; \theta^*). \tag{13}$$

3.1 "Re-read" Mechanism

The idea behind the mechanism is very intuitive, where we imitate humans to compress sentences. People first read the original sentence to master the sentence's global semantic information. Then we read the original sentence again, to re-understand the semantics of each word. At the same time, we make the decision to delete or reserve the words one by one according to the global sentence semantics. The concrete process is as follows: Original sentence is input into a BiLSTM model, thereby we get the hidden states H and the global sentence semantic representation vector h^1. H and h^1 assist secondary semantic modeling, with x input into a LSTM model again. In this way, the semantic representation of important words is highlighted, while the semantic representation of redundant words is weakened. What's more, we get the compressed sentences by a decoder (LSTM) based on a simple attention mechanism. The outline is shown in Fig. 2(a).

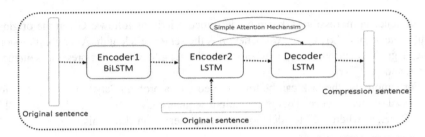

Fig. 2(a). The "Re-read" mechanism outline

Our "Re-read" mechanism consists of three parts: Bidirectional Semantic Encoder, "Re-read" Semantic Encoder and Decoder based on Simple Attention Mechanism. Its working principle is shown in Fig. 2(b).

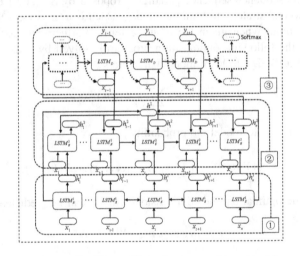

Fig. 2(b). The "Re-read" mechanism working principle

3.1.1 Bidirectional Semantic Encoder

The encoder includes a BiLSTM model, as indicated by the dotted line part ① in Fig. 2 (b). Supposing that the input sequence is $x = (x_1, x_2, \ldots, x_n)$, for time t:

$$h_t^1 = \left[LSTM_E^1(x_t, h_{t-1}^1), LSTM_E^1(x_t, h_{t+1}^1) \right]. \tag{14}$$

At time t, the hidden layer outputs can be expressed as the concatenation of the output of each hidden layer of BiLSTM. With the help of this encoder, the hidden layer outputs $H^1 = (h_1^1, h_2^1, \ldots, h_n^1)$ and $h^1 = \left[\vec{h}_n^1, \overleftarrow{h}_1^1\right]$, which are all input into the "Re-read" Semantic Encoder next.

3.1.2 "Re-read" Semantic Encoder

This encoder is a LSTM, as shown by the dotted line part ② in Fig. 2(b). Unlike conventional LSTM, this encoder's inputs involve four parts, i.e. x_t, h_{t-1}^2, h_t^1, h^1,

$$h_t^2 = LSTM_E^2\left(x_t, h_{t-1}^2, h_t^1, h^1\right). \tag{15}$$

The calculation process of $LSTM_E^1$ and $LSTM_E^2$ refers to the formulas (1)–(6). We use the hidden layer output h_n^2 at final time as the original sentence's ultimate semantic information input into the following decoder.

3.1.3 Decoder Based on Simple Attention Mechanism

Conventional attention mechanisms are computationally complex and redundant in attention information. In this paper, the sentence compression task has been transformed into a problem of labeling "0" or "1" for each word. The input and output sequence is strictly aligned, furthermore the "Re-read" semantic encoder has involved the global context information. So the conventional attention mechanism seems unnecessary in this case. We propose that the hidden state of "Re-read" semantic encoder at time t is directly used as the attention information for the decoder at time t. That is, only the context information most relevant to the predicted word is focused on, rather than considering all components of the sentence. Thereby, we can effectively remove redundant information, shown by the dotted line part ③ in Fig. 2(b).

$$s_t = f\left(x_t, y_{t-1}, h_t^2, s_{t-1}\right). \tag{16}$$

The final output layer is a Softmax classifier. It predicts the corresponding label (3-dimensional one-hot vector) for every word. If the word is the end of the sentence character, the third dimension is 1, indicating that the decoding prediction begins. If the word is retained, only the first dimension is 1, labelled "1". If the word is deleted, the second dimension is 1, labelled "0".

3.2 Bayesian Combination Model

In order to make full use of the prior information and information provided by the "Re-read" model, we establish the Bayesian combination classification model to obtain final compression sentences.

Firstly, we construct a vocabulary V_{train} based on all training corpus, and count the frequency p_r of each word which represents the word's reservation rate. The corresponding estimated probability is regarded as the prior information for word in the testing vocabulary V_{test}. So we get a uniform prior distribution $\pi(\theta_{w_i})$, where θ_{w_i} is the distribution's parameter, w_i represents the i-th word in the testing vocabulary V_{test}. For words that are not in V_{train}, we assume that the probability of being retained is 0.5.

$$\pi(\theta_{w_i}) = \begin{cases} 2 \cdot (1 - p_r), & \theta_{w_i} \in [0, 0.5] \\ 2 \cdot p_r, & \theta_{w_i} \in (0.5, 1] \end{cases} \qquad (17)$$

Secondly, the training data are equally divided into five parts randomly. With that we train five "Re-read" models under the same experimental configuration, obtaining five testing results R_1, R_2, R_3, R_4 and R_5. That is to say we have done five random tests, observing the number of the word labeled as "1". We call it the sample information. For the word w_i,

$$p(x|\theta_{w_i}) = \frac{R_1(w_i) + R_2(w_i) + R_3(w_i) + R_4(w_i) + R_5(w_i)}{5}. \qquad (18)$$

Finally, the posterior distribution is calculated from the prior distribution together with the sample information.

$$\pi(\theta_{w_i}|x) \propto h(x, \theta_{w_i}) = p(x|\theta_{w_i})\pi(\theta_{w_i}). \qquad (19)$$

We compute integrals for $\pi(\theta_{w_i}|x)$ in the interval (0.5,1) as an estimate of the probability of each word being retained.

$$p_{i,r} = \int_{0.5}^{1} \pi(\theta_{w_i}|x)d\theta_{w_i}. \qquad (20)$$

If $p_{i,r} > 0.5$, the word is retained, otherwise, the word is deleted. The model is shown in Fig. 3.

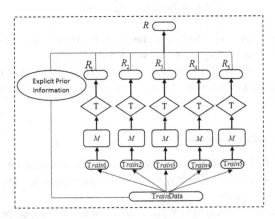

Fig. 3. The Bayesian combination model. M and T denotes the "Re-read" model and testing set, respectively. R_1, R_2, R_3, R_4 and R_5 represent the sample information, and R denotes the posterior result.

4 Experiments

4.1 Data and Preprocessing

We use Google Newswire sentence compression dataset as our experimental data, which includes 210,000 "original sentence-compression sentence" pairs. Among them, we use 200,000 as TrainData, the remaining 10,000 pairs as testing set. TrainData is divided into five parts equally and randomly: Train1, Train2, Train3, Train4 and Train5.

Prior to experiments, we preprocess our data. The original sentences were tokenized by NLTK's[1] word tokenizer. And then we use word2vec[2] for pre-training, accessing 97-dimensional word embedding vectors. After that, we build the shortlist of 8,000 (for 40,000 training pairs) and 20,000 (for 200,000 training pairs) most frequent words as a vocabulary for training our models. Any word not included in the vocabulary was replaced by a special token "unk" (unknown). At the end of a sentence, we added another special token "eos" (end-of-sentence) as a starting indication for decoding. Based on the above processing, the gold label sequences are constructed. We add one of the following three labels for each word in the sentence: 1 if this word is retained in the compression, 0 if this word is deleted, or 2 if it is the "eos". In addition, we count the retaining frequency of all words in the training set as prior information.

4.2 Experimental Configurations

In our experiments, each layer has 100 hidden units. The input is a 100-dimensional vector, where the first 97 dimensions are the current input word embedding vector. The last 3 dimensions are different for the encode and decode stage. In encoding, it is a zero vector. In decoding, there are two cases. For training, it is a one-hot vector of the gold label of the previous word. For testing, it is the one-hot vector of the previous predicted word.

We use the longest length of sentence in the data (210) for unfolding all the LSTM networks. According to Greff et al. [15] on the experience of parameters setting research, the learning rate was initialized to 0.001 (40,000 training pairs) and 0.01 (200,000 training pairs) according to the data volume. The decay rate [16] per 1,000 training steps is set to 0.9. In every case, instead of using a fixed number of epochs, we apply an early stop technique [17] in which the system stops training whenever the F1 score of the development set does not increase after 5 epochs. The specific model parameters are set as shown in Table 1.

---

[1] http://www.nltk.org/.

[2] http://code.google.com/p/word2vec/.

Table 1. Model parameters configuration

Parameters	Values
Max len	210
Vocabulary size	8000, 20000
Embedding size	97
Hidden size	100
Max epoch	50
Batch size	1000, 2000
Decay rate	0.9
Learning rate	0.001, 0.01

4.3 Results

Similar to Tran et al. [12], we use two metrics to evaluate the baseline and our systems: F1 score and per-sentence accuracy (abbreviated as Acc). This accuracy is computed by the number of compressions that could be fully reproduced. Based on the above experimental setup, we train baseline models and the Bayesian combination model with different training data, respectively.

Table 2 shows the baseline models' performance with random 40,000 training pairs (from Train1, Train2, Train3, Train4 or Train5). 3LSTM and BiLSTM-tA are the models proposed by Fillippova et al. [11] and Tran et al. [12], respectively. R-BiLSTM-tA is our "Re-read" model. The results show that "Re-read" model performs better than BiLSTM-tA or 3LSTM over a single training set, which highlights the positive role of "Re-read" mechanism in text semantic modeling.

Table 2. Baseline models

Model	F1	Acc
3LSTM [7]	0.7445	0.225
BiLSTM-tA [8]	0.7681	0.315
R-BiLSTM-tA	**0.7882**	**0.319**

Table 3 shows the results obtained using the 200,000 training pairs. CR-BiLSTM-tA is the Bayesian combination model. The results of the previous three models indicate that the increasement of training data only has a significant enhancement for 3LSTM model, with little effect on BiLSTM-tA and "Re-read" model. At the same time, it can be seen that the Bayesian combination model has a significant effect on the compression, which highlights the importance of the explicit prior information and the generalization capability of the combination model.

Table 3. Bayesian combination model

Model	F1	Acc
3LSTM [7]	0.7565	0.232
BiLSTM-tA [8]	0.7723	0.317
R-BiLSTM-tA	0.7909	0.320
CR-BiLSTM-tA	**0.8002**	**0.325**

4.4 Discussion

The twice semantic modeling of this paper makes the sentence semantic representation more focused. The first semantic modeling result assists to adjust the secondary one, highlighting the semantic contribution from important words, and weakening the redundant words' effects. The simple attention mechanism avoids the redundancy attention information and high computational complexity, enhancing efficiency and accuracy in decoding. In the case of single model with single training data (40,000 training pairs), our "Re-read" model's performance has exceeded other models. In contrast, when single model with all training data (200,000 training pairs), we may loss some global features, or can not mine more potential influence factors due to the difference of training data and the difficulty in hyperparameter adjustment. So the model generalization effect is poor. By the Bayesian combination model, five different classifiers are trained on different training sets. Each classifier can capture the potential correlation between different training data and testing data. It further uses the explicit prior information. Moreover, we reduced hardware requirements without increasing training costs.

5 Conclusion

In this paper, we use a "Re-read" mechanism to model the semantics of original sentence twice under the encoder-decoder framework for the first time, which is to address the problem in generalization of the current sentence compression model and insufficient utilization of explicit prior information. The first modeling result, as global information, adjusts the second one, obtaining a more comprehensive semantic vector. What's more, we use a simple attention mechanism to achieve more accurate and efficient decoding. On the basis of "Re-read" mechanism, we adopt a Bayesian combination model involving with the deep learning model to make full use of explicit prior information, and achieve better compression on Google Newswire sentence compression dataset. In the future, we will explore to generate compressed sentences directly based on the "Re-read" mechanism.

References

1. Dorr, B., Zajic, D., Schwartz, R.: Hedge trimmer: a parse-and-trim approach to headline generation. In: Proceedings of the HLT-NAACL 03 on Text Summarization Workshop, vol. 5, pp. 1–8. Association for Computational Linguistics (2003)
2. Jing, H.: Sentence reduction for automatic text summarization. In: Proceedings of the Sixth Conference on Applied Natural Language Processing, pp. 310–315. Association for Computational Linguistics (2000)
3. Knight, K., Marcu, D.: Summarization beyond sentence extraction: a probabilistic approach to sentence compression. Artif. Intell. **139**(1), 91–107 (2002)
4. Corston-Oliver, S.: Text compaction for display on very small screens. In: Proceedings of the NAACL Workshop on Automatic Summarization, pp. 89–98. Association for Computational Linguistics (2001)

5. Clarke, J., Lapata, M.: Global inference for sentence compression: an integer linear programming approach. J. Artif. Intell. Res. **31**, 399–429 (2008)
6. Filippova, K., Altun, Y.: Overcoming the lack of parallel data in sentence compression. In: EMNLP, pp. 1481–1491 (2013)
7. Filippova, K., Alfonseca, E., Colmenares, C.A., et al.: Sentence compression by deletion with LSTMs. In: Conference on Empirical Methods in Natural Language Processing, pp. 360–368 (2015)
8. Tran, N.T., Luong, V.T., Nguyen, N.L.T., et al.: Effective attention-based neural architectures for sentence compression with bidirectional long short-term memory. In: Proceedings of the Seventh Symposium on Information and Communication Technology, pp. 123–130. ACM (2016)
9. Klerke, S., Goldberg, Y., Søgaard, A.: Improving sentence compression by learning to predict gaze. arXiv preprint arXiv:1604.03357 (2016)
10. Hochreiter, S., Schmidhuber, J.: Long short-term memory. Neural Comput. **9**(8), 1735–1780 (1997)
11. Graves A. Supervised Sequence Labelling with Recurrent Neural Networks. Springer, Heidelberg (2012)
12. Sutskever, I., Vinyals, O., Le, Q.V.: Sequence to sequence learning with neural networks. In: Advances in Neural Information Processing Systems, pp. 3104–3112 (2014)
13. Bengio, Y., Simard, P., Frasconi, P.: Learning long-term dependencies with gradient descent is difficult[J]. IEEE Trans. Neural Netw. **5**(2), 157–166 (1994)
14. Bahdanau, D., Cho, K., Bengio, Y.: Neural machine translation by jointly learning to align and translate. arXiv preprint arXiv:1409.0473 (2014)
15. Greff, K., Srivastava, R.K., Koutník, J., et al: LSTM: a search space odyssey. IEEE Trans. Neural Netw. Learn. Syst. (2016)
16. Srivastava, N., Hinton, G., Krizhevsky, A., et al.: Dropout: a simple way to prevent neural networks from overfitting. J. Mach. Learn. Res. **15**(1), 1929–1958 (2014)
17. Raskutti, G., Wainwright, M.J., Yu, B.: Early stopping and non-parametric regression: an optimal data-dependent stopping rule. J. Mach. Learn. Res. **15**(1), 335–366 (2014)

Terminology Translation Error Identification and Correction

Mengyi Liu, Jian Tang, Yu Hong[(✉)], and Jianmin Yao

School of Computer Science and Technology, Soochow University,
Suzhou 215006, Jiangsu, China
mengyiliu22@gmail.com, JohnnyTang1120@gmail.com,
tianxianer@gmail.com, jyao@suda.edu.cn

Abstract. Statistical machine translation (SMT) system requires homogeneous training data in order to get domain-sensitive terminology translations. If the data is multi-domain mixed, it is difficult for SMT system to learn translation probability of context-sensitive terminology. However, terminology translation is important for SMT. The previous work mainly focuses on integrating terminology into machine translation systems and heavily relies on domain terminology resources. In this paper, we propose a back translation based method to identify terminology translation errors from SMT outputs and automatically suggest a better translation. Our approach is simple with no external resources and can be applied to any type of SMT system. We use three metrics: tree-edit distance, sentence semantic similarity and language model perplexity to measure the quality of back translation. Experimental results illustrate that our method improves performance on both weak and strong SMT systems, yielding a precision of 0.48% and 1.51% respectively.

Keywords: Statistical machine translation · Domain terminology · Post-processing · Back translation

1 Introduction

In general, the performance of the SMT heavily relies on the scale and quality of the training corpora [1]. High-quality and large-scale corpora tends to include richer linguistic phenomena. As a result, the training effect of the statistical model (translation model, language model, and reordering model) in translation system will be improved.

However, applying a generic SMT system to technical documents often leads to wrong results, especially in the translation of domain-specific terminology. This is mostly due to the lack of domain-specific parallel data from which the SMT system can learn translation knowledge. The importance of domain-specific terminology for SMT has been mentioned in several previous work [2, 3]. Most of the work handles the case how to integrate the terminology tightly into the translation system. This requires not only a large amount of in-domain parallel corpora which is often difficult to obtain, especially for low-resourced domains or languages, but also a good expertise in SMT. We look upon the problem from a different perspective where we post-process the

© Springer Nature Singapore Pte Ltd. 2017
X. Cheng et al. (Eds.): SMP 2017, CCIS 774, pp. 141–152, 2017.
https://doi.org/10.1007/978-981-10-6805-8_12

terminology translation instead of modifying the model. We propose a back translation based method to identify the terminology translation errors and suggest a better translation.

Given a sentence, machine translation system will not output an appropriate translation unless the sentence is logical, according with common sense and contextual semantic consistent. In order to facilitate the understanding of the above linguistic phenomena, two pairs of translation examples are given below (Table 1).

Table 1. Two pairs of translation examples

Source: 智能 工厂 用 传感器 和 **执行器** 赢得 管理 操作 和 知识 驱动 优化 的实时 信息 。

Target: *Intelligent plants use sensors and **actuators** to gain real-time information on management operations and knowledge-driven optimization.* **[Sample1]**

Source: 智能 工厂 用 传感器 和 **演员** 赢得 管理 操作 和 知识 驱动 优化 的 实时 信息 。

Target: *Intelligent factory with sensors and **actors** to win management operations and knowledge-driven optimization of real-time information.* **[Sample2]**

The source sentence in sample1 is normal statements, smooth and fluent on the whole; but in sample2 the source sentence is abnormal statements, phrase "*actor*" is contextual semantic inconsistent obviously. We use Google Translator[1] to translate two source language sentences, and two translation results show difference in syntactic structure and semantic. In the two source sentences, phrases "*管理操作*" and "*知识驱动优化*" are used to modify the phrase "*实时信息*". From the target sentence in sample1, we can see that phrases "*management operations*" and "*knowledge-driven optimization*" are used to modify the phrase "*real-time information*", the same as source sentence. But in sample2's target sentence, phrase "*real-time information*" is used to modify "*knowledge-driven optimization*", which is deviated from the meaning expressed by the source sentence. We further analyze this linguistic phenomenon and consider this is resulted from the translation mechanism. The system has translated "*演员*" as "*actors*", then it prefers "*win management operations*" as next translation rather than "*gain real-time information*" according with comprehensive score (language model et al.).

As can be seen from the above analysis, the irrationality of individual phrase in a sentence can affect the translation of the whole sentence. If the irrational element in the sentence is a term, this phenomenon will become more obvious. The reason for this is

---

[1] http://translate.google.cn.

that term conveys concepts of a text, term translation becomes crucial when the text is translated from its original language to another language [4].

In this paper, we aim to propose a method to identify terminology translation errors of the SMT outputs and suggest a better translation. Compared with integrating terminology into SMT models and building a sophisticated system, our method is simple and do not rely on domain resources. Our method is based on back translation, and we propose three metrics to measure the quality of back translation: (1) tree-edit distance; (2) sentence semantic similarity; (3) language model perplexity. Experimental results illustrate that they are all able to achieve improvements of precision on both weak and strong translation systems.

The remainder of the paper is organized as follows. Section 2 overviews the related work. We present the methodology and detail the metrics in Sect. 3. Section 4 shows the experimental settings and results. Section 5 draws conclusions and describes the future work.

2 Related Work

In this section, we briefly introduce related work and highlight the differences between our work and previous studies.

There has been a growing interest for terminology integration into SMT models recently. [5] investigate that bilingual terms are important for domain adaptation of machine translation. Direct integration of terminology into the SMT model has been considered, either by extending SMT training data [2], or via adding an additional term indicator feature into the translation model [3, 5]. [6] propose a binary feature to indicate whether a bilingual phrase contains a term pair. [4] investigate three issues of term translation in the context of document-informed SMT and integrate the three models into hierarchical phrase-based SMT. However, none of the above is possible when we deal with an external black-box SMT system.

[7] employ bilingual term bank as a dictionary and propose a post-processing step for a SMT system, where a wrongly translated term is replaced with a user-provided term translation. [8] propose a demonstration of a multilingual terminology verification/correction service, which detects the wrongly translated terms and suggest a better translation of the terms.

Our work is also related to machine translation error identification. [9] combine syntax feature, vocabulary feature and word posterior probability feature, which are extracted based on LG parsing, and use the binary classifier based on Maximum Entropy Model to predict the label of each word in machine translation. [10] rely on a random forest classifier and 16 features to predict the label of a word. [11] train two classifier models by using bidirectional long short-term memory recurrent neural networks and CRF to complete word level QE Task.

Our work departs from the previous work in two major respects.

- We focus on the terminology translation error identification and correction, and our method do not rely on external resources such as bilingual domain-specific terminology. This can be seen as post-editing focused on domain terminology.
- Our method is based on back translation, so we just need to compare the same language. This can avoid crossing-language comparison which is complicated.

3   Methodology

We propose a method to identify terminology translation errors and automatically suggest better translations. First of all, we present the methodological framework. Then we introduce the crucial part of comparing back translation and original sentence. Finally, we list preprocessing methods for collecting and processing raw data.

3.1   Back Translation Based Terminology-Checking Method

The method proposed in this paper does not modify the model of the translation system, but is used as the post processing of the existing translation system. Figure 1 shows the framework of back translation based terminology-checking method (BTTC).

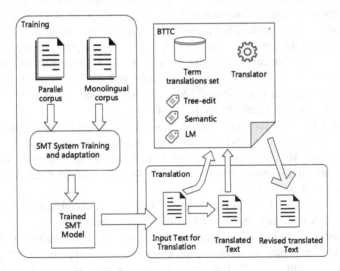

Fig. 1. Framework of BTTC

The left of the framework is the initial SMT system. Model training phase includes phrase table generation, translation model training, reordering model training, and language model training, et al. When these models have been trained, they are combined in a log-linear model. To obtain the best translation $\hat{e}$ of the source sentence f, log-linear model uses the following equation, in which h_m and λ_m denote the mth feature and weight.

$$\widehat{e} = \arg\max_{e} p(e|f)$$
$$= \arg\max_{e} \sum_{m=1}^{M} \lambda_m h_m(e,f)$$

(1)

Once we obtain a trained SMT system, given a sentence containing terminology, we can translate it into target language. The terminology translation may be correct or wrong and we don't know. To solve this problem, we propose a post-edit processing which contains several steps as follows:

- Locating the terminology translation. To identify the terminology translation errors, the first step is locating its position in the target sentence. Fortunately, we have access to the internal sub-phrase alignment provided by Moses[2], thus we know the exact location of the terminology translation. We just need to add parameters "-print-alignment-info" when decoding. Specific examples are shown below (Table 2):

Table 2. An example of internal sub-phrase alignments

Source: *It has great influence on the time complexity of the join algorithms to locate positions on **tertiary storage [16 17]**.*
Target: 它 可以 对 时间 复杂性 影响 连接 定位 算法 职位 **高等教育 仓库** . ‖‖ 0-0 0-1 6-3 7-4 3-5 10-6 11-8 13-7 14-9 **16-10 17-11** 18-12

The position of phrase "*tertiary storage*" in the source sentence is 16 and 17, and we can know the position of its translation in target sentence is 10 and 11 according to the alignment information, exactly the phrase "**高等教育 仓库**".

- Replacing terminology translation with other translations. The terminology we marked in the source sentence may have several translations in training data, and SMT system chooses the translation which has the highest probability score. Therefore, the translation which has more occurrences is more likely to be chosen. Differently, our method treats each translation equally and judge them from semantic perspective. In order to obtain all translation candidates for the terminology, we search the phrase table. The size of phrase table is usually very large, so we do hash operation on the phrase table and query terminology to improve efficiency. Then we obtain all terminology translations and filter some meaningless items.

---

[2] http://www.statmt.org/moses.

- Back translation. A back translation can be defined as the translation of a target sentence back to the original source language. In order to ensure the quality of the back translation, we call Youdao Translate API[3] interface instead of the reversed translation system constructed by ourselves. The input of the API is the text to be translated. In our case, it's a sentence which is the translation of the test sentence. The results returned from the API is the xml data structure.
- Selecting the best translation. For a test sentence, we have obtained several pseudo similar sentences. What we should do is to select the most similar sentence semantically and syntactically. We will detail this in the next section.

3.2 Compare Back Translation with Original Text

In this section, we will introduce three metrics to compare back translation with the original text. We think that terminology translation is more reliable when the similarity is higher between the back translation and the original sentence.

- Tree edit distance. Trees are among the most common and well-studied combinatorial structures in computer science. An optimal edit script between two trees is an edit script between them of minimum cost and this cost is the tree edit distance [12]. A tree edit model can be used to identify whether two sentences convey essentially the same meaning. In this paper, we use [13] 's method to calculate the tree edit distance between the dependency trees of two sentences. The smaller the distance, the greater the similarity of two sentences. We obtain dependency trees of sentences by Standford NLP toolkit[4]. We assume that we will get a bad translation when the source sentence includes an inappropriate terminology in it, even the dependency structure of the translation will be different from the original sentence.
- Sentence semantic similarity. Sentences that share semantic and syntactic properties are thus mapped to similar vector representations [14]. In [14]'s work, they propose a model called skip-thought vectors which encode a sentence to predict the sentences around it. The results of experiments on the SemEval 2014 Task 1 show that skip-thought vectors learn representations that are well suited for semantic relatedness. Sentence similarity refers to the matching extent in semantics of two sentences which is a real number, the greater the value, the greater the similarity of the two sentences. We use the cosine similarity here.
- Language model perplexity. [10] use language model perplexity feature to estimate the quality of machine translation at sentence level. Inspired by them, we use this metric to measure the quality of back translation.

3.3 Corpus Acquisition

To perform our method, we need the test set which consists of sentences and the terminology in each sentence should be marked.

---

[3] http://fanyi.youdao.com/openapi?path=data-mode.

[4] https://nlp.stanford.edu/software/nndep.shtml.

We find that journals on the web are good resources, we just need to click on the title of the paper with no downloading and then we can obtain keywords and abstracts both in Chinese and English. We crawl the keywords and abstracts by using urllib[5] which is a python package that collects several modules for working with URLs. On the basis, we use another python package BeautifulSoup[6] to extract keywords and abstracts from the structured source files of the crawled web pages.

The next step is to obtain the sentences which the keywords are in. We detect sentence boundaries on English abstracts by using OpenNLP[7] which is a machine learning based toolkit for the processing of natural language text. For Chinese abstracts, we write rules to detect sentence boundaries. We use a rough but simple way to extract parallel sentences which the keywords are in. Each article has about four keywords, for each keyword, we locate the sentence containing this keyword in the Chinese abstract, and then check the corresponding index sentence in English abstract with extending two sentences window at most. This is because English abstract is not translated by Chinese abstract sentence by sentence in many articles. Besides, we make all English keywords and abstracts lowercase to avoid case matching problems.

4 Experiments

We conduct a pilot study for verifying whether back translation based strategy is useful for the identification and correction of terminology translation errors in the SMT system outputs.

4.1 Setup

Our training data consists of 16M mix-domain sentence pairs extracted from web by [15]'s acquisition method. We randomly choose 2k sentences as tuning [16] set from CWMT09. The test set consists of 1657 sentences in English from the abstracts of a computer science's journal. We collect 11, 224 bilingual terms from the keywords of the journal.

The word alignments were obtained by running fast-align [17] on the corpora in both directions and using the "grow-diag-final-and" balance strategy [18]. We adopted KEN Language Modeling Toolkit [19] to train a 5-gram language model with modified Kneser-Ney smoothing on the Xinhua portion of the Chinese[8]/English[9] Gigaword corpus.

We use [13]'s method to calculate the tree edit distance between dependency trees of two sentences. We obtain dependency trees of sentences by Standford NLP toolkit.

---

[5] https://docs.python.org/3/library/urllib.html.

[6] https://www.crummy.com/software/BeautifulSoup/.

[7] http://opennlp.apache.org/.

[8] LDC2003T09 Gigaword Chinese Text Corpus Second Edition.

[9] LDC2009T13 Xinhua News Portion of English Gigaword Second Edition.

While the traditional sentence representation using mean pooled Word2Vec discards word order, SkipThoughts use a Recurrent Neural Network to capture the underlying sentence semantics. We use the pretrained model by [14] to compute a 4800 dimensional sentence representation.

We build several translation systems as follows:

- **Baseline:** We use Moses to construct English to Chinese translation system as our baseline system. The features used in baseline system include: (1) four translation probability features; (2) one language model feature; (3) distance-based and lexicalized distortion model feature; (4) word penalty; (5) phrase penalty.
- **Baseline+BiTerm:** [20] prove that concatenating the training data and the terms perform better than more complex techniques. We take the bilingual terms as parallel sentence pairs and add them into the training corpus.
- **Baseline+BTTC:** Performing our method on the outputs of the Baseline system.
- **Baseline+BiTerm+BTTC:** Performing our method on the outputs of the Baseline+ BiTerm system.

For the original terminology translation in the SMT system outputs, we think it may be wrong if it satisfies the following two conditions at the same time: (1) the result of the highest language model perplexity minus the original terminology translation's perplexity score is greater than the threshold value which we empirically set as 0.015; (2) its semantic similarity is lower than the highest score.

As for translation suggestion, we use three methods: (1) selecting the translation candidate whose back translation is the most similar to the test sentence semantically; (2) selecting the translation candidate whose back translation has the lowest tree-edit distance; (3) selecting the translation candidate whose back translation has the maximum difference between semantic similarity and tree-edit distance.

4.2 Evaluation Metrics

We conduct our method on the test set, with the aim to verify whether back translation based terminology-check method is able to identify the wrongly translated terminology and suggest a better translation. The basic evaluation metric is the precision rate (PR). Precision rate is defined as the percentage of the terms that are correctly translated as follow:

$$PR = \frac{\# \text{ of correctly translated terms}}{\text{Total} \ \# \text{ of terms}} \tag{2}$$

5   Results

Table 3 gives our experiment results. From this table, we can see that three suggestion methods all have positive effects, and semantic similarity method works better than the tree-edit distance method. For Baseline system, the tree-edit method achieves 0.36% precision improvement and the semantic method achieves 0.42% precision improvement. Baseline+BiTerm system also gives an evidence of this, the tree-edit method achieves 1.09% precision improvement and the semantic method achieves 1.21% precision improvement. Combing two metrics works best, which achieves 0.48% and 1.51% precision improvement on two systems respectively. The results also show that the BTTC can work better on the strong translation system. This is mainly because the strong translation system is trained from the higher quality corpora which contains more useful translation information. Therefore, our method is more likely to retrieve the correct terminology translation and make corrections.

Table 3. Performance of BTTC on different systems

Methods	Precision (%)
Baseline	25.05
Baseline+BTTC (tree-edit)	25.41
Baseline+BTTC (semantic)	25.47
Baseline+BTTC (semantic + tree-edit)	25.53
Baseline+BiTerm	54.19
Baseline+BiTerm+BTTC (tree-edit)	55.28
Baseline+BiTerm+BTTC (semantic)	55.40
Baseline+BiTerm+BTTC (semantic+tree-edit)	55.70

In order to know in what respects our method improve performance of translation, we manually analyze some test sentences and give some examples in Table 4. The back translations of all three sentences' original translations are semantically deviated from the source sentences. However, the replaced translation with the right terminology translation is more contextual consistent and their back translation is semantically similar to the source sentences.

Table 4. Translation examples

Src: *therefore the problem of quantum grammars generating quantum **regular languages** is solved.*

Baseline+BiTerm: 经常语言

Biseline+Biterm+BTTC: 正规语言

Src: *in order to optimize **data locality** and communication overhead, this paper proposes a novel alternate tiling stencil algorithm on distributed memory machines by exploiting the property of the iterative algorithm.*

Baseline+BiTerm: 数据区域性

Biseline+Biterm+BTTC: 数据局部性

We find that although many wrongly translated terminologies are corrected by BTTC, but the overall performance is not obvious. The reason is that some correct terminology translations are wrongly revised by BTTC. Considering a scenario where the user is dissatisfied with the outputs of the translation system, more specifically, he or she think the terminology translation is wrong. In such case, we get the feedback and know which terminology need to be corrected. Table 5 shows the better performance of our method in such situation. We perform our post-editing method on those true mistakes. The results show that BTTC achieves 0.96% and 3.38% precision improvement on Baseline system and Baseline+BiTerm system respectively.

Table 5. Performance of BTTC on true mistakes

Methods	Precision (%)
Baseline+BTTC	26.01
Baseline+BiTerm+BTTC	57.57

In addition, we find the sentence vector causes some mistakes. Table 6 shows an example. Obviously, the True_backtran is more similar with the Gold sentence, but the semantic similarity of True_backtran is 0.848 and lower than False_backtran's score, which is 0.972.

Table 6. Inappropriate scored examples

Gold: *The bus network is more suitable when you want to connect to some computers.*

True: 当 你 想 连接 一些 电脑 的 时候 , **总线 网络** 更 合适 。
True_backtran: *When you want to connect to some computers, the **bus network** is more appropriate.*

False: 当 你 想 连接 一些 电脑 的 时候 , **公交 网络** 更 合适 。
False_backtran: *When you want to connect to some computer time, **public transportation network** more suitable.*

6   Conclusion and Future Works

We propose a back translation based method to automatically identify terminology translation errors in the SMT system outputs and suggest a better translation. Our method relies on an external generic reversed MT engine and needs to know which is the terminology in the test sentence. We propose three metrics to measure the quality of back translation. Experimental results show that our method can suggest better terminology translations for both weak and strong translation systems. The performance of our method is better when the training data contains more translation information such as domain terminology. Besides, the performance can be further improved when the identification precision improves.

However, the strategies of measuring back translation are roughly simple and coarse in this paper. Complicated approach should be taken into account during identifying the true mistakes. In future work, we also consider representing the semantic of a sentence more accurately. In addition, acquiring terminology dictionary is also meaningful for our work, and each item in the dictionary corresponds to many possible translations.

Acknowledgments. This research work is supported by National Natural Science Foundation of China (Grants No. 61373097, No. 61672367, No. 61672368, No. 61331011), the Research Foundation of the Ministry of Education and China Mobile, MCM20150602 and the Science and Technology Plan of Jiangsu, SBK2015022101. The authors would like to thank the anonymous reviewers for their insightful comments and suggestions.

References

1. Axelrod, A., He, X., Gao, J.: Domain adaptation via pseudo in-domain data selection. In: Proceedings of the Conference on Empirical Methods in Natural Language Processing, pp. 355–362. Association for Computational Linguistics, Edinburgh (2011)
2. Carl, M., Langlais, P.: An intelligent terminology database as a pre-processor for statistical machine translation. In: Second International Workshop on Computational Terminology COLING-02 on COMPUTERM 2002, vol. 14, pp. 1–7. Association for Computational Linguistics (2002)
3. Skadiņš, R., Pinnis, M., Gornostay, T., Vasiļjevs, A.: Application of online terminology services in statistical machine translation. In: Proceedings of the XIV Machine Translation Summit, MT Summit XIV, France, pp. 281–286 (2013)
4. Meng, F., Xiong, D., Jiang, W., Liu, Q.: Modeling term translation for document-informed machine translation. In: Proceedings of the Conference on Empirical Methods in Natural Language Processing, pp. 546–556. Association for Computational Linguistics, Doha (2014)
5. Pinnis, M., Skadiņš, R.: MT adaptation for underresourced domains–what works and what not. In: Proceedings of the 5th International Conference Baltic HLT, p. 176. IOS Press (2012)
6. Ren, Z., Lu, Y., Cao, J., Liu, Q., Huang, Y.: Improving statistical machine translation using domain bilingual multiword expressions. In: Proceedings of the Workshop on Multiword Expressions: Identification, Interpretation, Disambiguation and Applications, pp. 47–54. Association for Computational Linguistics, Suntec (2009)
7. Itagaki, M., Aikawa, T.: Post-MT term swapper: supplementing a statistical machine translation system with a user dictionary. In: Proceedings of the 6th International Conference on Language Resources and Evaluation. European Language Resources Association, Marrakech (2008)
8. Bosca, A., Nikoulina, V., Dymetman, M.: A lightweight terminology verification service for external machine translation engines. In: Proceedings of the Demonstrations at the 14th Conference of the European Chapter of the Association for Computational Linguistics, pp. 49–52. Association for Computational Linguistics, Gothenburg (2014)
9. Xiong, D., Zhang, M., Li, H.: Error detection for statistical machine translation using linguistic features. In: Proceedings of the 48th Annual Meeting of the Association for Computational Linguistics, pp. 604–611. Association for Computational Linguistics, Uppsala (2010)

10. Wisniewski, G., Pécheux, N., Allauzen, A.: LIMSI submission for WMT'14 QE task. In: Proceedings of the 9th Workshop on Statistical Machine Translation, pp. 348–354. Association for Computational Linguistics, Baltimore (2014)

11. José, G.C., de Souza, J.G.-R., Buck, C., Turchi, M., Negri, M.: FBK-UPV-UEdin participation in the WMT14 quality estimation shared-task. In: Proceedings of the 9th Workshop on Statistical Machine Translation, pp. 322–328. Association for Computational Linguistics, Baltimore (2014)

12. Bille, P.: A survey on tree edit distance and related problems. Theoret. Comput. Sci. **337**(1), 217–239 (2005)

13. Yao, X., Van Durme, B., Callison-Burch, C., Clark, P.: Answer extraction as sequence tagging with tree edit distance. In: Proceedings of North American Chapter of the Association for Computational Linguistics, pp. 9–14. Association for Computational Linguistics Atlanta (2013)

14. Kiros, R., Zhu, Y., Salakhutdinov, R.R., Zemel, R., Urtasun, R., Torralba, A., Fidler, S.: Skip-thought vectors. In: Advances in Neural Information Processing Systems, pp. 3294–3302 (2015)

15. Liu, L., Hong, Y., Lu, J., Lang, J., Ji, H., Yao, J.M.: An iterative link-based method for parallel web page mining. In: Proceedings of the Conference on Empirical Methods in Natural Language Processing, pp. 1216–1224. Association for Computational Linguistics, Doha (2014)

16. Och, F.J.: Minimum error rate training in statistical machine translation. In: Proceedings of the 41st Annual Meeting on Association for Computational Linguistics, vol. 1, pp. 160–167. Association for Computational Linguistics, Sapporo (2003)

17. Dyer, C., Chahuneau, V., Smith, N.A.: A simple, fast, and effective reparameterization of IBM model 2. In: Proceedings of the Conference of the North American Chapter of the Association for Computational Linguistics: Human Language Technologies, pp. 644–649. Association for Computational Linguistics, Atlanta (2013)

18. Koehn, P., Och, F.J., Marcu, D.: Statistical phrase-based translation. In: Proceedings of the Conference of the North American Chapter of the Association for Computational Linguistics on Human Language Technology, vol. 1, pp. 48–54. Association for Computational Linguistics, Edmonton (2003)

19. Heafield, K., Pouzyrevsky, I., Clark, J.H., Koehn, P.: Scalable modified Kneser-Ney language model estimation. In: Proceedings of the 51st Annual Meeting of the Association for Computational Linguistics, pp. 690–696. Association for Computational Linguistics, Sofia (2013)

20. Bouamor, D., Semmar, N., Zweigenbaum, P.: Identifying bilingual multi-word expressions for statistical machine translation. In: Proceedings of the 8th International Conference on Language Resources and Evaluation, pp. 674–679. European Language Resources Association, Istanbul (2012)

Opinion Target Understanding in Event-Level Sentiment Analysis

Suyang Zhu, Shoushan Li[✉], and Guodong Zhou

School of Computer Science and Technology, Soochow University,
No. 1, Shizi Street, Gusu District, Suzhou 215021, Jiangsu, China
syzhu@stu.suda.edu.cn, {lishoushan,gdzhou}@suda.edu.cn

Abstract. In this paper, we focus on a critical subtask in *event-level* sentiment analysis, namely opinion target understanding, with the goal to determine which opinion target a comment talks about in an event description. Unlike traditional *aspect-level* sentiment analysis, opinion target understanding needs to not only recognize the opinion target in a comment, but also align the target to the corresponding opinion target in an event description. To address this problem, we propose a neural 2-sequences-to-1-sequence framework to jointly leverage both texts in an event description and an comment, and apply a word-by-word attention mechanism to capture the alignment between the two texts. Experimental results prove the effectiveness of the proposed approach to opinion target understanding in *event-level* sentiment analysis.

Keywords: Opinion target understanding · Sentiment analysis · 2-sequences-to-1-sequence framework · Word-by-word attention

1 Introduction

Sentiment analysis (aka opinion mining) mainly addresses the problem of determining the sentiment orientation (e.g. *positive, negative,* or *neutral*) in a piece of text [1,2]. Due to the rapid expansion of various types of social media, researches on sentiment analysis have been growing explosively during the past decade.

Related studies in early years focus on classifying a whole review towards a product. For convenience, in this paper, we refer to such task as *product-level* sentiment analysis [3,4]. In the literature, we usually recast it as a classification problem. A classifier accepts a review of a product as the input, and outputs the sentiment orientation of the review in order to determine whether the review is *positive* or *negative* for this product.

In recent years, researchers focus on identifying the sentiment orientation in a certain aspect of a product. We refer to such task as *aspect-level* sentiment analysis [5–8]. For instance, in a review of a restaurant *"the dishes are good but the waitresses are unskilled."*, *"dishes"* and *"waitresses"* are two aspects of the restaurant. A model takes a review as input, and outputs involved aspects alone with their sentimental orientations. In essence, such task is divided into

© Springer Nature Singapore Pte Ltd. 2017
X. Cheng et al. (Eds.): SMP 2017, CCIS 774, pp. 153–165, 2017.
https://doi.org/10.1007/978-981-10-6805-8_13

two subtasks: (1) aspect detection, which is usually recast as an information extraction problem; (2) sentiment classification with regard to each aspect.

In this paper, we turn our attention to sentiment analysis towards an event. For clarity, we refer to such task as a kind of *event-level* sentiment analysis. In such task, a model takes both texts in the event description and the comment as input, and outputs the opinion target of the comment in the corresponding event description with the sentiment orientation of the comment. Figure 1 gives an instance, translated from a Chinese news, of an event description involving two opinion targets, i.e., *"Hillary Clinton"* and *"Donald Trump"*, and two comments on it. The goal of *event-level* sentiment analysis is to find the sentiment orientation of a comment toward one certain opinion target. In principle, this task could be divided into two subtasks: (1) determining which opinion target a comment talks about in an event description; (2) determining the sentiment orientation of the comment. In this study, we address the first subtask. We call it opinion target understanding. Although this subtask looks similar to the subtask of aspect detection in *aspect-level* sentiment analysis, opinion target understanding in *event-level* sentiment analysis is much more challenging than aspect detection in *aspect-level* sentiment analysis due to following two reasons.

Event Description:
Hillary Clinton and Donald Trump met on October 9 for their second presidential debate. ... Trump stated that if elected, he would appoint a special prosecutor to investigate Secretary Clinton in relation to the matter. ... Public opinion generally believes that Clinton has a chance for being the first Madam President in the history of the US.

opinion targets: *H. Clinton, D. Trump*

Comment 1:
I'm looking forward to celebrating for the madam president.

opinion target: *H. Clinton*   sentiment orientation: *positive*

Comment 2:
This man is so naive that he thinks the real world is basically a Hollywood action movie.

opinion target: *D. Trump*   sentiment orientation: *negative*

Fig. 1. An example for two comments express different opinions on different targets in the event with different sentiment orientations

On the one hand, the input in aspect detection often contains only one kind of text, i.e., the comment, while the input in opinion target understanding contains two kinds of texts, i.e., event description and the comment. In order to leverage the hidden knowledge in both texts, a more sophisticated learning model is desperately needed.

On the other hand, since aspect detection recognizes the opinion target in a text, it is more likely to be an information extraction problem [7,8]. For instance, in Comment 1, *"madam president"* can be extracted as a aspect candidate using an information extraction model. However, only knowing *"madam president"* as an opinion target is not enough for opinion target understanding in *event-level* sentiment analysis. More importantly, we need to align *"madam president"* in the comment to the opinion target *"Hillary Clinton"* in the event description to know this comment is talking about *"Hillary Clinton"*. Therefore, opinion target understanding needs to "understand" the opinion target rather than to simply extract it.

In this paper, we tackle the first challenge by proposing a 2-sequences-to-1-sequence neural framework. In this framework, both texts in event description and the comment are modeled as character sequences. In order to align the opinion target in the comment to the mentions of the corresponding opinion target in the event description, a word-by-word attention mechanism is employed.

Furthermore, we tackle the second challenge by leveraging various mentions of one opinion target in the event description with an intermediate output introduced and represented as a vector where all the mention positions are valued as 1 and other positions are valued as 0. Taking Fig. 1 as an instance, four mentions of the opinion target *"Hillary Clinton"*, i.e., *"Hillary Clinton"*, *"Secretary Clinton"*, *"Clinton"* and *"Madam President"* appear in position 0, 27, 39, and 47 in the event description. We can represent the intermediate output as a one-hot vector of a fixed size, in which the 0th, 27th, 39th, and the 47th elements are 1, and the rest are 0. The final result is obtained by analyzing the predicted mention positions with hand-written rules. Experimental studies demonstrate that the proposed model performs better than some strong baselines.

The reminder of this paper is organized as follows. Section 2 gives a brief review of the collected data used in this paper. Section 3 proposes 2-sequences-to-1-sequence framework to opinion target understanding. Section 4 introduces the word-by-word attention mechanism. Section 5 describes the experiments and results for evaluating the proposed approach. Section 6 gives the conclusion and the future work.

2   Data Collection

The corpus we use are collected from Tencent News[1], a Chinese news portal site. We choose and crawl texts from 12 news articles, denoted as *Article i* ($i \in [1, 12]$) respectively, on this site. As a preliminary study, in this paper, we only concentrate on those news which have exactly 2 opinion targets in conflict because a comment of an conflict event often contains obvious sentiment tendency. The news articles are regarded as event descriptions along with 5,139 comments in total submitted by readers.

We annotate all mentions of opinion targets in each news article, and the opinion target of each comment. In this corpus, each comment is annotated with

---

[1] http://news.qq.com/.

one of the three categories: *target 1*, *target 2*, and *both*, and those comments talk about neither of two opinion targets are abandoned. The comments labeled with *both* category means these comments talk about both opinion targets. For instance, a comment like *"I think Trump and Clinton played fair."* for the event showed in Fig. 1 is categorized to *both*.

For each news article, P_{t1} and P_{t2} are two gold-annotated mention position vectors. These two vectors can be represented in the form of two sets: $S_{t1} = \{i \mid p_i = 1, p_i \in P_{t1}\}$, and $S_{t2} = \{i \mid p_i = 1, p_i \in P_{t2}\}$. For instance, in the event given in Fig. 1, let *"Hillary Clinton"* be *target 1*. Then we have $S_{t1} = \{0, 27, 39, 47\}$. For category *both*, we have vector $P_{t1} \vee P_{t2}$ and the label set $S_{t1} \cup S_{t2}$. Since one opinion target may refers to a group of people, only the plural personal pronouns which really refer to both targets are annotated as the mention of both *target 1* and *target 2*.

The comments of three categories are nearly balanced (1.25:1:1.25, approximately) in the whole corpus, but are severely unbalanced in most news article. Two annotators annotate the whole corpus, respectively. The annotation of this corpus achieves a consistency with 0.7663 Kappa value on three categories. The final labels are judged by a third annotator.

3   The 2-sequences-to-1-sequence Framework

In this section, we introduce the 2-sequences-to-1-sequence framework which leverage the information from the event description and the comment. **Model-basic** in Fig. 2 is our basic model, which consists of two LSTM-based encoders and one RNN-based decoder.

3.1   Input

Model-basic takes raw texts as input, and convert them to word embedding representations. Each word in a text is represented as a real-valued row vector. For instance, in **Model-basic**, given the raw text of event description consisting of l words $T = \{x_1, x_2, ..., x_l\}$, for a certain word x_i, we convert it to its word embedding e_i as follows:

$$e_i = E_e w_i \tag{1}$$

where the matrix $E_e \in \mathbb{R}^{d|V|}$ is the embedding lookup table for the event descriptions to be learned, d is the dimension of word embeddings, V is the vocabulary, and w_i is the one-hot representation of word x_i in the row vector form. The raw text of event description is then converted to a word embedding matrix $M_e \in \mathbb{R}^{ld}$. Likewise, we have table $E_c \in \mathbb{R}^{d|V|}$ for the comments, and the word embedding matrix $M_c \in \mathbb{R}^{kd}$ in **Model-basic**, where k is the length of the comment.

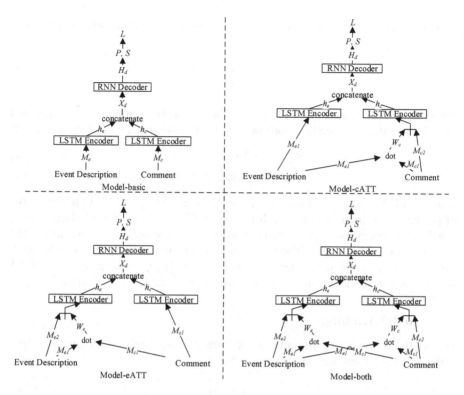

Fig. 2. Four models based on the proposed 2-sequences-to-1-sequence framework. **Model-basic** is the basic model. **Model-cATT** and **Model-eATT** apply word-by-word attention solely over the input of the comment and the event description, respectively. **Model-both** applies attention over both inputs

3.2 Encoder

We use LSTM [9,10] as the encoder to map the input sequence to a vector in a fixed size with mean pooling. In **Model-basic**, we denote the vectorized outputs of two encoders as follows:

$$h_e = LSTM(M_e) \tag{2}$$
$$h_c = LSTM(M_c) \tag{3}$$

where h_e and h_c refer to two encoded vectors generated from event description and the comment, respectively.

3.3 Decoder

We apply an RNN layer followed by a hidden layer as the decoder, which map the vectorized outputs of encoders to the mention position vector. Our model concatenates two encoded vectors, and duplicate the concatenated vector l (the length of the event description) times to obtain the input sequence for the decoder:

$$X_d = [x_1\ x_2\ ...\ x_l]^{\mathrm{T}} \tag{4}$$
$$x_i = h_e \oplus h_c,\ i \in [0, l] \tag{5}$$

where $\oplus$ is the concatenating operator. The duplicated sequence X_d is then fed into the decoder followed by a softmax function to predict the mention positions:

$$H_d = RNN(X_d) \tag{6}$$
$$P = softmax(WH_d + b) \tag{7}$$

where H_d is the sequential output of the decoder, $P = [p_1\ p_2\ ...\ p_l]$ is the vector of predicted probabilities, in which p_i ($i \in [1, l]$) refers to the probability of ith word in the event description being a mention of the opinion target which the comment talks about. W is the weight matrix of the hidden layer to be learned, and b is the bias term. We further have set $S = \{i\ |p_i > 0.5, p_i \in P, i \in [1, l]\}$ containing the predicted mention positions is used for label generation.

3.4   Model Training

Our model is trained to minimize a categorical cross-entropy loss function. Specially, the loss functions are defined as follows:

$$loss = -\frac{1}{n}\sum_{i=1}^{n}\sum_{j=1}^{l} y_{ij} \log p_{ij} \tag{8}$$

where n is the total number of samples, y_{ij} is the gold annotation indicates whether the jth word in the event description is truly a mention of the opinion target talked about by the comment in the ith sample, and p_{ij} refers to the corresponding predicted probability.

In this study, we apply Adadelta [11] as the optimizing algorithm. All the matrix and vector parameters in neural network are initialized with uniform samples in $[-\sqrt{6/(r + c)}, \sqrt{6/(r + c)}]$, where r and c are the numbers of rows and columns in the matrices [12].

Moreover, we apply dropout technique [13] to drop some dimensions of LSTM layers and RNN layers with a fixed probability to relieve the influence of overfitting. Early stopping technique is also applied to avoid overfitting.

The length of each event description and comment is set to two fixed sizes by zero padding. Zero-masking is applied for the flexibility on processing variable-length sequences.

3.5   Rules for Label Generation

Finally, we get the category label by analyzing the mention positions generated by the proposed model with hand-written rules. For each mention position vector and its corresponding event description, the rules are described as Algorithm 1.

Algorithm 1. Generating the category label.

Input:
    The mention position set, S;
    The set of gold-annotated mention position of $target\ 1$, S_{t1};
    The set of gold-annotated mention position of $target\ 2$, S_{t2};
Output:
    The category label of the mention position vector, L;
1: If $\frac{|S \cap S_{t1}|}{|S_{t1}|} \geq 0.5$ & $\frac{|S \cap S_{t2}|}{|S_{t2}|} \geq 0.5$, or $\frac{|S \cap S_{t1}|}{|S_{t1}|} = \frac{|S \cap S_{t2}|}{|S_{t2}|}$, Then $L = both$;
2: Else, If $\frac{|S \cap S_{t1}|}{|S_{t1}|} > \frac{|S \cap S_{t2}|}{|S_{t2}|}$, Then $L = target\ 1$;
3: Else, If $\frac{|S \cap S_{t1}|}{|S_{t1}|} < \frac{|S \cap S_{t2}|}{|S_{t2}|}$, Then $L = target\ 2$;
4: **return** L.

4   Word-by-Word Attention

The attention mechanism is widely used in various NLP tasks (e.g. machine translation [14]). Basing on **Model-basic**, we further propose **Model-cATT** and **Model-eATT** which apply word-by-word attention mechanism over the input of the comment and the event description, respectively, to capture the alignment between the opinion targets in comment and their corresponding mentions in the event description. Finally, we propose **Model-both** which applies word-by-word attention over both inputs. The attention mechanism used in this paper is similar to that in [14]. Figure 2 gives the structures of the three models.

For these models, we firstly define four lookup tables: $E_{e1} \in \mathbb{R}^{d|V|}$, $E_{c1} \in \mathbb{R}^{d|V|}$, $E_{e2} \in \mathbb{R}^{k|V|}$, and $E_{c2} \in \mathbb{R}^{l|V|}$. E_{e2} and E_{c2} are additionally learned in the need of dimension fitting. The corresponding word embedding matrices are $M_{e1} \in \mathbb{R}^{ld}$, $M_{c1} \in \mathbb{R}^{kd}$, $M_{e2} \in \mathbb{R}^{lk}$, and $M_{c2} \in \mathbb{R}^{kl}$, respectively.

We calculate the weight matrix via the dot product between two word embedding matrices of two inputs followed by the softmax function. In **Model-cATT**, the weight matrix over the comment is calculated as follows:

$$W_c = softmax(M_{c_1} M_{e_1}^{\mathrm{T}}) \qquad (9)$$

where $W_c \in [0,1]^{kl}$ is the probability matrix containing alignment probabilities over the input of comment. For instance, in W_c, a line in this matrix is a probability vector containing the alignment probabilities over one word in the comment and all words in the event description. The weight matrix is then added onto the additional embedding matrix M_{c_2}, and the result is fed into the LSTM encoder as its input. The vectorized outputs of two encoders in **Model-cATT** are:

$$h_e = LSTM(M_{e_1}) \qquad (10)$$
$$h_c = LSTM(M_{c_2} + W_c) \qquad (11)$$

Similarly, in **Model-eATT**, we first calculate the weight matrix over the event description:

$$W_e = softmax(M_{e_1} M_{c_1}^{\mathrm{T}}) \qquad (12)$$

And $W_e \in [0,1]^{lk}$ is then added onto the additional M_{e_2}. The vectorized outputs of two encoders in **Model-eATT** are:

$$h_e = LSTM(M_{e_2} + W_e) \tag{13}$$
$$h_c = LSTM(M_{c_1}) \tag{14}$$

Model-both applies word-by-word attention over both inputs. The W_e and W_c are both calculated. The vectorized outputs of two encoders in **Model-both** are:

$$h_e = LSTM(M_{e_2} + W_e) \tag{15}$$
$$h_c = LSTM(M_{c_2} + W_c) \tag{16}$$

5   Experimentation

In this section, we extensively evaluate the proposed models based on the 2-sequences-to-1-sequence framework for the opinion target understanding task.

5.1   Experimental Settings

Corpus: We conduct experiments on the annotated corpus introduced in Sect. 2. Since the data are nearly balanced, we keep the natural proportion of each category in our experiments. 80% of the data from each news article are used for model training, and 20% of the data are used for testing. Besides, 10% of the training set are set aside as the validation data for parameter tuning. Note that we don't train several models for each news article, but train one model for the whole corpus.

Table 1. Lists of hyper parameters during training

Parameters	Value
Word embedding dimension (d)	128
Learning rate	0.50
Learning rate decay	0.95
LSTM encoder output dimension	64
Batch size	128
Epoch (upper limit for early stopping)	300
Early stopping patience	20

Word Embeddings: The embedding lookup tables are initialized together with other layers described in Sect. 3.4, and are updated during model training. The word embedding dimensions d for E_e, E_c, E_{e1} and E_{c1}, is showed in Table 1. The dimensions of E_{e2} and E_{c2} equal to the padded length of comment and event description, respectively.

Hyper Parameters: The hyper parameters for the proposed models are well tuned on the validation data by the grid search method. Most important hyper parameters are shown in Table 1.

5.2   Experimental Results on Mention Position Output

In this subsection, we evaluate the intermediate output of mention positions of our models. Since the intermediate output S can be seen as a label set, we evaluate the result of mention positions in the view of multi-label classification.

Table 2. Accuracy and F_1-measure of each model in the view of multi-label classification

Methods	Acc. (%)	F_1 (%)
Model-basic	66.89	76.96
Model-cATT	67.65	77.20
Model-eATT	71.80	78.59
Model-both	**72.91**	**80.36**

Table 2 illustrates the results on the output of mention positions. Accuracy and F_1-measure for multi-label classification task proposed by [15] are applied as two evaluation metrics. From this table, we can find that **Model-both** overcomes the other three models with a relatively high accuracy of 72.91%, and a F_1-measure of 80.36%. Comparing to **Model-basic**, the word-by-word attention mechanism improves the ability of finding the corresponding mentions of the opinion target by achieving a significant promotion of 6.02% on accuracy and 3.40% on F_1-measure. These results justify that the proposed models are capable for finding the mentions of opinion target which a comment talks about in an event description, and the word-by-word attention mechanism can further improves this capability.

The improvement of **Model-eATT** over **Model-basic** is 4.15% higher in accuracy and 1.39% higher in F_1-measure than the improvement of **Model-cATT** over **Model-basic**. This result shows that the attention over the event description is more effective for the opinion understanding task in the view of multi-label classification.

5.3   Experimental Results on Category Label Output

In this subsection, we evaluate the output of category labels of our models. Besides the four proposed models, we further implement the following baselines for comparison:

Coreference resolution (CR): Since opinion target understanding is in some way similar to coreference resolution, we implement a coreference resolution system as a baseline. This baseline first determine whether each annotated opinion target mention in the news articles is the antecedent of the anaphor in the corresponding comment. Anaphors in the comment are assumed to be all the subjects/objects, except those first person pronouns which likely refer to the news readers. According to the results, this baseline outputs the final label for each instance with the similar algorithm to Algorithm 1. Features used in this baseline follow those in [16].

SVM classifier (SVM): This baseline uses BOW features for training. It takes the concatenated text of the event description and the comment as its input.

LSTM-based classifier (LSTM): This baseline accepts the same inputs as those of **Model-basic**. The difference between this baseline and **Model-basic** is that the concatenated vector $h_e \oplus h_c$ is directly fed into a hidden layer, followed by a softmax function, for the 3-way classification.

The results of 3-way classification of all approaches are given in Table 3. We use the standard accuracy as the evaluation metric. From these results, we find that the CR performs poorly in the opinion target understanding task due to the fact that a number of texts of comments from news portal sites may omit some sentence constituents. For instance, a reader may only write *"Good job!."* to express a positive sentiment towards one of the opinion target (mostly the agent of the event in this case) in the event. In such case, the performance of traditional feature-based coreference resolution system on the opinion target understanding task is likely to be limited.

SVM and LSTM are two strong baselines which perform better than **Model-basic** and **Model-cATT**. However, the proposed **Model-both** achieves the best performance among all baselines with an accuracy of 66.52%. Compared to **Model-basic**, **Model-both** has a higher improvement, i.e., 6.17% on accuracy, which justifies the noteworthy effectiveness of word-by-word attention mechanism in the view of 3-way classification. **Model-both** also overcomes SVM and LSTM with at least a promotion of 4.60% in accuracy. Significance test with t-test shows that **Model-both** significantly outperforms these two baselines (p-value < 0.05).

Similar to the results in Table 2, compared to **Model-basic**, **Model-eATT** improves 3.75% more than **Model-cATT** does in accuracy. This result also shows the higher effectiveness of the attention over the event description even in the view of 3-way classification task.

5.4   Error Analysis

The understanding of plural personal pronouns (e.g. "他们 *(they)*") performs relatively poor. The pronoun *"they"* can ambiguously refers to one target which consists of more than one person, or both targets. For instance, in *Article 7*, *target 1* is one person while *target 2* is a group of family members. A comment uses *they* may refers to either *target 2* or *both*. Such ambiguity can't be addressed

Table 3. Accuracy of each method in the view of 3-way classification

Methods	Acc. (%)
CR	48.02
SVM	61.23
LSTM	61.92
Model-basic	60.35
Model-cATT	61.08
Model-eATT	64.83
Model-both	**66.52**

well even in the coreference resolution task. Ambiguity of pronouns also occurs in some event descriptions of which the two opinion targets have the same gender. Since approximately 13% of the annotated mentions in all event descriptions are pronouns, and more pronouns occur in comments, the ambiguity of pronouns mainly limits the performance of our models.

There are some minor error cases in the results. Some less used aliases in comments can be hardly recognized because their low occurrence frequency. The unbalanced mentions of opinion targets in some events also affects the performance. For example, in *Article 12*, *target 1* has only 1 mention while *target 2* has 6. Some comments are miscategorized into *target 1* or *both* because the percentage of recognized mentions of *target 1* is easy to surpass the borderline of 50% defined in the rules described in Algorithm 1.

6 Conclusion

In this paper, we propose an opinion target understanding subtask in *event-level* sentiment analysis. Specifically, we propose a 2-sequences-to-1-sequence framework to leverage the information in both event description and comment, and further employ the word-by-word attention mechanism to capture the alignment of the opinion target and its corresponding mentions within two kinds of texts. Experimental results support the effectiveness of our approach to opinion target understanding task with a noteworthy promotion in performance.

Our future work will concentrate on seeking better modification for further improvement. We would also attempt to expand our approach into the multi-sequences-to-1-sequence, or even the multi-sequences-to-multi-sequences form.

References

1. Pang, B., Lee, L.: Opinion mining and sentiment analysis. Found. Trends Inf. Retrieval **2**(1–2), 1–135 (2007). http://dx.doi.org/10.1561/1500000011
2. Liu, B., Zhang, L.: A survey of opinion mining and sentiment analysis. In: Mining Text Data, pp. 415–463 (2012). http://dx.doi.org/10.1007/978-1-4614-3223-4_13

3. Pang, B., Lee, L., Vaithyanathan, S.: Thumbs up? Sentiment classification using machine learning techniques. CoRR cs.CL/0205070 (2002). http://arxiv.org/abs/cs.CL/0205070
4. Li, S., Huang, C., Zhou, G., Lee, S.Y.M.: Employing personal/impersonal views in supervised and semi-supervised sentiment classification. In: Proceedings of the 48th Annual Meeting of the Association for Computational Linguistics, ACL 2010, Uppsala, Sweden, 11–16 July 2010, pp. 414–423 (2010). http://www.aclweb.org/anthology/pp.10-1043
5. Lipenkova, J.: A system for fine-grained aspect-based sentiment analysis of Chinese. In: Proceedings of the 53rd Annual Meeting of the Association for Computational Linguistics and the 7th International Joint Conference on Natural Language Processing of the Asian Federation of Natural Language Processing, ACL 2015, Beijing, China, 26–31 July 2015. System Demonstrations, pp. 55–60 (2015). http://aclweb.org/anthology/P/P15/pp.15-4010.pdf
6. Ruder, S., Ghaffari, P., Breslin, J.G.: A hierarchical model of reviews for aspect-based sentiment analysis. In: Proceedings of the 2016 Conference on Empirical Methods in Natural Language Processing, EMNLP 2016, Austin, Texas, USA, 1–4 November 2016, pp. 999–1005 (2016). http://aclweb.org/anthology/D/D16/D16-1103.pdf
7. Jakob, N., Gurevych, I.: Extracting opinion targets in a single and cross-domain setting with conditional random fields. In: Proceedings of the 2010 Conference on Empirical Methods in Natural Language Processing, EMNLP 2010, MIT Stata Center, Massachusetts, USA, 9–11 October 2010. A Meeting of SIGDAT, a Special Interest Group of the ACL, pp. 1035–1045 (2010). http://www.aclweb.org/anthology/D10-1101
8. Liu, K., Xu, L., Zhao, J.: Extracting opinion targets and opinion words from online reviews with graph co-ranking. In: Proceedings of the 52nd Annual Meeting of the Association for Computational Linguistics, ACL 2014, Baltimore, MD, USA, 22–27 June 2014. Long Papers, vol. 1, pp. 314–324 (2014). http://aclweb.org/anthology/P/P14/pp.14-1030.pdf
9. Hochreiter, S., Schmidhuber, J.: Long short-term memory. Neural Comput. 9(8), 1735–1780 (1997). http://dx.doi.org/10.1162/neco.1997.9.8.1735
10. Graves, A.: Generating sequences with recurrent neural networks. CoRR abs/1308.0850 (2013). http://arxiv.org/abs/1308.0850
11. Zeiler, M.D.: ADADELTA: an adaptive learning rate method. CoRR abs/1212.5701 (2012). http://arxiv.org/abs/1212.5701
12. Glorot, X., Bengio, Y.: Understanding the difficulty of training deep feedforward neural networks. In: Proceedings of the Thirteenth International Conference on Artificial Intelligence and Statistics, AISTATS 2010, Chia Laguna Resort, Sardinia, Italy, 13–15 May 2010, pp. 249–256 (2010)
13. Hinton, G.E., Srivastava, N., Krizhevsky, A., Sutskever, I., Salakhutdinov, R.: Improving neural networks by preventing co-adaptation of feature detectors. CoRR abs/1207.0580 (2012). http://arxiv.org/abs/1207.0580
14. Bahdanau, D., Cho, K., Bengio, Y.: Neural machine translation by jointly learning to align and translate. CoRR abs/1409.0473 (2014). http://arxiv.org/abs/1409.0473

15. Godbole, S., Sarawagi, S.: Discriminative methods for multi-labeled classification. In: Dai, H., Srikant, R., Zhang, C. (eds.) PAKDD 2004. LNCS (LNAI), vol. 3056, pp. 22–30. Springer, Heidelberg (2004). doi:10.1007/978-3-540-24775-3_5
16. Kong, F., Ng, H.T.: Exploiting zero pronouns to improve Chinese coreference resolution. In: Proceedings of the 2013 Conference on Empirical Methods in Natural Language Processing, EMNLP 2013, Grand Hyatt Seattle, Seattle, Washington, USA, 18–21 October 2013. A meeting of SIGDAT, a Special Interest Group of the ACL, pp. 278–288 (2013). http://aclweb.org/anthology/D/D13/D13-1028.pdf

Topic Enhanced Word Vectors for Documents Representation

Dayu Li[1], Yang Li[1], and Suge Wang[1,2(✉)]

[1] School of Computer & Information Technology, Shanxi University,
Taiyuan 030006, China
wsg@sxu.edu.cn
[2] Key Laboratory of Computational Intelligence and Chinese Information Processing
of Ministry of Education, Shanxi University, Taiyuan 030006, China

Abstract. The words representation, as basic elements of documents representation, plays a crucial role in natural language processing. Topic models and Word embedding models have made great progress on words representation. There are some researches that combine the two models with each other, most of them assume that the semantics of context depends on the semantics of the current word and topic of the current word. This paper proposes a topic enhanced word vectors model (TEWV), which enhances the representation capability of word vectors by integrating topic information and semantics of context. Different from previous works, TEWV assumes that the semantics of the current word depends on the semantics of context and the topic, which is more consistent with common sense in dependency relationship. The experimental results on the 20NewsGroup dataset show that our approach achieves better performance than state-of-the-art methods.

Keywords: Words representation · Documents representation · Topic · Text categorization

1 Introduction

The words representation, as basic elements of documents representation, plays a crucial role in natural language processing. The conventional one-hot model is simple and effective, but ignores the semantic similarity between words.

Latent Dirichlet Allocation (LDA) model [1] is a probabilistic generative model, which the generation of a document depends on the document-topics distribution and the topic-words distribution. When the model is convergent, each word is assigned a topic label, the words with same topic have semantic similarity.

Word embedding model [2] has made significant progress both in words representation capability and algorithm efficiency. It makes full use of the semantic information of words, however, topic information is not taken into account. In fact, the topic information of words is important to many tasks, such as multi-class text classification.

© Springer Nature Singapore Pte Ltd. 2017
X. Cheng et al. (Eds.): SMP 2017, CCIS 774, pp. 166–177, 2017.
https://doi.org/10.1007/978-981-10-6805-8_14

There are some researches that combine LDA model and word embedding model with each other [3–5]. TWE model [3] and LTSG model [4] assume that the semantics of context depends on the semantics of the current word and topic of the word, which may not consistent with common sense in dependency relationship, we usually think that the semantics of the current word depends on the semantics of context and the topic. GTW model [5] is a generative model, it may have higher complexity due to its complex structure.

In this paper, we propose a topic enhanced word vectors (TEWV) model, which integrates topic information into word vectors based on CBOW model. The semantics of the current word depends on the semantics of context $\{w(t-2), w(t-1), w(t+1), w(t+2)\}$, and the topic $z(t)$ of the current word.

As shown in Fig. 2, according to the difference of input layer structure, we propose three sub-structures, TEWV-1, TEWV-2 and TEWV-3 respectively. For purposes of explanation, the size of the context window in each model here is 2, In experiments, the size of the window is a random number of $1 \sim X$, where X is empirical optimal value.

After the TEWV model is trained and convergent, the low dimensional vector representation of the word and topic can be obtained. In the experiment, we apply the model to multi-class text classification task, the experimental results show that word representation capability has been enhanced by integrating the topic information. In addition, we also investigate the running time of the algorithm, the experimental results show that the model can complete the training of large-scale corpus in a short time, and it is completely unsupervised.

The main contribution of this work is that, (1) the topic information and context information are embedded into the low dimensional vector representation of the word, which improves the representation capability of the word vector. (2) The proposed TEWV model is completely unsupervised, and the algorithm is fast and reliable, suitable for training on a large scale corpus. (3) In addition, the topic is also embedded into a low dimensional vector space.

We made the code for training the word and topic vectors based on the techniques described in this paper available as an open-source project[1].

2   Related Models

2.1   LDA Topic Model

LDA topic model is a probabilistic generative model, which assumes that each document is a probability distribution over an implicit topic, and that each implicit topic is a probability distribution over a word. The brief generative process of a word is as follows:

(1) Choose $\theta \sim \text{Dir}(\alpha)$;
(2) For each of the N words w_n:
   (a) Choose a topic $z_n \sim \text{Multinomial}(\theta)$;

---

[1] http://github.com/LoveMercy/Topic-Enhanced-Word-Vector.

(b) Choose a word w_n from $p(w_n|z_n, \beta)$, a multinomial probability conditioned on the topic z_n.

where, θ is document-topics distribution, and Dir() is Dirichlet distribution. Both α and β are hyperparameters of the Dirichlet distribution.

As we can see from the generative process, the generation of a word depends on the topic. Document-topics distribution and topic-words distribution can be obtained by Gibbs sampling, the former can be used as a vector representation of document, and the latter is words that represent the implicit topic. After the process converges, a topic label z_w is assigned to each word in the document. This topic label will be utilized to enhance the representation capability of word vector in our work.

2.2   CBOW Model

As a fast and efficient word embedding algorithm, CBOW model has been widely used in IR and NLP tasks in recent years. The basic idea of CBOW model is to predict the current word with the context of the current word. As shown in Fig. 1, each word in the vocabulary is mapped to a unique vector, the context vectors is used as input, the projection layer is the average of the input layer vectors, the projection layer is then used to predict the current word by a softmax function.

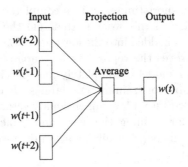

Fig. 1. CBOW model, the semantics of a word depends on context.

Given a sequence of words, $w_1, w_2, ..., w_T$, the objective function of the CBOW model is to maximize the average log probability $\mathcal{L}_{\text{CBOW}}$:

$$\mathcal{L}_{\text{CBOW}} = \frac{1}{T} \sum_{t=1}^{T} \log p(w_t|w_{t-k}, ..., w_{t+k}). \quad (1)$$

where T is the number of words in the corpus, w_t is the current word, $w_{t-k}, ..., w_{t+k}$ is the context of the current word, k is the size of the context window. In experiments, the size of the window is a random number of $1 \sim X$, X is empirical optimal value.

The prediction task is typically done via a multi-class classifier, such as soft-max:

$$p(w_t|w_{t-k}, ..., w_{t+k}) = \frac{e^{y_{w_t}}}{\sum_i e^{y_i}}. \tag{2}$$

Each of y_i is un-normalized log-probability for each output word i, computed as:

$$y = b + Uh(w_{t-k}, ..., w_{t+k}). \tag{3}$$

where U, b are the softmax parameters. h is constructed by a average of word vectors.

CBOW model utilized Hierarchical Softmax and Negative Sampling in output layer to speed up model training, and used stochastic gradient descent and back propagation algorithms to update the model parameters. After the training converges, words with similar meaning are mapped to a similar position in the vector space.

3 TEWV Model

We propose the topic enhanced word vectors (TEWV) model based on the LDA topic model and the CBOW model. As shown in Fig. 2, the structure of TEWV model consists of three layers: input layer, projection layer and output layer. The context vectors and the topic vector are input to the input layer, the projection layer is the average of the input layer or the input layer itself, The output layer is constructed with a Huffman tree.

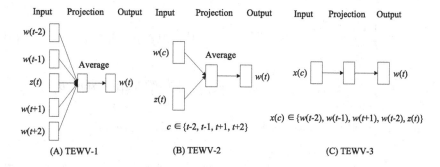

Fig. 2. TEWV model, the semantics of a word depends on context and its topic.

The basic idea of the TEWV model is that the semantics of a word depends on both the semantics of context and the topic. Model aims at maximizing the objective function $\mathcal{L}_{\text{TEWV}}$.

$$\mathcal{L}_{\text{TEWV}} = \frac{1}{T} \sum_{t=1}^{T} \log p(w_t|w_{t-k}, ..., w_{t+k}, z_t). \tag{4}$$

where T is the number of words in the corpus, w_t is the current word, $w_{t-k}, ..., w_{t+k}$ is the context of the current word, k is the size of the context window. z_t is the topic of current word w_t.

Hierarchical Softmax:

The probability $p(w_t|w_{t-k}, ..., w_{t+k}, z_t)$ in the objective function is constructed by the Hierarchical Softmax algorithm.

In the output layer, a Huffman tree is constructed based on the word frequency of each word in the vocabulary, each leaf node in the tree corresponds to a word in the vocabulary, the larger the word frequency, the closer the leaf node is from the root node. We define the conditional probability as:

$$p(w_t|w_{t-k}, ..., w_{t+k}, z_t) = \prod_{j=1}^{J_{w_t}} \{[\sigma(\boldsymbol{x}_{w_t}^\mathrm{T}\boldsymbol{\theta}_j^{w_t})]^{1-d_j^{w_t}} \cdot [1 - \sigma(\boldsymbol{x}_{w_t}^\mathrm{T}\boldsymbol{\theta}_j^{w_t})]^{d_j^{w_t}}\}. \quad (5)$$

where, J_{w_t} represents the number of non-leaf nodes on the path of the leaf node of current word w_t to the root node. $\sigma()$ is the logistic function, $\sigma(\boldsymbol{x}_{w_t}^\mathrm{T}\boldsymbol{\theta}_j^{w_t}) = \frac{1}{1+e^{-\boldsymbol{x}_{w_t}^\mathrm{T}\boldsymbol{\theta}_j^{w_t}}}$. $\boldsymbol{x}_{w_t}$ is the projection layer vector. $\boldsymbol{\theta}_j^{w_t}$ is the weight of projection layer node to non-leaf node j. $d_j^{w_t}$ is the Hoffman encoding of non-leaf node j, $d_j^{w_t} \in \{0,1\}$, if the non-leaf node j is the left node, the code is 1, whereas the code is 0.

Then the objective function $\mathcal{L}_\mathrm{TEWV}$:

$$\mathcal{L}_\mathrm{TEWV} = \frac{1}{T} \sum_{t=1}^{T} \log p(w_t|w_{t-k}, ..., w_{t+k}, z_t)$$

$$= \frac{1}{T} \sum_{t=1}^{T} \log \prod_{j=1}^{J_{w_t}} \{[\sigma(\boldsymbol{x}_{w_t}^\mathrm{T}\boldsymbol{\theta}_j^{w_t})]^{1-d_j^{w_t}} \cdot [1 - \sigma(\boldsymbol{x}_{w_t}^\mathrm{T}\boldsymbol{\theta}_j^{w_t})]^{d_j^{w_t}}\} \quad (6)$$

$$= \frac{1}{T} \sum_{t=1}^{T} \sum_{j=1}^{J_{w_t}} \{(1 - d_j^{w_t}) \cdot \log[\sigma(\boldsymbol{x}_{w_t}^\mathrm{T}\boldsymbol{\theta}_j^{w_t})] + d_j^{w_t} \log[1 - \sigma(\boldsymbol{x}_{w_t}^\mathrm{T}\boldsymbol{\theta}_j^{w_t})]\}.$$

The content in the braces in Eq. 6 denoted as $\mathcal{L}_\mathrm{TEWV}(w_t, j)$.

$$\mathcal{L}_\mathrm{TEWV}(w_t, j) = (1 - d_j^{w_t}) \cdot \log[\sigma(\boldsymbol{x}_{w_t}^\mathrm{T}\boldsymbol{\theta}_j^{w_t})] + d_j^{w_t} \log[1 - \sigma(\boldsymbol{x}_{w_t}^\mathrm{T}\boldsymbol{\theta}_j^{w_t})]. \quad (7)$$

Maximizing the objective function $\mathcal{L}_\mathrm{TEWV}$ is equivalent to maximizing the new objective function $\mathcal{L}_\mathrm{TEWV}(w_t, j)$. We use gradient ascent algorithm to maximize the objective function, take the derivative of objective function $\mathcal{L}_\mathrm{TEWV}(w_t, j)$ with respect to $\boldsymbol{x}_{w_t}$ and $\boldsymbol{\theta}_j^{w_t}$:

$$\frac{\partial \mathcal{L}_\mathrm{TEWV}(w_t, j)}{\partial \boldsymbol{x}_{w_t}} = [1 - d_j^{w_t} - \sigma(\boldsymbol{x}_{w_t}^\mathrm{T}\boldsymbol{\theta}_j^{w_t})]\boldsymbol{\theta}_j^{w_t}. \quad (8)$$

$$\frac{\partial \mathcal{L}_\mathrm{TEWV}(w_t, j)}{\partial \boldsymbol{\theta}_j^{w_t}} = [1 - d_j^{w_t} - \sigma(\boldsymbol{x}_{w_t}^\mathrm{T}\boldsymbol{\theta}_j^{w_t})]\boldsymbol{x}_{w_t}. \quad (9)$$

The propagation error e_{w_t} of the output layer to the projection layer is updated as:

$$e_{w_t} \leftarrow e_{w_t} + [1 - d_j^{w_t} - \sigma(x_{w_t}^{\mathrm{T}} \theta_j^{w_t})] \theta_j^{w_t}. \tag{10}$$

The weight $\theta_j^{w_t}$ of the projection layer to the output layer is updated as:

$$\theta_j^{w_t} \leftarrow \theta_j^{w_t} + \alpha [1 - d_j^{w_t} - \sigma(x_{w_t}^{\mathrm{T}} \theta_j^{w_t})] x_{w_t}. \tag{11}$$

We design three sub-structures, TEWV-1, TEWV-2 and TEWV-3. They have different projection layer vector x_{w_t} due to the difference of input layer structure.

3.1 TEWV-1

As shown in Fig. 2(A), the vectors corresponding to $w_{t-k}, ..., w_{t+k}, z_t$ are averaged as the projection layer vector,

$$x_{w_t} = \text{Average}\left(v\left(w_{t-k}\right), ..., v\left(w_{t+k}\right), v\left(z_t\right)\right). \tag{12}$$

The pseudo code implementation of the algorithm TEWV-1 is shown in Algorithm 1.

Algorithm 1. TEWV-1

Input: sequence of words $\{w_1, w_2, ..., w_T\}$ and topics $\{z_1, z_2, ..., z_T\}$ of each words
Output: vector representation of words and topics
1: Initialize the words vector $v(w)$ and topics vector $v(z)$ randomly
2: **for** each $w_t \in \{w_1, w_2, ..., w_T\}$ **do**
3:     $e_{w_t} = 0$
4:     $x_{w_t} = \text{Average}\left(v\left(w_{t-k}\right), ..., v\left(w_{t+k}\right), v\left(z_t\right)\right)$
5:     **for** $j = 1 \rightarrow J_{w_t}$ **do**
6:         Update e_{w_t} as in Eq. 10
7:         Update $\theta_j^{w_t}$ as in Eq. 11
8:     **end for**
9:     **for** each $c \in \{w_{t-k}, ..., w_{t+k}, z_t\}$ **do**
10:         $v(c) \leftarrow v(c) + e_{w_t}$
11:     **end for**
12: **end for**

3.2 TEWV-2

As shown in Fig. 2(B), the vectors corresponding to w_c, z_t are averaged as the projection layer,

$$x_{w_t} = \text{Average}\left(v\left(w_c\right), v\left(z_t\right)\right), c \in \{t - k, ..., t + k\}. \tag{13}$$

The pseudo code implementation of the algorithm TEWV-2 is shown in Algorithm 2.

Algorithm 2. TEWV-2

Input: sequence of words $\{w_1, w_2, ..., w_T\}$ and topics $\{z_1, z_2, ..., z_T\}$ of each words
Output: vector representation of words and topics
1: Initialize the words vector $v(w)$ and topics vector $v(z)$ randomly
2: **for** each $w_t \in \{w_1, w_2, ..., w_T\}$ **do**
3:    **for** $c \in \{t - k, ..., t + k\}$ **do**
4:       $e_{w_t} = 0$
5:       $x_{w_t} = \text{Average}(v(w_c), v(z_t))$
6:       **for** $j = 1 \rightarrow J_{w_t}$ **do**
7:          Update e_{w_t} as in Eq. 10
8:          Update $\theta_j^{w_t}$ as in Eq. 11
9:       **end for**
10:      $v(w_c) \leftarrow v(w_c) + e_{w_t}$
11:      $v(z_t) \leftarrow v(z_t) + e_{w_t}$
12:   **end for**
13: **end for**

3.3 TEWV-3

As shown in Fig. 2(C), the vectors corresponding to $w_{t-k}, ..., w_{t+k}, z_t$ are directly used as projection layer,

$$x_{w_t} = v(c), c \in \{w_{t-k}, ..., w_{t+k}, z_t\}. \tag{14}$$

The pseudo code implementation of the algorithm TEWV-3 is shown in Algorithm 3.

Algorithm 3. TEWV-3

Input: sequence of words $\{w_1, w_2, ..., w_T\}$ and topics $\{z_1, z_2, ..., z_T\}$ of each words
Output: vector representation of words and topics
1: Initialize the words vector $v(w)$ and topics vector $v(z)$ randomly
2: **for** each $w_t \in \{w_1, w_2, ..., w_T\}$ **do**
3:    **for** $c \in \{w_{t-k}, ..., w_{t+k}, z_t\}$ **do**
4:       $e_{w_t} = 0$
5:       $x_{w_t} = v(c)$
6:       **for** $j = 1 \rightarrow J_{w_t}$ **do**
7:          Update e_{w_t} as in Eq. 10
8:          Update $\theta_j^{w_t}$ as in Eq. 11
9:       **end for**
10:      $v(c) \leftarrow v(c) + e_{w_t}$
11:   **end for**
12: **end for**

---

4 Experiments

In this section, we evaluate related models on multi-class text classification, then we present the running time of our algorithm, and we also compare the difference of topic-words between LDA model and our TEWV model.

4.1 Text Classification

Here we investigate the effectiveness of TEWV model in multi-class text classification task. We learn the vector representation of each word on the training corpus, then build the vector of a document by averaging all the word vectors in this document and train a linear classifier using Liblinear [6].

Compared Methods. We compare the performance of our methods against seven methods:

- **BOW:** The BOW model represents each document as a bag of words and the weighting scheme is TFIDF.
- **LDA:** LDA model [1] represents each document as its inferred topic distribution.
- **CBOW:** CBOW model [2] represents each document as the average of each word vectors.
- **TWE:** Topical Word Embedding model [3], which represents each document as the average of all the concatenation of word vectors and topic vectors.
- **GTE:** Generative Topic Embedding model [5], which the topic embeddings and the topic mixing proportions jointly represent the document.
- **SLRTM:** Sentence Level Recurrent Topic Model [15], which models the sequential dependency of words and topic coherence within a sentence using Recurrent Neural Networks.
- **LTSG:** Latent Topical Skip-Gram model [4], which makes topic models and vector representations mutually improve each other within the same corpus.

Datasets. We run the experiments on the dataset 20NewsGroup[2] [7]. The 20 Newsgroups data set is a collection of approximately 20,000 newsgroup documents, partitioned evenly across 20 different newsgroups. Data is preprocessed by lowercasing, removing stop-words, and removing low-frequency words before training.

Experiment Setting. Same as Liu et al. [3], we set the dimensions of both word and topic embeddings as $K = 400$, the number of topics as $T = 80$. And we set the window size of TEWV-1 as $X = 8$, the window size of remaining model as $X = 15$, they are the optimal values in the experiment.

Evaluation metrics. We adopted accuracy, macro-averaged precision, macro-averaged recall and macro-averaged F-measure as the evaluation metrics.

Evaluation Results. Table 1 presents the performance of the different methods on the classification task. The highest scores were highlighted with boldface.

---

[2] http://qwone.com/~jason/20Newsgroups/.

It can be seen that TEWV-3 obtained the best performance. As compared to the BOW model, we incorporate the semantic information into the vector representation of words, and reduce the dimension of vector representation of text significantly. As compared to the LDA model and CBOW model, we incorporate the topic and the context into the vector representation of words simultaneously, which means the semantics of words is more abundant. Moreover, TEWV-3 model performed slightly better than TWE and LTSG, which indicates that TEWV-3 model fully embedding the topic and context information into the representation of words and topics, and the structure of TEWV-3 model is more reasonable on this text classification task.

Table 1. Performance on multi-class text classification. Best score is in boldface.

Model	Accuracy	Precision	Recall	F-measure
BOW	79.7	79.5	79	79.2
LDA	72.2	70.8	70.7	70.7
CBOW	80.1	79.4	79.0	79.0
TWE	81.5	81.2	80.6	80.9
GTE	-	72.1	71.9	71.8
SLRTM	73.9	-	-	-
LTSG	82.8	82.4	81.8	81.8
TEWV-1	80.6	80.0	79.6	79.6
TEWV-2	82.4	82.5	81.2	81.2
TEWV-3	**83.7**	**83.5**	**82.7**	**82.7**

Among the three submodels we proposed, TEWV-3 performs best, TEWV-2 is in the middle, TEWV-1 is the worst. This shows that when parameters are updating, although averaging operations lead to computational convenience, part of the semantic information is lost. TEWV-3 directly uses the word vector or topic vector as the projection layer to update, and perform the best, which is consistent with our expectations.

4.2 Running Time

Algorithm running time is also an indicator of the quality of an algorithm, which is particularly important in the era of big data. In this section, we discuss the model training time for word vectors and topic vectors on the dataset 20News-Group.

Our experiments were carried out on a 12 core processor, The running time of CBOW, TEWV-1, TEWV-2 and TEWV-3 are 25 s, 29 s, 5 min1 s and 5 min18 s respectively (their iterations are set to 15). As can be seen from the experimental results, the training time of TEWV-1 model is almost the same as CBOW,

TEWV-2 model and TEWV-3 model are a little bit longer. In general, the training time is less than 6 min, and completely unsupervised, which means our model is suitable for training on large scale corpora.

4.3   Topic-Words

After the model training, not only do we represent words as vectors, but we also represent topics as vectors. In this section, we will explore how the topic-words distribution obtained by the LDA topic model differs from the our model.

The correlation between topic and word is measured by cosine similarity, several words that are most relevant to the topic are displayed. As shown in Table 2, the topic-words of LDA model are more general, while topic-words of TEWV model tend to describe specific information because the context is incorporated. Take topic 16 as an example, it can be seen that both are related to the topic of "power", the words "power", "input" and "audio" presented by the LDA model are more abstract and general, the words "amplifier", "analog" and "amp" presented by the TEWV-3 model are more specific and explicit.

Table 2. Nearest words for each topic.

Topic	LDA	TEWV-1-3
Topic_2	god, jesus, bible, christ, christian, church, christians	jesus, christ, doctrine, lord, bible, ephesians, verse, scriptures, sola
Topic_6	key, encryption, chip, clipper, keys, government, security	encryption, clipper, wiretap, chip, cryptography, crypto, nsa, enforcement
Topic_16	radio, power, output, lines, audio, current, input, tv	amplifier, analog, amps, amp, voltage, capacitor, dmm, vdc, diode, ic

5   Related Work

In recent years, the low dimensional vector representation of words and documents has made great progress in Information Retrieval and Natural Language Processing. The bag of words (BOW) model treats the document as a collection of words, ignoring word order and syntax. Because of simplicity and effectiveness [8], BOW is widely used in various tasks of IR and NLP. However, the BOW model suffers from curse of dimensionality and sparse.

Blei et al. [1] proposed Latent Dirichlet Allocation (LDA) topic model to overcome the shortcomings of the conventional BOW model, LDA model document mapping to a low dimensional space theme, the document a low dimensional vector space representation in the theme, but the LDA model is still the bag of words model assumption, and that the theme of information to generate a word depends only on the document.

Bengio et al. [9] proposed a neural probabilistic language model (NPLM), which uses a feedforward neural network with a linear projection layer and a nonlinear hidden layer to learn the vector representation of words. The word vector representation based on the neural probabilistic language model significantly improves the effectiveness of various Natural Language Processing tasks, However there are problems of long training times [10–12].

CBOW model is proposed to deal with the training on large scale corpus [2,13,14], CBOW model removed the hidden layer of NPLM, and uses the hierarchical softmax and negative sampling in the output layer to improve algorithm efficiency. CBOW model can complete training on a large corpus in a short time, and the semantic and syntactic information of the word is embedded into the low dimensional vector representation of the word. CBOW model is widely used in the NLP in recent years. However, the CBOW model holds that the semantics of a word is only related to the context of the current word, without considering the global topic information.

Liu et al. [3] proposed topical word embeddings (TWE) model, The TWE model holds that the same words have different vector representations under different topics. It has achieved remarkable results in tasks such as text categorization by concatenating the topic vector and the word vector as the new topical word vector.

In addition, there are several works that combine topic model with neural network language model. They all made full use of advantages of two models, progress has been made in their respective tasks.

Li et al. [5] proposed a generative model combining word embedding and LDA, Tian et al. [15] proposed a novel topic model called Sentence Level Recurrent Topic Model (SLRTM), which models the sequential dependency of words and topic coherence within a sentence using Recurrent Neural Networks.

Law et al. [4] proposed an algorithm framework that makes topic models and vector representations mutually improve each other.

6   Conclusion and Future Work

In this paper, we propose three topic enhanced word vectors model, which incorporate context and topic information simultaneously, and fully embedding the topic and context information into the representation of words and topics. We evaluate our model on text classification and running time, the experimental results show that our model can learn high-quality document representations, and efficient in running time.

We consider the following future research directions: (1) The topic numbers and window size in TEWV model must be pre-defined, we will investigate non-parametric TEWV model in future. (2) The topic vectors obtained by TEWV model has potential applications in various scenarios, such as document retrieval, document generation and summarization. (3) In the future, we plan to integrate more information into our model, such as sentiment and emotion to accommodate more complex tasks.

Acknowledgments. We thank the anonymous mentor provided by SMP for the careful proofreading. This work was supported by: National Natural Science Foundation of China (61632011, 61573231, 61672331, 61432011).

References

1. Blei, D.M., Ng, A.Y., Jordan, M.I.: Latent Dirichlet allocation. J. Mach. Learn. Res. **3**(1), 993–1022 (2003)
2. Mikolov, T., Chen, K., Corrado, G., et al.: Efficient estimation of word representations in vector space. arXiv preprint arXiv:1301.3781 (2013)
3. Liu, Y., Liu, Z., Chua, T.S., et al.: Topical word embeddings. In: 29th AAAI Conference on Artificial Intelligence, pp. 2418–2424. AAAI Press, California (2015)
4. Law, J., Zhuo, H.H., He, J., et al.: LTSG: Latent Topical Skip-Gram for mutually learning topic model and vector representations. arXiv preprint arXiv:1702.07117 (2017)
5. Li, S., Chua, T.S., Zhu, J., et al.: Generative topic embedding: a continuous representation of documents. In: Meeting of the Association for Computational Linguistics, pp. 666–675 (2016)
6. Fan, R.E., Chang, K.W., Hsieh, C.J., et al.: LIBLINEAR: a library for large linear classification. J. Mach. Learn. Res. **9**(8), 1871–1874 (2008)
7. Lang, K.: Newsweeder: learning to Filter Netnews. In: 12th International Conference on Machine Learning, pp. 331–339 (1995)
8. Manning, C.D., Raghavan, P., Schütze, H.: Introduction to information retrieval. J. Am. Soc. Inf. Sci. Technol. **43**(3), 824–825 (2008)
9. Bengio, Y., Ducharme, R., Vincent, P., et al.: A neural probabilistic language model. J. Mach. Learn. Res. **3**(2), 1137–1155 (2003)
10. Collobert, R., Weston, J.: A unified architecture for natural language processing: deep neural networks with multitask learning. In: 25th International Conference on Machine Learning, pp. 160–167. ACM Press, New York (2008)
11. Collobert, R., Weston, J., Bottou, L., et al.: Natural language processing (Almost) from scratch. J. Mach. Learn. Res. **12**(8), 2493–2537 (2011)
12. Turian, J., Ratinov, L., Bengio, Y.: Word representations: a simple and general method for semi-supervised learning. In: 48th Annual Meeting of The Association for Computational Linguistics, pp. 384–394. Association for Computational Linguistics, Pennsylvania (2010)
13. Mikolov, T., Sutskever, I., Chen, K., et al.: Distributed representations of words and phrases and their compositionality. In: Advances in Neural Information Processing Systems, pp. 3111–3119 (2013)
14. Le, Q., Mikolov, T.: Distributed representations of sentences and documents. In: 31st International Conference on Machine Learning, pp. 1188–1196 (2014)
15. Tian, F., Gao, B., He, D., et al.: Sentence level recurrent topic model: letting topics speak for themselves. arXiv preprint arXiv:1604.02038 (2016)

Text Mining and Sentiment Analysis

Social Annotation for Query Expansion Learning from Multiple Expansion Strategies

Yuan Lin, Bo Xu, Luying Li, Hongfei Lin[✉], and Kan Xu

DaLian University of Technology, No 2 LingGong Road GanJingZi District,
DaLian 116023, China
{zhlin, hflin, xukan}@dlut.edu.cn,
xubo2011@mail.dlut.edu.cn, liluying_dlut@foxmail.com

Abstract. User-generated content, such as web pages, is often annotated by users with free-text labels, called annotations, which can be an effective source of information for query formulation tasks. The implicit relationships between annotations can be important to select expansion terms. However, extracting such knowledge from social annotations presents many challenges, since annotations are often ambiguous, noisy, and uncertain. Besides, most research uses a single query expansion method for query expansion tasks, and never considers the annotations attributes. In contrast, in this paper, we proposed a novel framework that optimized the combination of three query expansion methods used for expansion terms from social annotations in three strategies. Furthermore, we also introduce learning to rank methods for phrase weighting, and select the features from social annotation resource for training ranking model. Experimental results on three TREC test collections show that the retrieval performance can be improved by our proposed method.

Keywords: Social annotation · Query expansion · Learning to rank

1 Introduction

Query expansion technologies are developed to solve the problem that the queries users submit can't describe their truly information need and have been proved to be effective in many information retrieval tasks. Early query expansion technologies are mainly treat the retrieval documents itself as the source of extensions terms. Nowadays, many kinds of external extension resources are introduced as the source of the query expansion, such as the online social network developed (e.g., Flikr, Del.icio.us, Bibsonomy, etc. [1, 2]), anyone can easily annotate objects (sites, pages, media, and so on) that someone else authored. Social annotations have become a popular way to allow users to contribute descriptive metadata for Web information, such as Web pages and photos. The so-called folksonomy [3, 4], based on free-form tag annotation, then rises. In folksonomy systems, the online web resources are defined as resources, the tags attached resources are defined as annotations, and the people who carry out the tagging behaviors are defined as users. With the popularity of folksonomy, social annotations attract the attention of many researchers on various fields [5, 6]. Some researchers explore the possibility of the social annotation serving as the source of expansion

© Springer Nature Singapore Pte Ltd. 2017
X. Cheng et al. (Eds.): SMP 2017, CCIS 774, pp. 181–192, 2017.
https://doi.org/10.1007/978-981-10-6805-8_15

terms. Lin et al. [7] has proved that social annotations could improve the performance for query expansion through mining better expansion terms, they also propose a method to weight expansion terms by term ranking model from learning to rank method. Their work expose some truth: (1) annotations are good resource for query expansion; (2) learning to rank method is an effective technology to merge valid features thus getting better rank and improve the performance in the information retrieval field.

However, annotations are always informal and irregular in folksonomy. In other words, not all the annotations can be treated as good expansion terms. Therefore, although the good performance of the social annotation based query expansion method has been verified, there is still some room to improve the effectiveness. So it is essential to achieve more appropriate annotations as good expansion terms. There are two important problems for select good terms from the annotations. One is how to select the candidate expansion terms; the other is how to evaluate the quality of expansion terms.

To solve the problem mentioned above, we propose a novel framework based on three query expansion methods in folksonomy system to select candidate expansion terms, calling Query Expansion Learning from Multiple Strategies. They are the term co-occurrence based query expansion method [7] and two approaches which are based on the term co-occurrence and considering user quality information proposed by Guo et al. [8].

In this paper we choose the three query expansion approaches to achieve the expansion terms set which contains more appropriate expansion terms according to multiple expansion strategies. From our initial results of this method, we introduce multiple learning to rank methods to weighting the candidate terms according to their potential impact on retrieval effectiveness. Once the ranking list is obtained, the top ranked terms will be selected to expand the original query. Actually, the framework for query expansion this paper proposed is flexible and expandable. Once we discover better expansion approaches or features, the framework can be expanded to merge more effective factor to achieve better results. Therefore, our work is to provide some approaches and procedure to conduct query expansion research to a higher level.

The contributions of this paper can be summarized as follows: (1) we propose a novel framework to select candidate terms from social annotations by merging the three effective query expansion methods in five patterns; (2) In order to improve the performance of term ranking, we introduce different kinds of learning to rank methods for ranking models, furthermore, we also include the features from social annotation, which are extracted from resources, users and annotations. (3) We explore the relationship between the results of term ranking model and final retrieval model, which help select better learning to rank methods for expansion terms.

2   Related Work

More recently, some studies focus on using an external resource for query expansion. They found one of the query expansion failure reasons is the lack of relevant documents in the local collection, so researchers started to examine the benefits of external information sources for pseudo-relevance feedback. Therefore, the performance of query expansion can be improved by using a large external collection. Diverse sources such as Wikipedia [9, 10], large web and news corpora [9] were found to be beneficial for document retrieval on both newswire and web corpora. At the same time, several external collection enrichment approaches have been proposed. Inspired by previous work that demonstrates that query expansion using external corpora is highly effective, our work follows this strategy of a query expansion approach using an external collection as a resource of query expansion terms. With a rapid development of the social networks, more and more researches shift their perspective towards social annotations application.

Learning to rank [12] is an effective method to apply the machine learning approaches to improve the performance of information retrieval. Learning to rank can be applied to a wide variety of applications in information retrieval and natural language processing. Some techniques have been devised to improve this kind of approaches for better performance. One of such novel ideas is to construct new samples to improve the ranking accuracy. Lin et al. [13] proposed a novel group ranking approach based on cross entropy loss (GroupCE) to improve the ranking performance. In this paper, GroupCE is the main ranking model for ranking the expansion terms.

Recently Lin et al. [7] indicate that annotations are good resources for query expansion. They also attempt to use learning to rank method to rank the candidate terms according to their potential impact on retrieval effectiveness. Guo et al. [10] propose a novel algorithm to give each user a reasonable quality score according to his tagging behavior and the mutual reinforcement relationship; Based on their work, in this paper we also explore to apply query expansion methods effectively to social annotation resource for better retrieval performance. We propose a novel framework based on several query expansion approaches to achieve more appropriate expansion terms. And we also distinguish the different importance of these terms by learning to rank methods according to their potential impact on retrieval effectiveness.

3 Methodology

In this section, we briefly introduce our query expansion framework for candidate terms. At first, we will introduce the three expansion methods which are the basis of our query expansion framework. Then we apply three strategies working on the candidate expansion terms obtained from the three expansion methods to achieve an optimized expansion term set. In addition, we also review the ranking algorithms to optimize term set ranking list.

3.1 Expansion for Candidate Terms

There are three query expansion approaches being considered in our framework, one is the term co-occurrence approach without user information [7], the other two are proposed by Guo et al. [8], which are based on high quality users approach focusing on mining user quality to achieve better expansion terms calling filtering resources through user quality and adding user quality into co-occurrence metric.

Term Co-occurrence Approach. The term co-occurrence is usually used to measure how often terms appear together in a text window. In our experiment, the text window is defined as the whole content of the annotated resource r. For a query term qj and a candidate term ti in the social annotation sample S, the co-occurrence value is defined as follows:

$$coo\left(a_i, q_j | R_j\right) = \frac{\sum_{r \in R_j} \log(tf(a_i, r) + 1.0) \times \log(tf(q_{j,r}) + 1.0)}{\log(n)} \tag{1}$$

where N is the sum of articles in social annotation sample S, $tf(.|f)$ is frequency of term appears in annotate resource r.

Filtering Resource through User Quality. In folksonomy system, each user can be weighted by a score which measures the contribution of user's tagging behavior, and we call it Quality Score (QS for short) [8]. The query expansion method is to apply QS is to re-rank the resources returned from initial retrieval, from which we extract resources tagged by high quality users. Each resource is tagged by many users using different annotations, and each user is weighted by QS for specific annotation. Therefore, each resource can be measured as Eq. (2), which is called Resource Score (RS):

$$RS(r) = \sum_{a \in A(r)} \sum_{u \in U(r,a)} QS(u, a) \tag{2}$$

where $U(r, a)$ is the user set for users who annotate resource r using annotation a; $A(r)$ is the annotation set for annotations which are used to tag the given resource r.

Hence, the returned resources from the initial retrieval are re-ranked by RS. The top N resources are filtered from the re-ranked resources to conduct the further experiment according to formula 1 (SATUSER1).

Weighting Annotations through User Quality. By means of integrating QS to modify the traditional co-occurrence metric, we apply user quality selecting better expansion terms, which is the second way to use user quality and the third way to implement the query expansion. We have tried a number of metric to observe the result, and the best one resulted from the metric below:

$$coo\left(a_i, q_j | R_j\right) = \frac{\sum_{r \in R_j} \log\left(tf(a_i, r) \times \left(1.0 + e^{2QS(u,a_i)}\right) + 1.0\right) \times \log\left(tf(q_{j,r}) + 1.0\right)}{\log(n)} \tag{3}$$

where $QS(u, a)$ aims at weighting the annotation used by high quality user higher than the one used by other users when we select expansion terms(SATUSER2).

Then, based on the assumption that terms in the document are independent from each other, the expansion terms are selected according to the equation below.

$$coo\left(a_i, Q | R_j\right) = \sum_{q_j \in Q} idf\left(q_j, R_j\right) idf\left(a_i, R_j\right) \log\left(coo\left(a_i, q_j | R_j\right) + 1.0\right) \tag{4}$$

where $idf(*, R_j)$ shares the similar meaning with the inverse document frequency in traditional information retrieval model; $coo(a_i, q_j|R_j)$ is computed by the three query expansion method mentioned above. Finally, top K expansion terms selected according to Eq. (4) compose the new expanded query Q_{exp} with the corresponding original query Q. The proportional distribution of the expansion terms and the corresponding original query Q is control by a parameter α.

All of the three methods are proved effective for extracting expansion terms. In order to obtain better candidate terms for query expansion, we collect more appropriate annotation into the optimized expansion term set according to some strategies working on the three approaches, and obtain features for each expansion term according to different expansion methods ranges from their parameters and annotation attribute. In this way, we will finally design a framework to take all the advantage of three methods into account.

3.2   Strategies for Expansion Terms Selection

In our framework, there are three strategies working on three types of query expansion methods to fuse the expansion terms of the query expansion methods. We choose 150 candidate expansion terms per each method respectively. For the convenience of description, $Mi\{t1,...,ti,...,tn\}$ stands for one of query expansion method; ti is one of candidate expansion term from Mi.

Strategy 1. Selecting the terms which appear in the result term list of all the query expansion methods serve as the expansion terms.

For example, there are three query expansion methods: $M1\{t1, t2\}$, $M2\{t3, t1\}$, M3 $\{t1, t4\}$. t1 is selected as expansion term by M1, M2 and M3 at the same time, it is one of candidate query expansion terms selected by strategy 1. Then, we will get the feature vector of t1: M1(t1); M2(t1); M3(t1).

Strategy 2. Selecting the terms based on the optimal method of query expansion word serve as the expansion terms.

We examine the performance of the methods on training set to obtain the best performance method of all the three query expansion method for extracting the candidate terms. For example, three expansion methods: $M1\{t1, t2\}$, $M2\{t3, t1\}$, $M3\{t1, t4\}$, assume that M1 is the best performance method of all, and then we choose expansion terms t1 and t2 into the optimized expansion term set. The feature vector can be represented as: t1: M1(t1); M2(t1); M3(t1) and t2:M1(t2); M2(t2); M3(t2).

Strategy 3. Selecting the terms by merging top-k terms on results list sorted by different methods serve as the expansion terms.

For example, three expansion methods: $M1\{t1, t2\}$, $M2\{t3, t1\}$, $M3\{t1, t4\}$, we set k=1, then expansion terms t1 and t3 can be selected into the optimized expansion term set. The total number of list is 150 for each method respectively. We select top-50 terms from the each method to construct the final term list of strategy 3.

The three strategies aim at obtaining the better candidate term collection for further processing. For each expansion term, we will get its features from different expansion methods ranges from their parameters and annotation attribute. For details, we can obtain the fraction of the feature vector according to the score from each expansion method, and treat annotation's own attributes as features, too. Therefore, we infer that the combination of results may lead to obtain the optimized candidate term set.

3.3 Learning to Rank Expansion Terms

In this section, we will give a general description of learning to rank a set of expansion terms according to their effectiveness.

Term Labeling. In order to rank terms in machine learning method, each candidate term should be given a ground truth label which reveals the relationship between the expansion term and original query. In our experiment, we use three TREC collections, among which each query is divided into five groups with ratio 3:1:1 (training set, validating set, test set). In the training dataset, each query has a term list which ranks the candidate expansion terms according to $chg(t)$. To generate the development dataset, we label each term with binary relevance judgments relevant or irrelevant (1 or 0).

Term Ranking Model. Learning to rank method is the foundation of term ranking model. Different ranking method show the different performance. So the method can benefit the query expansion based on social annotations a lot. Learning to rank is grouped into three approaches: the pointwise approach, the pairwise approach and the listwise approach. We construct term ranking model by following four ranking methods: Regression, RankSVM [14], ListNet [15], LambdaMART [16], GroupCE [13].

We apply the four ranking methods to examine which type of learning to rank method is most effective to rank the query expansion terms from social annotation.

Features for Term Ranking. Feature is an important factor of learning to rank model, which decides the performance of ranking model directly. In this paper, the feature selection is a crucial issue, which is divided into two parts: co-occurrence with query terms and useful statistical attributes based on user and resource.

Co-occurrence with query. The co-occurrence features are used to estimate the relationship between high frequency terms more reliably. Therefore, we define the co-occurrence feature as formula 1. These features do not include the user information. By different parameter setting, we can obtain 10 features in this way.

Filtering Resources through User Quality. These features come from the first method to add user quality; we get M returned resources for the initial retrieval from the delicious dataset. Then the M resources are re-ranked, from which we pick up N top most resources as the final returned resources to select the candidate expansion terms. The top K annotations are selected according to traditional co-occurrence metric as the final expansion terms. After conversion of the parameters, we can also obtain 10 features.

Adding User Quality into Co-occurrence Metric. These features come from the second method adding user quality; the top K annotations are selected from the N resources in the initial retrieval according to the Eq. (4) and then compose the expanded query Q_{exp}. Different parameter settings can produce 10 features.

Different from previous research, we also take the tag attributes into account to construct the features for ranking model. The features can be very effective to qualify the expansion terms, since they depend on the manual annotations, which provide a wealth of information of relevant terms.

Terms in a same web page. The features are statistics of expansion term and query term which appear in the same social annotation web page: maximum, minimum and average frequencies.

Terms annotating the same web page. The features are statistics of expansion term and query term which are used to annotate the same social annotation web page: maximum, minimum and average frequencies.

Term annotating the web pages. The features are statistics of the same expansion term annotating the different social annotation web pages: maximum, minimum and average frequencies.

User quality of term. The tag can be qualified by the user who annotated it to a web page. The features can be obtained by user quality: maximum, minimum and average values.

Resource quality of term. The quality of resource which is annotated can be also a important factor for query expansion terms from social annotation. The features can be obtained by user quality: maximum, minimum and average values.

The total number of our basic features is 45.

4 Experiments

In our experiment, the dataset is collected from Del.icio.us and consists of 4414 users, 41204 resources and 28733 annotations in total, which is the same with [17]. The data is processed in the two steps: first, stem all of the annotations, making the annotations which share the same stem get together; second, some annotations are made up by several words, such as *"java/programming"*, *"java_programming"*, and so on. We divide these annotations into several words with the help of the delimiters.

The expansion procedure is conducted on the delicious data set, and retrieval procedure is conducted on three standard TREC collections, AP88-90(Associated Press), WSJ87-90 (Wall St. Journal) and Robust 2004 (the dataset of TREC Robust Track started in 2003) are also needed in our experiment. In the term ranking experiments, for each collection, we divide its topics by query numbers into three parts: the training set, the validation set and the testing set. We conduct 5-fold cross validation experiments, each one using 3/5 of the queries for training a term ranking model, 1/5 for estimating the parameters and 1/5 for predicting on new queries. Four learning to rank methods are investigated for term ranking.

Inspired by the work of Lin et al. [7], we will test each of these terms to see its impact on the retrieval effectiveness. We measure the performance change due to the expansion term t from social annotation by the ratio:

$$\text{chg}(t) = \left(\frac{MAP(Q \cup t) - MAP(Q)}{MAP(Q)} \right) \quad (5)$$

where MAP(Q) and MAP(Q $\cup$ t) are respectively the MAP of the original query and expanded query (expanded with t). It means good (or bad) expansion term which can improve (or hurt) the effectiveness should produce a performance change such that $|chg(t)| > 0.005$.

4.1   Performance of Term Selected Strategies

In this experiment, we use MAP as the evaluation method. We give the results of ListNet model to rank the expansion terms. Table 1 shows the comparing ListNet approach with the three approaches in three TREC data sets.

From Table 1, we can see that the strategy 1 enhances the retrieval performance over the other two strategies. This indicates that the strategy 1 provided more effective terms closely related to the query expansion terms based on social annotations. The terms which are selected by three query expansion methods at the same time may be more relevant than the others. Because strategy 1 can produce more good terms, we adopt this strategy for following experiments. In this section, we only choose ListNet to examine effectiveness of the term selection strategy. The following section will examine the performance of ranking models.

Table 1. MAP of query expansion for different strategies on AP, WSJ, Robust collections

Collection	Strategy 1	Strategy 2	Strategy 3
WSJ	0.2496	0.2244	0.2149
AP	0.2084	0.2062	0.1817
Robust	0.2118	0.1855	0.1675

4.2   Performance of Term Ranking Models

We obtain a set of candidate expansion terms from social annotations by strategy 1 and then extract features described in Sect. 3.3 **Features for Term ranking** with respect to the terms to constitute term feature vectors. We will test performance of different ranking models. For pointwise approach, we select Regression approach for term ranking model, and for pairwise approach, we select RankSVM approach, LambdaMart is also used to learn listwise ranking model compared with ListNet approach evaluated by MAP.

Table 2. Performance of ranking model evaluated by MAP

Collection	ListNet	Regression	Ranksvm	LambdaMart
WSJ	0.5333	0.5278	0.5383	0.5558
AP	0.5326	0.5182	0.5334	0.5448
Robust	0.3418	0.2395	0.3248	0.3621

The results on Table 2 reveal the term ranking performance of each ranking model. MAP can evaluate the effectiveness of ranking models, which can provide good terms and set the terms higher weight to rank good terms on the top of list for query expansion. Based on the features we extract from social annotation, we can see the listwise approaches achieve better performance for providing the good terms.

4.3   Relationship Between Ranking Models and Retrieval Results

Listwise method can effectively select relevance terms for the original query. Compared with the listwise methods, the GroupCE method performs much better on ranking [13]. So we also use GroupCE method to learn the term ranking model for expansion words. We list results on Table 3.

Table 3. Performance of listwise based ranking model

Collection	ListNet	LambdaMart	GroupCE
WSJ	0.5333	0.5558	0.5790
AP	0.5326	0.5448	0.5580
Robust	0.3418	0.3621	0.3888

Table 4 shows that GroupCE achieves the best term ranking performance, for this method, top k terms with high weight scores are selected to expand the query. The weight score could be used to reflect the importance of the expansion term. Using the expansion terms extracted from social annotation collection based on different term ranking model, we will examine which method can also achieve the best performance on query expansion.

Table 4. MAP of query expansion for different ranking model

Collection	ListNet	LambdaMart	GroupCE
WSJ	0.2496	0.2693	0.2787
AP	0.2084	0.2181	0.2212
Robust	0.2118	0.2132	0.2197

From Table 4, we can see GroupCE also achieve best performance on the retrieval results based on query expansion as its performance on term ranking. The better performance the term ranking model achieves, the more accurate the retrieval result is. So GroupCE, which is an improved listwise method, can obtain the best retrieval results. The results list on Table 4 reveals that the feature space which we construct for ranking model is also effective.

4.4   Performance of Expansion

After above experiments, we choose the strategy 1 as the method for providing candidate terms, and apply the GroupCE to learn a ranking model for ranking the expansion terms, and set them weights according to their relevance to the query. For each expansion method we choose 150 terms for strategy 1, and there are 10 expansion terms selected by term ranking model. Finally we set the weights to 0.8, 0.2 for query terms and expansion terms in the expansion model. We have tried a number of parameters to observe the result, and the best one resulted from the setting above. We show the experimental results on Tables 5, 6 and 7 on three TREC datasets.

Table 5. Performance of different methods on WSJ collection

WSJ	p@5	p@10	p@20	MAP
Original	0.4861	0.4569	0.4106	0.2694
F_MAX	0.4935	0.4592	0.4135	0.2763
QELMS	0.4970	0.4604	0.4192	0.2800

Table 6. Performance of different methods on AP collection

AP	p@5	p@10	p@20	MAP
Original	0.4227	0.4012	0.3756	0.2132
F_MAX	0.4246	0.4126	0.3779	0.2178
QELMS	0.4352	0.4158	0.3808	0.2223

Table 7. Performance of different methods on Robust2004 collection

Robust	p@5	p@10	p@20	MAP
Original	0.4359	0.4014	0.3479	0.2114
F_MAX	0.4409	0.4067	0.3485	0.2160
QELMS	0.4731	0.4193	0.3555	0.2235

We compare two baseline methods: Original and F_MAX with our algorithm. Original denotes the query likelihood model retrieval using original query; F_MAX means the best performance one of the methods, which used to construct the features for ranking model. The methods include the retrieval after the query expansion based on social annotations using PRF method and the user based query expansion methods and etc.

As we have seen, QELMS gets improvement comparing with Original and F_MAX. Our method shows relatively obvious improvement in all the terms of evaluations on AP and WSJ collections, however, on Robust 2004 collection, the increasing is outstanding. All the results reveal that our methods are effective to improve the query expansion performance. And it can raise retrieval results obviously.

5 Conclusions

In this paper, we have explored the feasibility to improve the social annotation query expansion by candidate terms selection and term weighting methods. We propose three strategies to obtain expansion terms and further experiments demonstrate that the strategy 1 is valid and works best compared with the others. Besides, learning to rank models with the features proposed in this work is also effective to rank the candidate terms and set weights according to their relevance with respect to queries. We also show that our ranking approach works satisfactorily on the different TREC collections.

This study suggests several research avenues for our future investigations. For the ranking model selection, because of performance of ranking model operates on the final retrieval results. The better performance ranking model can obtain better retrieval results. This means that there is still much room to improve the retrieval performance by applying better learning to rank approach to query expansion. As the development of social annotation, there will be much room to extract novel features to indicate the relevance of expansion terms and query terms to improve the expansion procedure. In addition, we will explore more applications of social annotation based on learning to rank methods.

Acknowledgments. This work is partially supported by grant from the Natural Science Foundation of China (No. 61402075, 61602078, 61572102, 61632011), the Ministry of Education Humanities and Social Science Project (No. 16YJCZH12), the Fundamental Research Funds for the Central Universities (No. DUT17RC(3)016).

References

1. Borrego, A., Fry, J.: Measuring researchers' use of scholarly information through social bookmarking data: a case study of BibSonomy. J. Inf. Sci. **38**(3), 297–308 (2012)
2. Heymann, P., Koutrika, G., Molina, H.G.: Can social bookmarking improve web search? In: Proceedings of the International Conference on Web Search and Data Mining, pp. 195–205 (2008)
3. Bouadjenek, M.R., Hacid, H., Bouzeghoub, M., Vakali, A.: Using social annotations to enhance document representation for personalized search. In: Proceedings of the 36th International ACM SIGIR Conference on Research and Development in Information Retrieval, pp. 1049–1052. ACM (2013)
4. Laura, G.M.: Social bookmarking, folksonomies, and web 2.0 tools. Searcher **14**(6), 26–38 (2006)
5. Kiu, C., Tsui, E.: TaxoFolk: a hybrid taxonomy-folksonomy structure for knowledge classification and navigation. Expert Syst. Appl. **38**(5), 6049–6058 (2011)
6. Xu, S., Bao, S., Fei, B., Su, Z., Yu, Y.: Exploring folksonomy for personalized search. In: Proceedings of the 31st Annual International ACM SIGIR Conference on Research and Development in Information Retrieval, pp. 155–162. ACM (2008)
7. Lin, Y., Lin, H., Jin, S., Ye, Z.: Social annotation in query expansion: a machine learning approach. In: SIGIR 2011: Proceedings of the 34st Annual International ACM SIGIR Conference on Research and Development in Information Retrieval, pp. 405–414. ACM (2011)
8. Guo, Q., Liu, W., Lin, Y., Lin, H.: Query expansion based on user quality in folksonomy. In: Hou, Y., Nie, J.-Y., Sun, L., Wang, B., Zhang, P. (eds.) AIRS 2012. LNCS, vol. 7675, pp. 396–405. Springer, Heidelberg (2012). doi:10.1007/978-3-642-35341-3_35
9. Bendersky, M., Fisher, D., Croft, W.B.: UMass at TREC 2010 web track: term dependence, spam filtering and quality bias. In: Proceedings of TREC-10 (2011)
10. Yuvarani, M., Lyengar, N., Kannan, A.: Improved concept-based query expansion using Wikipedia. Int. J. Commun. Netw. Distrib. Syst. **11**(17), 26–41 (2013)
11. Diaz, F., Metzler, D.: Improving the estimation of relevance models using large external corpora. In: Proceedings of the 29st Annual International ACM SIGIR Conference on Research and Development in Information retrieval, pp. 154–161. ACM (2006)
12. Liu, T.: Learning to rank for information retrieval. Found. Trends Inf. Retrieval **3**(3), 225–331 (2009)

13. Lin, Y., Lin, H., Wu, J., Xu, K.: Learning to rank with cross entropy. In: Proceedings of the 20th ACM International Conference on Information and Knowledge Management (CIKM 2011), pp. 2057–2060. ACM (2011)

14. Joachims, T.: Optimizing search engines using click through data. In: Proceedings of the 8th ACM SIGKDD International Conference on Knowledge Discovery and Data Mining, pp. 133–142. ACM (2002)

15. Cao, Z., Qin, T., Liu, T.Y., Tsai, M.F., Li, H.: Learning to rank: from pairwise approach to listwise approach. In: Proceedings of the ICML, pp. 129–136. ACM (2007)

16. Wu, Q., Burges, C., Svore, K., Gao, J.: Ranking, boosting and model adaptation. Technical report, Microsoft Technical Report MSR-TR-2008-109 (2008)

17. Lu, C., Hu, X., Chen, X., Park, J.R.: The Topic-perspective model for folksonomy systems. In: Proceeding of the 16th ACM SIGKDD International Conference on Knowledge Discovery and Data Mining, pp. 683–691 (2010)

Supervised Domain Adaptation for Sentiment Regression

Jian Xu, Hao Yin, Shoushan Li$^{(\boxtimes)}$, and Guodong Zhou

Natural Language Processing Lab, School of Computer Science and Technology,
Soochow University, Suzhou, China
{jxul017,hyin}@stu.suda.edu.cn,
{lishoushan,gdzhou}@suda.edu.cn

Abstract. Sentiment regression is a task of summarizing the overall sentiment of a review with a real-valued score. However, the regression model trained in one domain probably performs poorly in a different domain due to the distribution variety. Different from existing studies, domain adaptation in sentiment regression is more challenging because the rating range in one domain might be different from that in the other domain. In this study, we propose a novel approach to domain adaptation for sentiment regression. Specifically, our approach employs an auxiliary Long Short-Term Memory (LSTM) layer to learn the auxiliary representation from the source domain, and simultaneously join the auxiliary representation into the main LSTM layer for the target domain regression setting. In the learning process, the LSTM regression models for the source and target domains are jointly learned. Empirical studies demonstrate that our joint learning approach performs significantly better than several strong baselines.

Keywords: Sentiment analysis · Domain adaptation · Natural language processing

1 Introduction

This article discusses a subfield of sentiment analysis referred to as sentiment regression, which aims to summarize the overall sentiment of a review by labeling it with a real-valued score. Recently, sentiment regression gains significant popularity [1], and consequently much academic attention, as it plays a key role in many social applications, such as information retrieval [2], online advertising [3] and recommendation system [4]. For instance, in a recommendation system, one popular way to recommend a product is to sort all products according to their rating scores which are obtained from the review rating component.

Most previous studies on sentiment regression focus on machine learning approaches which leverage a large amount of annotated samples to train a regression model. However, the main criticism of such approaches is that the regression model trained in one domain probably isn't a good proxy for a different domain due to the distribution variety, thus performing dramatically bad in the other domain. Generally,

X. Cheng et al. (Eds.): SMP 2017, CCIS 774, pp. 193–205, 2017.
https://doi.org/10.1007/978-981-10-6805-8_16

sentiment is expressed differently in different domains, and annotating corpora for every possible domain of interest is difficult and too much time-consuming.

To solve the above problem, many domain adaptation approaches are proposed, e.g. SCL [5]. Domain adaptation approaches are mainly divided into two categories, supervised approaches and semi-supervised approaches. The former leverages both the labeled data from the source domain and a small amount of labeled data from the target domain to perform domain adaptation (e.g., [6]) while the latter leverages both the labeled data from the source domain and only unlabeled data from the target to perform domain adaption (e.g., [5]). This paper focuses on the former.

However, almost all existing domain adaptation studies in sentiment analysis focus on the tasks of sentiment classification. In sentiment regression, domain adaptation becomes more challenging. One major challenge is that the two sentiment regression tasks in the source and target domains might possess different label ranges. For example, in one domain, the rating score is from 1-star to 5-stars while in the other domain, the rating score is from 1-star to 10-stars. In such scenarios, existing approaches to classification-based domain adaptation cannot be directly applied.

Besides, although existing approaches to sentiment regression have achieved some success in the study of sentiment analysis, most of these approaches are built with shallow learning architectures. In recent years, learning methods with deep architectures have achieved significant success in many natural language processing (NLP) tasks, such as machine translation [7] and question answering [8]. It is a pressing need to extensively exploit the effectiveness of the deep learning method on the task of sentiment regression.

In this paper, we employ a popular deep learning method, named Long Short-Term Memory (LSTM) network, to perform sentiment regression with both the target and source domains. The main merits of the LSTM method lie in that it equips with a special gating mechanism that controls access to memory cells and it is powerful and effective at capturing long-term dependencies [9].

More importantly, based on the LSTM regression model, we propose a novel approach, namely cross-domain LSTM, to supervised domain adaptation for sentiment regression. Specifically, we separate the whole sentiment regression task into a main task (regression on the target domain) and an auxiliary task (regression on the source domain). An auxiliary representation learned from the auxiliary task with a shared LSTM layer is integrated into the main task for joint learning. Since our approach jointly learns the sentiment regression tasks in the source and target domains through sharing the auxiliary representations of the samples and thus does not need the two tasks have the same label range. With the help of the auxiliary task, our approach boosts the performance of the main task through incorporating the auxiliary representations learned from the auxiliary task. The experimental result demonstrates that our approach performs better than several strong baselines.

The remainder of this paper is organized as follows. Section 2 overviews related work on sentiment analysis and domain adaptation. Section 3 presents a basic LSTM approach to sentiment regression. Section 4 presents our cross-domain LSTM approach to domain adaptation for sentiment regression. Section 5 evaluates the proposed approach. Finally, Sect. 6 gives the conclusion and future work.

2   Related Work

In the last decade, sentiment analysis has become a hot research area in natural language processing [1]. In this area, sentiment regression is an important task and has attracted more and more attention since the pioneer work by Pang and Lee [10]. One major research line on sentiment analysis is to design effective features. Another major research line is to propose novel learning models. However, the main criticism of such approaches is that the classification or regression model trained in one domain probably performs poorly for a different domain.

To solve the above problem, many domain adaptation approaches are proposed. Domain adaptation approaches for sentiment analysis are mainly divided into two categories, supervised approaches and semi-supervised approaches. As for semi-supervised approaches, Blitzer et al. [5] extend to sentiment classification the structural correspondence learning (SCL) algorithm. Tan and Cheng [11] propose a weighted SCL model (W-SCL) which weights the feature as well as the instances. Tan et al. [12] propose Adapted Naïve Bayes (ANB), a weighted transfer version of Naïve Bayes Classifier, to gain knowledge from the new domain data. As for supervised approaches, there exist much fewer related studies in sentiment analysis. But in the whole natural language processing (NLP) community, a famous approach named feature augmentation approach has been proposed to perform supervised domain adaptation in several NLP tasks [6]. In their approach, three versions of the original feature vector are generated to integrate the classification knowledge from the source domain, the target domain and both domains.

Our work belongs to the setting of supervised domain adaptation in sentiment analysis. Unlike all the above domain adaptation approaches, our work is the first to solve the domain adaptation for sentiment regression where the label range of the target domain is different from that of the source domain.

3   Basic LSTM Model for Sentiment Regression

In this section, we describe a basic LSTM approach to sentiment regression. The first subsection introduces basic LSTM network. The second subsection delineates the LSTM approach to sentiment regression.

3.1   Basic LSTM Network

Long short-term memory network (LSTM) is proposed by Hochreiter and Schmidhuber [9] to specifically address this issue of learning long-term dependencies. The LSTM maintains a separate memory cell inside it that updates and exposes its content only when deemed necessary. A number of minor modifications to the standard LSTM unit have been made. In this study, we apply the implementation used by Graves [13] to map the input sequence of main task to a fixed-sized vector.

The LSTM unit consists of an input gate i, an output gate o, a forget gate f, a hidden state h, and a memory cell c. At time step t, LSTM unit is updated as follows:

$$i_t = \sigma(W_i x_t + U_i h_{t-1} + V_i c_{t-1}) \tag{1}$$

$$f_t = \sigma\left(W_f x_t + U_f h_{t-1} + V_f c_{t-1}\right) \tag{2}$$

$$o_t = \sigma(W_o x_t + U_o h_{t-1} + V_o c_{t-1}) \tag{3}$$

$$\tilde{c}_t = \tanh(W_c x_t + U_c h_{t-1}) \tag{4}$$

$$c_t = f_t \odot c_{t-1} + i_t \odot \tilde{c}_t \tag{5}$$

$$h_t = o_t \odot \tanh(c_t) \tag{6}$$

Where x_t denotes the input at time step t, σ denotes the logistic sigmoid function, $\odot$ denotes elementwise point multiplication. W, U and V represent the corresponding weight matrices connecting them to the gates. Intuitively, the forget gate controls how much the information is discarded in each memory unit, the input gate controls the amount of updated information in each memory unit, and the output gate controls the exposure of the internal memory state.

3.2 LSTM Model for Sentiment Regression

Figure 1 illustrates the model architecture of sentiment regression with a LSTM layer. We utilize T^{input} to represent the input, and the input propagates through the LSTM layer, yielding the high-dimensional vector, i.e.,

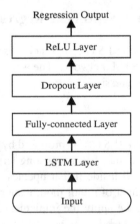

Fig. 1. LSTM sentiment regression model

$$h = LSTM(T^{input}) \tag{7}$$

Where h is the output from the LSTM layer.

Subsequently, the fully-connected layer which is similar to a hidden layer in the conventional multi-layer perceptron, accepts the output from the previous layer, weighting them and passing through a normally activation function as follows:

$$h^* = dense(h) = \phi(\theta^T h + b) \tag{8}$$

Where ϕ is the non-linear activation function, employed "ReLU" in our model. h^* is the output from the fully-connected layer.

The dropout layer has been very successful on feed-forward networks [14]. By randomly omitting feature detectors from the network during training, it can obtain less interdependent network units and achieve better performance, which is used as a hidden layer in our framework, i.e.,

$$h^d = h^* \cdot D(p^*) \tag{9}$$

Where D denotes the dropout operator, p^* denotes a tune-able hyper parameter (the probability of retaining a hidden unit in the network), and h^d denotes the output from the dropout layer.

The ReLU output layer is used for a regression task. The output from the previous layer is then fed into the output layer to get the predicted value, i.e.,

$$f = \text{ReLU}(W^d h^d + b^d) \tag{10}$$

Where f is the predicted value, which is a discrete variable, W^d is the weight vector to be learned, and b^d is the bias term.

For sentiment regression, we employ "mean squared error" for loss function. Specially, the loss function is defined as follows:

$$loss = \frac{1}{2m} \sum_{i=1}^{m} ||f_i - y_i||^2 \tag{11}$$

Where $loss$ is the loss function of sentiment regression, and y_i is the ground truth label of the i-th sample, and f_i indicates the predicted value of i-th sample, and m is the total number of the training samples.

4 Cross-Domain LSTM Model for Sentiment Regression

Figure 2 delineates the overall architecture of cross-domain LSTM model which contains a main LSTM layer and an auxiliary LSTM layer. In our study, we consider the sentiment regression task of the target domain as the main task and the sentiment regression task of the source domain as the auxiliary task. The goal of the approach is to employ the auxiliary representation to assist the regression performance of the main task. The main idea of our cross-domain LSTM approach lies in that the auxiliary LSTM layer is shared by both the main and auxiliary tasks so as to leverage the learning knowledge from both the target domain and the source domain.

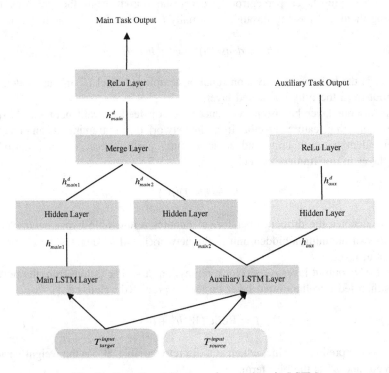

Fig. 2. Overall architecture of cross-domain LSTM

4.1   The Main Task

Formally, the target domain representation is generated from both the main LSTM layer and the auxiliary LSTM layer respectively:

$$h_{main1} = LSTM_{main}(T_{target}^{input}) \tag{12}$$

$$h_{main2} = LSTM_{aux}(T_{target}^{input}) \tag{13}$$

Where the output h_{main1} and h_{main2} respectively represent the representations for the target domain via the main LSTM layer and via the auxiliary LSTM layer.

Then h^d_{main1} and h^d_{main2} are fed into two different fully-connected layers (dense layer) respectively to obtain two target representations:

$$h^d_{main1} = dense_{main1}(h_{main1}) \tag{14}$$

$$h^d_{main2} = dense_{main2}(h_{main2}) \tag{15}$$

Then we concatenate the two target domain representations as the input of the hidden layer in the main task:

$$h^d_{main} = dense_{main}(h^d_{main1} \oplus h^d_{main2}) \tag{16}$$

Where h^d_{main} denotes the outputs of fully-connected layer in the main task, and $\oplus$ denotes the concatenate operator.

4.2 The Auxiliary Task

The source representation is also generated by the auxiliary LSTM layer, which is a reused LSTM layer and is employed to bridge across the target and source domains. The reused LSTM layer encodes source input sequence with the same weights:

$$h_{aux} = LSTM_{aux}(T^{input}_{source}) \tag{17}$$

Where h_{aux} represents the representation for the source domain via the reused LSTM layer.

Then a fully-connected layer is utilized to obtain a feature vector for source regression, which is the same as the hidden layer in the main task:

$$h^d_{aux} = dense_{aux}(h_{aux}) \tag{18}$$

Where h^d_{aux} denotes the output of fully-connected layer (dense layer) in the auxiliary task.

4.3 Joint Learning

Once we obtain the main representation h^d_{main} and the auxiliary representation h^d_{aux}, we feed them into the ReLU layers to get the predicted values of the main task and the auxiliary task respectively.

$$f^{main} = ReLU(W^m h^d_{main} + b^m) \tag{19}$$

$$f^{aux} = ReLU(W^a h^d_{aux} + b^a) \tag{20}$$

Where f^{main} is output of the main task and f^{aux} is the output of the auxiliary task.

Finally, we define our joint cost function for cross-domain LSTM model as a linear combination of the cost functions of both the main task and auxiliary task as follows:

$$loss_{joint} = loss_{main} + loss_{aux} \tag{21}$$

$$loss_{main} = \frac{1}{2m}\sum_{i=1}^{m}||f_i^{main} - y_i^{main}||^2 \tag{22}$$

$$loss_{aux} = \frac{1}{2m}\sum_{i=1}^{m}||f_i^{aux} - y_i^{aux}||^2 \tag{23}$$

5 Experimentation

In this section, we systematically evaluate the performance of our cross-domain LSTM model for supervised domain adaptation for sentiment regression.

5.1 Experimental Settings

Data Settings: We use two data sets in our experiment. The first data set consists of product reviews collected from Amazon[1] by Mcauley [15]. This data set contains 4 domains, e.g. Book, CD, Electronic and Kitchen. Each domain's ratings range from 1 star to 5 stars. The second data set is consist of movie reviews crawled from IMDB[2]. Movie reviews range from 1 star to 10 stars. We extract a balanced data set from the source domain and the target domain by selecting 2000 samples from each category. We use 80% of the source domain data and 10% of the target domain data in each review category as the training data and 20% of the target domain data as the test data. When the source domain data are from the first data set of product reviews, the target domain data are from the second data set of movie reviews (or otherwise).

Text Representation: For word representation, we employ skip-gram algorithm (gensim[3] implementation) by word2vec to pre-trained word embedding on the whole data. The length of each text is set to a fixed size.

Basic Prediction Algorithm: LSTM is employed as the basic prediction algorithm in our approach, which is implemented with the tool Keras[4]. The hyper parameters of LSTM are well tuned on the validation data by the grid search method, and most important hyper parameters are shown in Table 1.

---

[1] http://Amazon.com/.
[2] http://www.imdb.com/.
[3] http://radimrehurek.com/gensim/.
[4] https://github.com/fchollet/keras.

Table 1. Parameters setting in LSTM

Parameter description	Value
Dimension of embedding	100
Dimension of the LSTM layer output	128
Dimension of the full-connected layer output	64
Learning rate	0.01
Dropout probability	0.5
Epochs of iteration	30

Evaluation Metric: We employ the coefficient of determination R^2 to measure the performance. Coefficient of determination R^2 is used in the context of statistical models with the main purpose to predict the future outcomes on the basis of other related information. R^2 nearing 1.0 indicates that a regression line fits the data well. Formally, the coefficient of determination R^2 is defined as follows:

$$R^2 = 1 - \frac{SS_{err}}{SS_{tot}} \tag{24}$$

$$SS_{tot} = \sum_i \left(y_i - \bar{y}\right)^2 \tag{25}$$

$$SS_{err} = \sum_i (y_i - f_i)^2 \tag{26}$$

$$\bar{y} = \frac{1}{n}\sum_{i=1}^{n} y_i \tag{27}$$

Where y_i is the real value and f_i is the predicted value of each sample.

Significance Test: We randomly split the whole data into training and test data 10 times and employ two different learning approaches, namely *A1* and *A2*, to perform review rating. Then, we employ *t*-test to perform the significance test to test whether the learning approach *A1* performs better than *A2* (or otherwise).

5.2 Experimental Results

For a thorough comparison, we implement several domain adaptation approaches to sentiment regression. These approaches are introduced as follows.

- **Baseline:** The LSTM regression model which is trained with only a small number of labeled data from the target domain.

- **Label_Mapping:** The LSTM regression model which is trained with a small number of labeled data from the target domain and a large number of labeled data with mapping labels from the source domain. Note that since the source domain's range is different from the target domain, we map the ratings of source domain to the range of target domain. For example, if target domain ratings range from Lt to Ht and source domain ratings range from Ls to Hs, we change the rating $OldVal$ of source domain to rating $NewVal$ as follows:

$$NewVal = Lt + (OldVal - Ls) * \frac{Ht - Lt}{Hs - Ls} \qquad (28)$$

- **Feature_Augmentation:** This is an approach proposed by Hal Daumé III [6]. They augment the feature space of both the source and target data and use the augmentation feature space as the input to a standard learning algorithm. We also leverage the label mapping strategy as mentioned above in this feature augmentation approach.
- **Cross-domain_LSTM:** This is our approach which learns an auxiliary representation for joint learning, which has been described in Sect. 4 in detail. Note that, in our approach, we do not need to map labels of the source domain data to the range of target domain.

Figure 3 shows the R^2 results of different approaches. From the figure, we can see that Label_Mapping performs even worse than the baseline in several settings, such as Movie→Electronic and Kitchen→Movie. This result demonstrates that simply merging the data from the source and target domains with label mapping is not a good solution. Feature_Augmentation performs consistently better than the baseline and performs generally better than Label_labeling except in the setting of Movie→CD. Among all approaches, our approach cross-domain LSTM always performs best in all settings. In general, cross-domain LSTM outperforms the baseline with about 0.046 in R^2. Furthermore, significance test shows that cross-domain LSTM significantly outperforms Label_Mapping in all settings (p-value < 0.05). As for the Feature_Augmentation approach, cross-domain LSTM significantly outperforms it in 6 settings, i.e., Book→Movie, Movie→CD, Electroinc→Movie, Moive→Electronic, Kitchen→Movie, and Movie→Kitchen (p-value < 0.05). In the other 2 settings, our approach and Feature_Augmentation achieve no significantly different performance.

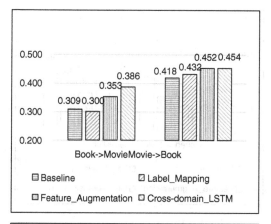

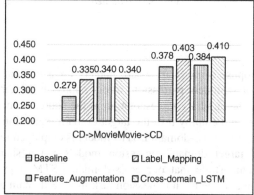

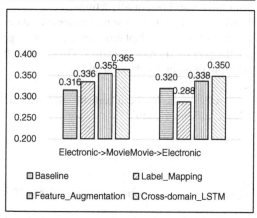

Fig. 3. R^2 Results of Different Domain Adaptation Approaches

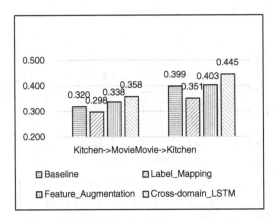

Fig. 3. (*continued*)

6   Conclusion

In this paper, we propose a novel approach, namely cross-domain LSTM model, to domain adaptation for sentiment regression. In our approach, we employ an auxiliary LSTM layer to learn the auxiliary representation in the source domain regression task (as the auxiliary task) and employ it in the target domain regression task (as the main task). To achieve this, our cross-domain LSTM model is employed to bridge across the source domain and target domain regression models via a shared LSTM layer. Empirical studies demonstrate that our cross-domain LSTM approach significantly boosts the performance of the main regression task in all 8 domain adaptation settings and performs better than three baselines in all settings.

In our future work, we would like to exploit the unlabeled data in the target domain to improve the performance of domain adaptation for sentiment regression. Moreover, we would like to apply the proposed cross-domain LSTM approach to the supervised domain adaptation problems in other NLP applications.

Acknowledgments. This research work has been partially supported by three NSFC grants, No. 61375073, No. 61672366 and No. 61331011.

References

1. Pang, B., Lee, L.: Opinion mining and sentiment analysis. Found. Trends Inf. Retreival **2**, 1–135 (2008)
2. Paltoglou, G., Thelwall, M.: A study of information retrieval weighting schemes for sentiment analysis. In: Proceedings of the 48th Annual Meeting of the Association for Computational Linguistics, pp. 1386–1395. Association for Computational Linguistics (2010)
3. Qiu, G., He, X., Zhang, F., et al.: DASA: dissatisfaction-oriented advertising based on sentiment analysis. Expert Syst. Appl. **37**(9), 6182–6191 (2010)

4. Yang, D., Zhang, D., Yu, Z., et al.: A sentiment-enhanced personalized location recommendation system. In Proceedings of the 24th ACM Conference on Hypertext and Social Media, pp. 119–128. ACM (2013)
5. Blitzer, J., Dredze, M., Pereira, F.: Biographies, bollywood, boom-boxes and blenders: domain adaptation for sentiment classification.ACL 7, 440–447 (2007)
6. Daumé III, H.: Frustratingly easy domain adaptation. arXiv preprint arXiv:0907.1815 (2009)
7. Bahdanau D, Cho K, Bengio Y. Neural machine translation by jointly learning to align and translate. arXiv preprint arXiv:1409.0473 (2014)
8. Iyyer, M., Boyd-Graber, J.L., Claudino, L.M.B., et al.: A neural network for factoid question answering over paragraphs. In: EMNLP, pp. 633–644 (2014)
9. Hochreiter, S., Schmidhuber, J.: Long short-term memory. Neural Comput. 9(8), 1735–1780 (1997)
10. Pang, B., Lee, L.: Seeing stars: exploiting class relationships for sentiment categorization with respect to rating scales. In: Proceedings of the 43rd Annual Meeting on Association for Computational Linguistics, pp. 115–124. Association for Computational Linguistics (2005)
11. Tan, S., Cheng, X.: Improving SCL model for sentiment-transfer learning. In: Proceedings of Human Language Technologies: The 2009 Annual Conference of the North American Chapter of the Association for Computational Linguistics, Companion Volume: Short Papers, pp. 181–184. Association for Computational Linguistics (2009)
12. Tan, S., Cheng, X., Wang, Y., Xu, H.: Adapting Naive Bayes to domain adaptation for sentiment analysis. In: Boughanem, M., Berrut, C., Mothe, J., Soule-Dupuy, C. (eds.) ECIR 2009. LNCS, vol. 5478, pp. 337–349. Springer, Heidelberg (2009). doi:10.1007/978-3-642-00958-7_31
13. Graves, A.: Generating sequences with recurrent neural networks. arXiv preprint arXiv: 1308.0850 (2013)
14. Hinton, G.E., Srivastava, N., Krizhevsky, A., et al.: Improving neural networks by preventing co-adaptation of feature detectors. arXiv preprint arXiv:1207.0580 (2012)
15. McAuley, J., Pandey, R., Leskovec, J.: Inferring networks of substitutable and complementary products. In: Proceedings of the 21st ACM SIGKDD International Conference on Knowledge Discovery and Data Mining, pp. 785–794. ACM (2015)

Dependency-Attention-Based LSTM for Target-Dependent Sentiment Analysis

Xinbo Wang$^{(\boxtimes)}$ and Guang Chen

Beijing University of Posts and Telecommunications, Beijing, China
xinbow@foxmail.com

Abstract. Target-dependent sentiment analysis is a fine-grained sentiment analysis and has received an increasing attention. For target-dependent sentiment analysis, the key issue is to capture the important context information according to the given target word. While some critical information in the context may be in a long distance from the target word, so it is significant to explore how to adequately and directly capture these long-range information. The dependency relation can connect words which are relevant in syntax but far in word order. Inspired by this, we propose Dependency-Attention-based Long Short-Term Memory Network (DAT-LSTM) and Segmented Dependency-Attention-based Long Short-Term Memory Network (Seg-DAT-LSTM) for target-dependent sentiment analysis. The dependency-attention mechanism utilizes dependency relation to fully capture long-range information for certain target. Experiments on the tweet dataset and SemEval 2014 dataset indicate that our models achieve state-of-the-art performance on target-dependent sentiment classification.

Keywords: Sentiment analysis · Attention · Dependency

1 Introduction

Sentiment analysis is a fundamental task of natural language processing; especially, it plays a critical role in the area of data mining [1]. Sentiment analysis aims at predicting user's sentiment through the generated text. It is significant to understand user's attitude in social networks and product reviews, and sentiment analysis attracts an increasing attention in academia and industry.

Our work focuses on the target-dependent sentiment analysis, which is a fine-grained sentiment analysis and has been extensively studied in the field of natural language processing. Given a sentence and a target, target-dependent sentiment analysis aims at predicting the sentiment polarity (e.g. positive, negative and neutral) of the sentence towards the target. For example,

"I bought a new camera. The picture quality is amazing but the battery life is too short."

The sentence polarity is positive if the target is *"picture quality"*, but negative when considering the target *"battery life"*.

© Springer Nature Singapore Pte Ltd. 2017
X. Cheng et al. (Eds.): SMP 2017, CCIS 774, pp. 206–217, 2017.
https://doi.org/10.1007/978-981-10-6805-8_17

Existing methods include feature-based Support Vector Machine [2–4] and neural network models [5–8]. Neural models are of increasing interest for their capacity to learn low dimensional representation without feature engineering, and neural models can capture semantic relation between target and context words in an elegant way.

Despite these advantages, existing neural networks have some limits. For target-dependent sentiment analysis, the key issue is to capture the important context information for certain target, and then to infer the sentiment polarity. Some critical context information may be in a long distance from target words. Existing neural models have limited capacity to capture these long-range information. Considering the sentence,

"I love the simplicity and respect which was given to the food, as well the staff was friendly and knowledgeable."

For the target *"food"*, it expresses positive polarity mainly through *"I love the simplicity and respect"*, and these information is far away from the target. ATAE-LSTM [7] acquired relative context information through attention mechanism, and gained promising result. Traditional attention model has limited capacity to capture long-range information [9], so ATAE-LSTM has finite ability to learn the context information which is important but in a long distance from the target. We believe it is significant for target-dependent sentiment analysis to adequately capture long-range information.

To pursuit this goal, we develop ATAE-LSTM for target-dependent sentiment analysis by introducing dependency information. We design a dependency-attention-based LSTM (DAT-LSTM) to learn long-range information for certain target. Dependency relation can capture information between words which are relevant in syntax but far in word order. DAT-LSTM combines these dependencies with traditional attention mechanism, so that it can acquire long-range information for the given target.

The main contributions of our work can be summarized as follows:

(1) We propose dependency-attention-based LSTM for target-dependent sentiment classification. This model introduces dependencies between words and is able to capture long-range information which is important for certain target. Results show this model is effective.

(2) Since long sentences lead to the problem of insufficiently capturing long-range information, we propose segmented dependency-attention-based LSTM to solve it. This model segregates sentences into two shorter parts by the position of target word and respectively processes them; because the two parts are both shorter than original sentences, they can be analyzed more fully. This model gains promising results.

The structure of this paper is as follows. In Sect. 2, we describe the related work. In Sect. 3, we introduce our models. In Sect. 4, we show and analyse the experiment results. In Sect. 5, we summarize this work and future direction.

2   Related Work

Target-dependent sentiment analysis is a widely studied field focused by many scholars. Most existing methods treat it as a text classification problem. Such methods take sentiment polarities as categories and apply machine learning algorithms to train classifiers. The performance of classifiers deeply depends on the extracted features [2–4].

Motivated by the success of deep neural networks in natural language processing [12–14], some neural network methods were proposed for target-dependent sentiment analysis. These methods can automatically learn low dimensional representation of texts without feature engineering [5–7,17,18].

ATAE-LSTM [7] captured the significant context information for certain target through attention mechanism with target embedding, and achieved state-of-the-art performance on target-dependent sentiment classification. Despite the advantage of ATAE-LSTM, traditional attention mechanism has limited capacity to capture long-range information [9]. After the size of attention window reaching a certain value, the performance of language model would not continue to improve though enlarging the window size. It indicated that attentive neural language model mainly utilized a memory of most recent history and failed to exploit long-range information. For target-dependent sentiment classification, the important context information for certain target may be far away from the target. Therefore we are motivated to design a neural network to capture long-range important information for certain target.

TC-LSTM [6] modeled the connection between target word and context words through a left LSTM and a right one. Inspired by this, we segregate sentences into two shorter parts and respectively learn the important information for targets. Since the two parts are both shorter than original sentences, they are not that seriously hindered by long-range information and can be fully analyzed.

3   Methods

3.1   Dependency-Attention-Based LSTM (DAT-LSTM)

For target-dependent sentiment analysis, the critical issue is to capture the significant information from the context according to the given target word. Intuitively, attention model is able to solve this problem, and ATAE-LSTM [7] does get a promising result. But it also has some limitation. Experiments [9] indicate that attention model has limited capacity to capture long-range information. For target-dependent sentiment analysis, the important context information may be far away from the target word.

To fully capture these distant context information, we propose dependency-attention-based LSTM (DAT-LSTM). This model captures long-range information by introducing the dependency relation, so as to learn the specific context information more adequately. The model structure is shown in Fig. 1.

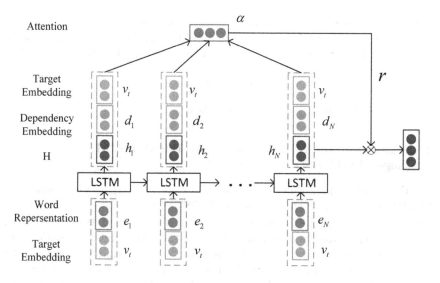

Fig. 1. The architecture of DAT-LSTM.

Given a sentence $s = \{w_1, w_2, ..., w_N\}$ and a target word w_t, we get the directly dependent term of each word through Stanford CoreNLP [19], and map each word to its embedding. v_t represents the target embedding. If target is a single word like *"food"* or *"service"*, target embedding is the embedding of the target word. If target is multi-word expression like *"battery life"*, target embedding is an average of its constituting word embedding. Let $H \in \mathbb{R}^{d_H \times N}$ be the matrix consisting of hidden vectors, $D \in \mathbb{R}^{d_D \times N}$ be the dependency matrix.

$$H = [h_1, h_2, ..., h_N] \tag{1}$$

$$D = [d_1, d_2, ..., d_N] \tag{2}$$

Where h_i represents the hidden vector of ith time step, d_i represents the embedding of the word which has the direct dependency relation with w_i. Let $M \in \mathbb{R}^{(d_H + d_D + d_{v_t}) \times N}$ be the matrix combining H, D and target embedding.

$$M = tanh \left[W_H H, W_D D, W_V v_t \cdot e_N \right]^T \tag{3}$$

Where e_N is a column vector with N 1s. The dependency-attention mechanism will produce a weight vector $\alpha \in \mathbb{R}^N$ and a weighted representation $r \in \mathbb{R}^{d_H}$.

$$\alpha = softmax \left(w^T M \right) \tag{4}$$

$$r = H \cdot \alpha^T \tag{5}$$

$W_H \in \mathbb{R}^{d_H \times d_H}$, $W_D \in \mathbb{R}^{d_D \times d_D}$, $W_V \in \mathbb{R}^{d_{v_t} \times d_{v_t}}$ and $w \in \mathbb{R}^{d_H + d_D + d_{v_t}}$ are projection parameters. The final sentence representation is given by,

$$h_* = tanh \left(W_r r + W_h h_N \right) \tag{6}$$

W_r, W_h are parameters to be learned during training. h_* is considered as the feature representation of the sentence, considering certain target. Then we utilize softmax to transfer the sentence vector to conditional probability distribution.

$$y = softmax\,(W_* h_* + b_*)$$ (7)

W_* and b_* are parameters of the softmax layer.

3.2 Segmented Dependency-Attention-Based LSTM (Seg-DAT-LSTM)

To further capture long-range information for certain target, we propose Seg-DAT-LSTM, inspired by the success of TC-LSTM [6].

On the base of DAT-LSTM, we segregate sentences into two parts by the location of target word and respectively learn the important information from the two parts. Since the two parts are both shorter than original sentences, they are not that seriously hindered by the long distance; thus they can be learned more adequately, compared with directly analyzing the original sentences. At the same time, taking the target word as the last unit can better utilize the target information. Finally the model merges the information captured from the two parts. The model structure is shown in Fig. 2.

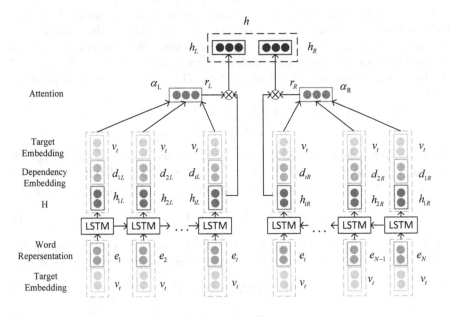

Fig. 2. The architecture of Seg-DAT-LSTM.

This model has two units, a left DAT-LSTM, DAT_L, and a right one, DAT_R. The input of DAT_L is the preceding context and target. Analogously, the input of DAT_R is the following context and target. We run DAT_L from left to right, and run DAT_R from right to left. As we believe that the last unit of LSTM contains most semantic information. Similar to DAT-LSTM, DAT_L and DAT_R combine the hidden output, dependency embedding and the target embedding, then utilize dependency-attention mechanism to get weighted representation.

$$h_L = tanh\left(W_{rL}r_L + W_{hL}h_{tL}\right) \qquad (8)$$

$$h_R = tanh\left(W_{rR}r_R + W_{hR}h_{tR}\right) \qquad (9)$$

h_L is the final representation of left part, and h_R is the final representation of right part. Then, the sentence representation is given by,

$$h = tanh\left(W_L h_L + W_R h_R\right) \qquad (10)$$

h is considered as the feature representation combined the target information. Finally we utilize the softmax to gain the probability distribution.

3.3   Model Training

We train DAT-LSTM and Seg-DAT-LSTM in an end-to-end way with supervised framework. The loss function is cross entropy, defined as follows,

$$loss = -\Sigma_{d=1}^{D}\Sigma_{c=1}^{C}p_c^g(d) \cdot log\left(p_c(d)\right) \qquad (11)$$

Where D is the training dataset, C is the number of sentiment categories, $p_c^g(d)$ represents the gold probability of sentiment category c, $p_c(d)$ represents the probability of predicting d as label c. We update parameters with stochastic gradient descent.

4   Experiments

In this section, we introduce the experiment settings and empirical results on the task of target-dependent sentiment analysis.

4.1   Dataset and Evaluation

We evaluate our models on two benchmark datasets, the tweet dataset [15], which consists of tweets with targets and corresponding polarities, and the restaurant dataset of SemEval 2014 task4 [16], which consists of customers reviews with targets and corresponding polarities. The detailed statistics of the dataset is given in Table 1. Evaluation metrics are accuracy over three categories (positive, negative and neutral) and two categories (positive and negative). The accuracy is defined as follows,

$$Accuracy = \frac{The\ count\ of\ correctly\ predicted\ samples.}{The\ count\ of\ all\ samples.} \qquad (12)$$

Table 1. Statistics of dataset.

Data	Positive	Negative	Neutral
Tweet-Train	1561	1560	3127
Tweet-Test	173	173	346
Restaurant-Train	2164	805	633
Restaurant-Test	728	196	196

4.2   Comparison with Other Methods

We compare our models with following baseline models,

LSTM [6], modeled the sentence through LSTM, without considering the target information.

TD-LSTM [6], consisted of a left LSTM and a right one to capture the target-dependent context information.

TC-LSTM [6], on the base of TD-LSTM, TC-LSTM connected the target word and each context word to capture the context information according to the target word more fully.

ATAE-LSTM [7], combined the attention mechanism with LSTM to capture the significant information according to the aspect, at the same time this model connected the aspect embedding to the hidden layer and input layer.

Experimental results of baseline models and our models are given in Table 2. Compared with the models we can find that our models are effective on target-dependent sentiment analysis.

Table 2. Experimental results.

Model	Tweet (Three)	Tweet (Two)	Restaurant (Three)	Restaurant (Two)
LSTM [6]	66.5	80.16	74.3	80.18
TD-LSTM [6]	70.8	82.09	75.6	85.27
TC-LSTM [6]	71.5	83.99	76.29	87.08
ATAE-LSTM [7]	73.1	84.21	77.2	90.9
DAT-LSTM	**74.01**	**85.13**	**78.11**	**91.81**
Seg-DAT-LSTM	73.87	84.96	77.84	91.49

4.3   Analysis

Target Information. TD-LSTM performed better than standard LSTM for its considering the information of target word. On the base of TD-LSTM, TC-LSTM connected the target word embedding with each word in the context and gained further promotion. Thus we can get the conclusion that taking the association of target word and context words into consideration is significant and effective for target-dependent sentiment analysis.

Long-Range Information. The crucial problem of target-dependent sentiment analysis is to learn the important information for certain target from context words and then to infer sentiment polarity of the target. ATAE-LSTM captured these information through attention mechanism, while attention mechanism had limited ability to capture long-range information [9]. For target-dependent sentiment analysis, it is possible that the crucial information is far away from the target word. ATAE-LSTM may be not able to fully capture these information. Dependency relation connects words which are relevant in syntax but far in word order; thus it is able to carry long-range information between words. DAT-LSTM combines dependencies to learn long-range information which is vital for certain target. Experiments show DAT-LSTM performs better than baseline models.

The Segmented Model. Large length of sentences leads to the problem of inadequately capturing long-distance information. From this point of view, separating sentences into two parts by the position of target words, and respectively analyzing them would be an effective way. Since each part is shorter than original sentences, the model can fully acquire important information from preceding context and following context. While experiments show it is not effective just as we expected. This may be because the coherence of sentences and some semantic information are lost during the segmentation processing.

4.4   Case Study

As we demonstrated, our models obtain promising results. In this section, we will show the advantages of our model through examples.

An Example of Long Sentences. We define the sentence as long sentences when it is longer than 20. Table 3 shows the polarities towards different targets of *"I love the simplicity and respect which was given to the food, as well the staff was friendly and knowledgeable."*

ATAE-LSTM predicts the sentiment polarity of *"food"* incorrectly, and DAT-LSTM gets the right label. Figure 3 shows the main dependency relation of this example. We can see the dependency relation combines *"food"* and *"simplicity"* through *"given"*, which are far apart but closely related. In this way, the significant information for *"food"* is directly connected to *"food"*. Changes of attention weights are shown as Fig. 4. In ATAE-LSTM, *"love"*, *"simplicity"* and *"respect"*, which are important for *"food"*, don't make much difference from other words. So it is hard to correctly predict the polarity. In DAT-LSTM, when considering the target *"food"*, weights of *"love"*, *"simplicity"* and *"respect"* are obviously

214   X. Wang and G. Chen

Table 3. The true labels and predicted labels from ATAE-LSTM and DAT-LSTM towards different targets of *"I love the simplicity and respect which was given to the food, as well the staff was friendly and knowledgeable."*

Target	True label	ATAE-LSTM	DAT-LSTM
food	Positive	Neutral	Positive
staff	Positive	Positive	Positive

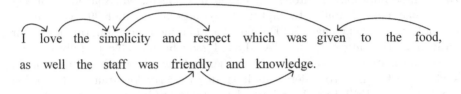

Fig. 3. The main dependency relation of *"I love the simplicity and respect which was given to the food, as well the staff was friendly and knowledgeable"*.

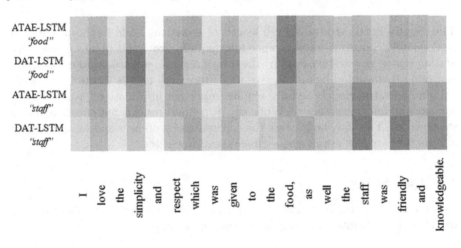

Fig. 4. Changes of attention weights in different models of the mentioned targets, *"food"* and *"staff"*.

larger than other context words. According to these information, DAT-LSTM predicts the polarity of *"food"* correctly. This indicates that combining tokens which have direct dependency relation is effective to capture important long-range information for certain target.

For the target *"staff"*, both in ATAE-LSTM and DAT-LSTM, *"friendly"* and *"knowledge"*, which are significant to *"staff"*, have obvious difference from other context words. And in DAT-LSTM, this difference is much greater. It indicates that introducing dependency relation can reinforce the role of relative information in short distance.

An Example of Short Sentences. We define the sentence as short sentences when it is shorter than 20. Table 4 shows the polarities with different models of *"The appetizers are ok, but the service is slow."*

Table 4. The true labels and predicted labels from ATAE-LSTM and DAT-LSTM towards different targets of *"The appetizers are ok, but the service is slow."*

Target	True label	ATAE-LSTM	DAT-LSTM
appetizers	Positive	Positive	Positive
service	Negative	Neutral	Negative

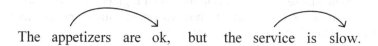

The appetizers are ok, but the service is slow.

Fig. 5. The main dependency relation of *"The appetizers are ok, but the service is slow."*

Figure 5 shows the main dependency relation of *"The appetizers are ok, but the service is slow."*, and Fig. 6 shows changes of attention weights in ATAE-LSTM and DAT-LSTM. In ATAE-LSTM, when considering the target *"service"*, *"slow"*, which expresses important information for *"service"*, has a weight similar to *"ok"* and *"appetizer"*. So ATAE-LSTM is not able to properly predict the sentiment polarity of *"service"*. Dependency relation combines *"service"* and *"slow"* directly, and in DAT-LSTM, *"slow"* has a more obvious difference from other words. This further indicates that introducing dependency relation can reinforce the role of relative information in short distance and is significant to infer the sentiment polarity of certain target in short sentence.

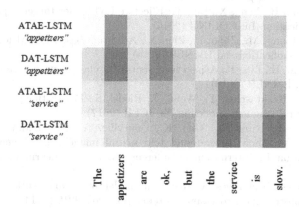

Fig. 6. Changes of attention weights in different models of the mentioned targets, *"appetizers"* and *"service"*.

5   Conclusions

In this paper, we develop the ATAE-LSTM for target-dependent sentiment analysis. For target-dependent sentiment classification, it is critical to learn the important information, which may be in a large distance from target word, for certain target from the context words. While existing methods have limited capacity to capture long-range information. DAT-LSTM solves this problem by combining dependency information. Experiments show it is effective. On this basis, Seg-DAT-LSTM separates sentences into two parts which are both shorter than original sentences, and respectively learns the information, thus to gain the important context information for certain target. Seg-DAT-LSTM doesn't perform as we expected, this may be because it loses the coherence of sentences and some important semantic information when segmenting the sentences. For further work we want to investigate how to retain important information for certain target when segmenting the sentences.

References

1. Pang, B., Lee, L.: Opinion mining and sentiment analysis. Found. Trends. Inf. Retr. **2**(1–2), 1–135 (2008)
2. Jiang, L., Mo, Y., Zhou, M., Liu, X., Zhao, T.: Target-dependent twitter sentiment classification. In: Proceedings of the 49th Annual Meeting of the Association for Computational Linguistics: Human Language Technologies-Volume 1, pp. 151–160. Association for Computational Linguistics (2011)
3. Kiritchenko, S., Zhu, X., Cherry, C., Mohammad, S.: NRC-Canada-2014: detecting aspects and sentiment in customer reviews. In: Proceedings of the 8th International Workshop on Semantic Evaluation, pp. 437–442 (2014)
4. Wagner, J., Arora, P., Cortes, S., Barman, U., Bogdanova, D., Foster, J., Tounsi, L.: DCU: aspect-based polarity classification for semeval task 4 (2014)
5. Zhang, M., Zhang, Y., Vo, D.-T.: Gated neural networks for targeted sentiment analysis. In: AAAI, pp. 3087–3093 (2016)
6. Tang, D., Qin, B., Feng, X., Liu, T.: Effective LSTMs for target-dependent sentiment classification. In: COLING (2016)
7. Wang, Y., Huang, M., Zhao, L., Zhu, X.: Attention-based LSTM for aspect-level sentiment classification. In: EMNLP (2016)
8. Tang, D., Qin, B., Liu, T.: Aspect level sentiment classification with deep memory network. In: EMNLP (2016)
9. Daniluk, M., Rocktaschel, T., Welbl, J., Riedel, S.: Frustratingly short attention spans in neural language modeling. In: ICLR (2017)
10. Xu, K., Ba, J., Kiros, R., Cho, K., Courville, A., Salakhudinov, R., Zemel, R., Bengio, Y.: Show, attend and tell: neural image caption generation with visual attention. In: International Conference on Machine Learning, pp. 2048–2057 (2015)
11. Mnih, V., Heess, N., Graves, A.: Recurrent models of visual attention. In: Advances in Neural Information Processing Systems, pp. 2204–2212 (2014)
12. Bengio, Y., Ducharme, R., Vincent, P., Jauvin, C.: A neural probabilistic language model. J. Mach. Learn. Res. **3**, 1137–1155 (2003)

13. Sutskever, I., Vinyals, O., Le, Q.V.: Sequence to sequence learning with neural networks. In: Advances in Neural Information Processing Systems, pp. 3104–3112 (2014)
14. Collobert, R., Weston, J., Bottou, L., Karlen, M., Kavukcuoglu, K., Kuksa, P.: Natural language processing (almost) from scratch. J. Mach. Learn. Res. **12**, 2493–2537 (2011)
15. Dong, L., Wei, F., Tan, C., Tang, D., Zhou, M., Ke, X.: Adaptive recursive neural network for target-dependent twitter sentiment classification. In: ACL, vol. 2, pp. 49–54 (2014)
16. Pontiki, M., Galanis, D., Pavlopoulos, J., Papageorgiou, H., Androutsopoulos, I., Manandhar, S.: Semeval-2014 task 4: aspect based sentiment analysis. In: Proceedings of SemEval, pp. 27–35 (2014)
17. Ruder, S., Ghaffari, P., Breslin, J.G.: A hierarchical model of reviews for aspect-based sentiment analysis. In: EMNLP (2016)
18. Lakkaraju, H., Socher, R., Manning, C.: Aspect specific sentiment analysis using hierarchical deep learning. In: NIPS Workshop on Deep Learning and Representation Learning (2014)
19. Manning, C.D., Surdeanu, M., Bauer, J., Finkel, J.R., Bethard, S., McClosky, D.: The stanford CoreNLP natural language processing toolkit. In: ACL, pp. 55–60 (2014)

A Novel Fuzzy Logic Model for Multi-label Fine-Grained Emotion Retrieval

Chu Wang[1], Daling Wang[1,2(✉)], Shi Feng[1,2], and Yifei Zhang[1,2]

[1] School of Computer Science and Engineering, Northeastern University,
Shenyang 110169, People's Republic of China
wangchu@research.neu.edu.cn,
{wangdaling, fengshi, zhangyifei}@cse.neu.edu.cn
[2] Key Laboratory of Medical Image Computing, Northeastern University,
Ministry of Education, Shenyang 110169, People's Republic of China

Abstract. The traditional opinion retrieval methods can acquire the topic-relevant and subjective documents or sentences with the issued query. However, these methods usually focus on the sentiment polarities in the retrieval results, but ignore the emotion intensities. In fact, the users may pay more attentions on the emotion similarity between the query words and retrieval results, i.e. the sentences with exactly the same fine-grained emotion labels and similar emotion intensities with the query should be ranked higher. To address the problem, we propose a new method based on fuzzy set theory. According to the theory, we build a model for multi-label fine-grained emotion retrieval, which utilizes fuzzy relation equation to calculate the value of sentiment words and then uses lattice close-degree to retrieval emotions and rank on their intensity. Extensive experiments are conducted on a well-known Chinese blog emotion corpus. Experimental results show that our proposed multi-label fine-grained emotion retrieval algorithm outperforms baseline methods by a large margin.

Keywords: Emotion retrieval · Sentiment analysis · Fuzzy relation equation

1 Introduction

The explosion of social media on Web has created unprecedented opportunities for people to publicly voice their opinions on trending events and commercial products. Therefore, mining opinions of the user generated content in social media has become a popular research for both academic and industrial communities in recent years [16].

The task of opinion retrieval is to find the topic-relevant and subjective documents/sentences in Web text such as blogs and tweets, which has been studied in depth in recent years [9]. However, the human's fine-grained emotions are much more complex than sentiment orientations such as positive and negative. For example,

The project is supported by National Natural Science Foundation of China (61370074, 61402091).

X. Cheng et al. (Eds.): SMP 2017, CCIS 774, pp. 218–231, 2017.
https://doi.org/10.1007/978-981-10-6805-8_18

Robert Plutchik defined eight basic emotions: *joy, sadness, anger, fear, trust, disgust, surprise* and *anticipation* [12]. Multiple basic emotions could be co-existing even in a short sentence, as the example sentence shown below: *"You are a fine person, Mr Baggins, and I'm very fond of you. But you are only quite a little fellow, in a wide world, after all."*—Gandalf, Hobbit.

In the above example, because multiple emotions often exist at the same time, a sentence with different sentiment words and contexts may show different sentiment and intensity. Traditional opinion retrieval methods can find the topic-relevant opinionative sentences in a dataset, but cannot address the general problem of how to retrieval emotion-relevant sentences in a large social media text collection.

Table 1 contains some examples. If an author expresses stronger emotion in a post, the emotion will have higher intensity. Because the sentimental word "expect" has stronger emotion than the word "OK", so the intensities value on the label of *joy* emotions in the third post are higher than those of the second post.

Table 1. The examples of multiple emotions with different intensities

Social text	joy	hate	love	sorrow	anxiety	surprise	anger	expect
I love my hometown, but its air condition is getting worse all these years	0	0.2	0.9	0.6	0	0	0	0
The dinner is OK, but I am looking forward to the party tomorrow!	0.4	0	0	0	0	0	0	0.9
Oh my god, I did not expect I can get such a good result in the final exam.	0.8	0	0	0	0	0.8	0	0
This laptop is rubbish! It cost me 1500$ but was broken in one month!	0	0.7	0	0	0	0	0.9	0

To tackle these challenges, in this paper, we regard the emotion analysis with different intensity as fuzzy problem, and use fuzzy theory to address the problem. In summary, the main contributions of this paper are summarized as follows:

(1) We present a new research problem of emotion retrieval. We discuss the differences between opinion retrieval and emotion retrieval.

(2) We propose a novel fuzzy logic model for the emotion retrieval task. The proposed model can not only identify the mixed emotion labels in the query and relevant sentences, but capture the emotion intensity similarity between the sentences.

(3) We conduct extensive experiments on a public available Chinese blog dataset with emotion intensity labels. The experimental results demonstrate that the proposed method achieves excellent performance in terms of information retrieval metrics.

The remainder of this paper is structured as follows: We introduce the related work in Sect. 2. Section 3 introduces fuzzy set theory and lattice close degree. In Sect. 4, we describe the proposed fuzzy theory based emotion retrieval algorithm and framework. In Sect. 5, we discuss a series of experiments conducted for emotion retrieval by using public blog dataset. Finally we highlight the conclusion of the paper and give the future work in Sect. 6.

2 Related Work

Firstly, opinion classification inferred extraction and analysis on various aspects of the content [3]. In recent years, some research focused on co-referencing area [1]. Moreover, currently more and more researches considered the sentiment categories such as *joy, hate, love, sorrow, anxiety, surprise, anger, expect* [13] called as fine-grained emotion. Classification algorithms were mainly based on traditional machine learning techniques like supervised and unsupervised with good efficiencies [7].

Secondly, in opinion retrieval, many sentiment retrieval methods were based on machine learning, such as Zhang, et al. [22] and Liu, et al. [10]. In the early days, some other approaches were based on capturing topic-related opinion expression, even built a query-specific opinion lexicon for retrieval [8]. For retrieval results ranking, some methods based on candidate antecedents got a target anaphoric expression in general texts [14]. Some researchers found linguistic discourse structures were not suitable for short messages, such as texts in microblogs due to words limit in a text [19].

Thirdly, it is more common to use fuzzy logic to deal with the retrieval problems. Gupta, et al. proposed a ranking function based on fuzzy theory in [5]. Some approach used fuzzy theory to solve the retrieval challenge by bringing the advantages of the annotations and feed them back to adjust the retrieval results [11]. Nowadays, some researchers use semantic network to solve emotion recognition and sentiment retrieval [2], and find out that social features and unsupervised opinionatedness can improve the performance of the opinion retrieval on tweets [4], and even used for big data [20].

However, there are few methods could retrieve the data based on emotional intensity. Therefore, we propose a fuzzy logic method which perfectly matched human emotional logic to deal with such a fine-granted opinion retrieval problem with emotional intensity.

3 Fuzzy Theory

3.1 Fuzzy Set and Fuzzy Relation Equation

Fuzzy set theory was proposed by Zadeh in 1965 [21]. U is a finite and non-empty set, and is seemed as universe, which is defined as:

$$\mu : U \to [0, 1] \tag{1}$$

where for each $x \in U$, $\mu_A(x)$ is called as the membership degree of x in A. The fuzzy power set is denoted by $F(U)$ [17] showing the set of all fuzzy sets in the universe U.

Suppose there are three finite and non-empty sets $U = \{u_1,...u_n\}$, $V = \{v_1,...v_n\}$, $W = \{w_1,...w_n\}$, for a given F matrix $A \in \mu_{m \times l}$, $B \in \mu_{n \times l}$, we calculate F matrix $X \in \mu_{n \times m}$ to meet the formula $X \circ A = B$ as:

$$
\begin{bmatrix} x_{11} & \cdots & x_{1m} \\ \vdots & \ddots & \vdots \\ x_{n1} & \cdots & x_{nm} \end{bmatrix} \circ \begin{bmatrix} a_{11} & \cdots & a_{1l} \\ \vdots & \ddots & \vdots \\ a_{m1} & \cdots & a_{mn} \end{bmatrix} = \begin{bmatrix} b_{11} & \cdots & b_{1l} \\ \vdots & \ddots & \vdots \\ b_{n1} & \cdots & b_{nl} \end{bmatrix} \tag{2}
$$

We solve $A \circ X = B$ the same as $X \circ A = B$. Fuzzy union and fuzzy intersection (FI) is a very widely used method in the fuzzy set methods. Fuzzy union is defined as: $(A \cup B)(x) = \max(A(x), B(x))$ for all $x \in X$, and fuzzy intersection is defined as: $(A \cup B)(x) = \min(A(x), B(x))$ for all $x \in X$.

3.2 Lattice Close-Degree

According to Fuzzy set theory [12], given an fuzzy set A and B, where $A = (a_1, a_2,..., a_n)$, $B = (b_1, b_2, b_3,..., b_n)$. Making fuzzy set B closer to A, will increase the inner product $A \circ B$ and reduction of outer product $A \overset{\wedge}{\circ} B$. In another words, when $A \circ B$ is larger and $A \overset{\wedge}{\circ} B$ is smaller, A and B are closer. Therefore, we take the fuzzy inner product and fuzzy outer product combination of "lattice close degree" to describe the close degree of the two fuzzy sets.

Since the emotion intensity between [0, 1] can be regarded as fuzzy degree, we can use lattice close-degree method to fix the interval problem. It represents the degree of similarity between two fuzzy sets. To define lattice closeness, we first introduce two definitions of inner product and outer product. Then let A and B be two finite fuzzy subsets on the universe, thus it can be shown as follows: Let $A, B \in F(U)$, $A = (a_1, a_2,..., a_n)$, $B = (b_1, b_2,..., b_n)$, we call $A \circ B = \overset{n}{\underset{i=1}{\vee}} (a_i \wedge b_i)$ as the A, B inner product of $F(U)$, and $A \overset{\wedge}{\circ} B = \overset{n}{\underset{i=1}{\vee}} (a_i \wedge b_i)$ as outer product of $F(U)$, which $\wedge$ means $min(a, b)$, $\vee$ means $max(a, b)$. And the lattice close degree can be defined as:

$$
N(A, B) = (A \circ B) \wedge (A \overset{\wedge}{\circ} B) \tag{3}
$$

It also can be regarded as:

$$
N(A, B) = (A \circ B) \wedge (A^c \circ B^c) \tag{4}
$$

where c is fuzzy complement of vector A, which means $A^c = 1 - A$.

4 Multi-labeled Fine-Grained Emotion Retrieval

The process includes modeling and retrieval stage, and whole model is shown in Fig. 1.

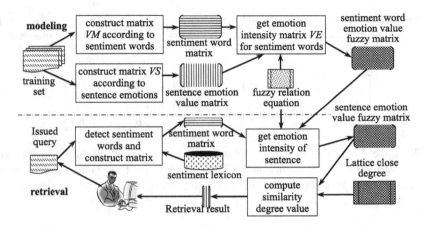

Fig. 1. Overall framework of multi-label emotion retrieval

4.1 Fuzzy Relation Equation Calculation for Modeling

In fuzzy relation modeling in Fig. 1, for our problem, the emotion intensity of training set we used has been labeled at sentence level, and we also know the embedded sentiment words (even unknown, we still can detect them with existing techniques and sentiment lexicon). So in modeling stage, especially in the training part, we use a fuzzy relation equation (Formula (2)) to get emotion intensity range of the sentiment words in the query with the training set.

To explain our method more specifically, we rewrite fuzzy relation equation shown in Formula (2) and represent it as Formula (5). VW corresponds to the matrix A in Formula (2), so that VE corresponds to X and VS corresponds to B.

$$
\begin{bmatrix} vw_{11} & \cdots & vw_{1m} \\ \vdots & \ddots & \vdots \\ vw_{n1} & \cdots & vw_{nm} \end{bmatrix} \circ \begin{bmatrix} ve_{11} & \cdots & ve_{18} \\ \vdots & \ddots & \vdots \\ ve_{n1} & \cdots & a_{n8} \end{bmatrix} = \begin{bmatrix} vs_{11} & \cdots & vs_{18} \\ \vdots & \ddots & \vdots \\ vs_{m1} & \cdots & vs_{m8} \end{bmatrix} \tag{5}
$$

In the equation above, the first item on the left $VW = [vw_{mn}]$ is sentiment word matrix, here assuming the training set T has n sentiment words and m sentences. If jth sentiment word w_j exists in ith sentence s_i, $vw_{ij} = 1$ else $vw_{ij} = 0$.

The item in right $VS = [vs_{m8}]$ is sentence emotion intensity matrix, here we consider eight emotions as the same as [13], i.e. $e_1 = joy$, $e_2 = hate$, $e_3 = love$, $e_4 = sorrow$, $e_5 = anxiety$, $e_6 = surprise$, $e_7 = anger$, $e_8 = expect$. For the ith sentence s_i, vs_{i1}, vs_{i2}, ..., vs_{i8} represent the emotion intensity value of joy, $hate$, $love$, $sorrow$, $anxiety$, $surprise$, $anger$, and $expect$, respectively. We can construct the matrix based on known multi-label emotion intensity of every sentence in T.

The second item in left $VE = [ve_{n8}]$ is emotion intensity matrix of all sentiment words. For the ith sentiment word w_i, ve_{i1}, ve_{i2}, ..., ve_{i8} represent the corresponding eight emotion intensity values of w_i, respectively. In the modeling stage, our goal is just to calculate it by solving the fuzzy relation equation in Formula (5).

The following Algorithm 1 describes the process for achieving *VE* matrix.

Algorithm 1: Modeling Algorithm for Multi-label Emotion Intensity

---

Input: Training Set T // In T, all sentiment words have been labeled with eight emotions intensities of every sentence are known.
Output: emotion intensity matrix *VE* about all sentiment words.
Description:
    1. For every sentence $s \in T$;
    2.   {Construct a row of *VW* matrix with all sentiment words in s;
    3.   Construct a row of *VS* matrix with eight emotion intensity values of s;}
    4. Solve and Return fuzzy matrix *VE* with fuzzy relation equation in Formula (5);

4.2 Similarity Measure Based on Lattice Close-Degree for Retrieval

In sentiment retrieval stage in Fig. 1, for a new sentence s, we apply *VE* matrix returned by Algorithm 1 to achieve a new emotion intensity matrix *VE'*. Because *VE* is a fuzzy matrix, *VE'* is also fuzzy. After we get the range of the emotional value of the sentiment words from the training set, we use lattice close degree method to calculate the similarity between the query and sentence in the dataset.

We use the fuzzy range of the emotional value of the sentiment words from *VE* on the lattice close degree to solve the problem. As we see, the value we get from the matrix is a range, but what lattice close degree needs is a certain value. So we make a change to fit the problem. When A, B belong to *VE'*, let A be one kind of emotional value of the sentence to be retrieved, which $A = \{A_1, A_2, A_3, ..., A_i\}$, A_i means the value of sentiment words (VE'_i), then A_i is a range which is calculated by Formula (5), in which each A_i is defined as $[a_{i(min)}, a_{i(max)}]$. We combine all the value of sentiment words together. Using fuzzy union and fuzzy intersection, given $a_{(min)} = \wedge a_{i(min)}$, and $a_{(max)} = \vee a_{i(max)}$, then the range of A is $[a_{(min)}, a_{(max)}]$. Similarly, we can see a kind of sentiment value of sentences B in the dataset can be seemed as $B_j = [b_{j(max)}, b_{j(min)}]$ (VE'_j). Then the range of B is $B = [b_{(max)}, b_{(min)}]$, which i, j are the sentiment words which contain the same kind of emotion. After that, the Formula (3) can be changed as

$$N_1(A, B) = (a_{(max)} \circ b_{(max)}) \wedge (a_{(max)} \overset{\wedge}{\circ} b_{(max)}) \tag{6}$$

$$N_2(A, B) = (a_{(min)} \circ b_{(min)}) \wedge (a_{(min)} \overset{\wedge}{\circ} b_{(min)}) \tag{7}$$

The average value $N(A, B)$ of $N_1(A, B)$ and $N_2(A, B)$ is the similarity measure of our retrieval work. Where $N(A, B)$ is larger, comes more relevance between A and B. According to the calculation, this part is described in the following Algorithm 2.

Algorithm 2: Similarity measure based on lattice close-degree for emotion retrieval

---

Input: a new sentence s, emotion intensity matrix VE from Algorithm 1
Output: 5 sentences as retrieval result
Description:
1. Find all sentiment words $w \in s$;
2. Construct sentiment word matrix VW' with all above $w \in s$;
3. For every $w \in VW'$
4.   Achieve emotion intensity value range of w from VE;
5. Compute the lattice close-degree of VE' between the query sentence and dataset as:

$$N_1(A,B) = (a_{(max)} \circ b_{(max)}) \wedge (a_{(max)} \hat{\circ} b_{(max)}) ; \quad N_2(A,B) = (a_{(min)} \circ b_{(min)}) \wedge (a_{(min)} \hat{\circ} b_{(min)}) ;$$

// where A corresponds to VE' and B corresponds to VE

6. Calculate the similarity value as: $N(A,B) = \dfrac{N_1(A,B) + N_2(A,B)}{2}$;

7. Rank $N(A, B)$ in descending orders and Return top 5 sentences with max N;

---

5   Experiments

5.1   Dataset and Evaluation Metric

At the experiment section, we choose Quan's Chinese blog dataset [13] to evaluate our method. There are 1,487 documents, 11,953 paragraphs, 38,051 sentences, and 971,628 Chinese words in this corpus. All the sentiment words are labeled, and every sentence and sentiment word are annotated by eight basic kinds of emotions with intensities between 0 and 1. To verify the method we proposed more effectively, we removed the sentences with negation words. An example is shown in Fig. 2.

Fig. 2.   An example sentence of the dataset with emotion label and intensity

In the paper, we ignore the corresponding emotion intensities in the word level. We use 10 folds cross validation for the experiments.

On average, the dataset is divided into 10 parts, and each time we get 9 training sets and 1 test set. We not only focus on the label classification accuracy, but also the intensity of each emotion. So during this experiment, the results of our experiments are the average of ten folds cross-validation. Note that we focus on the sentence level emotion retrieval.

The system measure is more complicated. To solve the problem, our measure approaches should take care both of the two sides, label and intensity. We decide to pick up 20 sentences, each of them can be seemed as a query. Then we give every query 5 retrieval results using the proposed algorithms, and 20 accurate answers using the gold standard labels, which are based on Euclidean distance between the query and the testing set. Here we introduce our evaluation metrics.

Subset Accuracy (S-A for short): It evaluates whether the most relevant sentence we retrieved is in the 20 accurate answers or not. x_i is the most relevant sentence, Y_i is accurate answers. p is the sum of number of queries, and in this paper, $p = 20$.

$$subsetacc_s(h) = \frac{1}{p} \sum_{i=1}^{p} [|h(x_i = Y_i)|] \qquad (8)$$

Total Precision (T-P for short): The total precision evaluation metric is different from the classic one, which is seemed as:

$$TP = f(y)/T, TP \in [0, 1] \qquad (9)$$

In this formula, $f(y)$ is the number of real related documents we detected, and T represents the sum of number of documents we find out. For example, in Table 2, $T = 20$, in Table 3, $T = 10$. As we all know, the value of precision is the larger, the better.

Average Precision (A-P for short): The average precision evaluates the average fraction of relevant sentences ranked higher than a particular sentence, $y \in Y_i$. For example, if there are 5 resulted sentences, the ranking order of the 5 results in the 20 answers is 1, 2, 5, 10 and 20. $average_s(h) = (1/1 + 2/2 + 3/5 + 4/10 + 5/20)/5 = 0.64$.

Binary Preference (B-P for short): B-pref (Binary preference) was introduced for the first time to TREC Terabyte in 2005. This evaluation metric is primarily concerned with the number of times an unrelated document appeared before the relevant document.

$$B - pref = \frac{1}{Y} \sum_y 1 - \frac{|\{n \; ranked \; higher \; then \; y, \; y \in Y_i\}|}{Y} \qquad (10)$$

Normalized Discounted Cumulative Gain: $nDCG$ is well suited to evaluation of recommendation system, as it rewards relevant items in the top ranked results more heavily than those ranked lower [6]. For a given user profile, the ranked results are examined top-down, where *nDCG* is computed as:

$$nDCG_i = Z_i \sum_{j=1}^{R} \frac{2^{r(j)} - 1}{\log(1+j)} \qquad (11)$$

where Z_i is a normalization constant calculated so that a perfect ordering would obtain *nDCG* of 1; and each $r(j)$ is an integer relevance level (for our case, $r(j) = 1$ and $r(j) = 0$ for relevant and irrelevant recommendations, respectively) of result returned at the rank j ($j = 1, \cdots, R$). Therefore, in this work, we use *nDCG@R* ($R = 5$ or 10) for evaluation where R is the number of top-R sentences returned by our proposed approaches.

Average Response Time (R-t for short)**:** In the opinion retrieval area, average response time is a very important metric, our experimental equipment is a computer with 8 Gb RAM and i5-4590T CPU, which contains 4 cores with 2.00 GHz.

5.2   Experiment Setup

Rarely researches were proposed for multi-label and fine-grained emotion retrieval problem. We will compare our method with the following methods that can be divided into following categories.

(1) **Calculating the sentence emotion intensity labels:** In this method, firstly, the comparing methods need to get the value of every sentence in the testing set, then we calculate the Euclidean distance between them and the query as the basis of the final ranking.

(2) **Classic retrieval method:** At this part, we use classic BM25 method [15] as a baseline. Then we rank them top-down.

$$Score(Q, d) = \sum_{i}^{n} IDF(q_i) \cdot \frac{f_i \cdot (k_1 + 1)}{f_i + k_1 \cdot (1 - b + b \cdot \frac{dl}{avgdl})} \qquad (12)$$

$$IDF(q_i) = \log \frac{N - n(q_i) + 0.5}{n(q_i) + 0.5} \qquad (13)$$

In Formula (13), n is the total number of documents. q_i is the sentiment words. n (q_i) is the number of documents that contain q_i. In Formula (12), *avgdl* is the average length of all documents. k_1 and b are the adjustment factor, usually set according to experience. In this experiment, we set $k_1 = 2$, $b = 0.75$. f_i is the frequency of the term in the document. *dl* is the length of the document d.

(3) **Multi-label emotion intensity analysis:** In this part, we use MBL method [18] as a comparing method. In this method, there are three steps: firstly, we use fuzzy matrix to calculate the intensity of the sentiment word value just as we do.

Secondly, we use the improving fuzzy-rough set to calculate the emotional intensities of every sentence in the testing set, which is shown as following:

$$\bar{F}(A)(e) = \vee_{w\in W}[F(e)(\bar{w}) \wedge \bar{A}(w)], e \in E \tag{14}$$

$$\underline{F}(A)(e) = \begin{cases} \wedge_{w\in W}[(1 - F(e)(\underline{w})) \vee (\underline{A})(w)], & e \in E, \ F(e)(\underline{w}) \in [0.5, 1] \\ \wedge_{w\in W}[F(e)(\underline{w}) \wedge (\underline{A})(w)], & e \in E, F(e)(\underline{w}) \in [0, 0.5) \end{cases} \tag{15}$$

The pair $(\bar{F}(A), \underline{F}(A))$ is referred to as a generalized fuzzy rough set, and $F(e)(w)$ is referred to as upper and lower generalized fuzzy rough approximation operators. A is strongest emotional intensity object. After that, finally we use Euclidean distance to calculate the similarity between the query and testing set.

(4) **Word2vector:** Word2vec is a group of related models that are used to produce word embeddings, it is very popular in recent years. In this paper, we use the classic method as a training algorithm which can be downloaded from Google. As we can see, classic word2vec is suitable for evaluating the similarity of words. So to fit our sentence problem, we choose to add the vectors of the keywords contained in this sentence as its sentence vector, then we can compare the distance between sentences.

5.3 Experiment Results

In this section, we compare our methods with all the other methods which have been mentioned above. Because our method uses cross-validation, the results shown in Tables 2 and 3 are averaged from 10 runs.

Table 2. 5 retrieval results for 20 standard answers of sentence emotion retrieval

	Fuzzy union	BM25	FI	MBL	Word2vec	Our method
S-A	0.578	0.26	0.58	0.846	0.553	0.822
T-P	0.382	0.12	0.43	0.827	0.430	0.803
B-P	0.305	0.28	0.41	0.691	0.391	0.573
A-P	0.206	0.10	0.29	0.626	0.192	0.584
nDCG	1.28	0.49	1.62	2.86	1.43	2.89
R-t (ms)	825	1917	886	264862	6962	2988

Table 3. 10 retrieval results for 10 standard answers of sentence emotion retrieval

	Fuzzy union	BM25	FI	MBL	Word2vec	Our method
S-A	0.441	0.227	0.461	0.789	0.506	0.726
T-P	0.255	0.108	0.347	0.759	0.389	0.669
B-P	0.213	0.186	0.36	0.705	0.23	0.65
A-P	0.182	0.105	0.26	0.630	0.16	0.61
nDCG	1.97	0.88	2.33	5.22	2.35	5.06
R-t (ms)	902	2013	996	302465	7604	3082

Here gives an example of query: 这不能不说是世界经济历史上的奇迹, 当然,中国仍然面临一个最大的难题, 那就是普通民众如何分享这样的经济成功。 (emotion label: *Joy* = 0.4, *Love* = 0.4, *Anxiety* = 0.6) There are the five retrieval results as follows:

(1) 温暖是我们每个人都需要的, 特别是别人处于危难之中。 (emotion label: *Joy* = 0.4, *Love* = 0.4, *Anxiety* = 0.4, *except* = 0.5)
(2) 无比自豪的同时, 又感到肩上重担的沉重。 (emotion label: *Joy* = 0.8, *Love* = 0.8, *Anxiety* = 0.6)
(3) 孩子, 感谢你的提醒, 但愿你没有学到一些不该学的东西! (emotion label: *Joy* = 0.5, *Love* = 0.7, *Anxiety* = 0.5, *except* = 0.8)
(4) 日本有车的人家太多了, 但是日本的车位挺贵的, 可能这也会制约人们买车的欲望, 不过日本的地铁这么发达,实用主义至上的日本人更多还是选择乘坐地铁上下班。 (emotion label: *Joy* = 0.4, *Love* = 0.7, *Anxiety* = 0.3)
(5) 吴老师说: "这个办法好是好, 可是她不听怎么办? (emotion label: *Joy* = 0.3, *Love* = 0.3, *Anxiety* = 0.3)

According to the experimental results shown in Table 3, under this new situation, which demands 10 answers and 10 results, our retrieval system performance is the most superior and appreciative of all.

Because the core of our approach is based on the evaluation of keywords, so although our approach is used for sentence emotion similarity retrieval, which means the query for input is a sentence, it can also be instead by a combination of emotional words with different emotional categories. As what is shown in Tables 4 and 5.

Table 4. 5 retrieval results for 20 standard answers of word based emotion retrieval

	Fuzzy union	BM25	FI	MBL	Word2vec	Our method
S-A	0.546	0.285	0.609	0.798	0.465	0.762
T-P	0.392	0.237	0482	0.759	0.489	0.727
B-P	0.267	0.264	0.396	0.625	0.359	0.609
A-P	0.229	0.152	0.349	0.580	0.301	0.523
nDCG	1.24	0.52	1.16	2.72	1.07	2.46
R-t (ms)	662	1819	539	258913	6840	2816

Table 5. 10 retrieval results for 10 standard answers of word based emotion retrieval

	Fuzzy union	BM25	FI	MBL	Word2vec	Our method
S-A	0.519	0.225	0.491	0.722	0.423	0.667
T-P	0.286	0.119	0.367	0.680	0.303	0.629
B-P	0.242	0.157	0.324	0.613	0.308	0.587
A-P	0.205	0.130	0.297	0.587	0.289	0.501
nDCG	2.06	0.75	2.15	4.96	1.93,	4.68
R-t (ms)	884	1982	739	284679	7280	2968

5.4   Discussion

Although the dataset we use labeled every sentiment word with emotion intensity, in this paper, our main purpose is to evaluate the result of the retrieval, so we do not use the emotion intensity which was labeled on the sentiment words in the dataset.

In Wang's comparing method, which is called MBL in Table 2 [18], at some of the evaluation metrics, it shows better. However, because it was not originally designed for retrieval problem, which means it needs to calculate and get every exact value of emotional words in advance, which costs much more response time. So it does not show any feasible at all.

The word2vec method is suitable for evaluating semantic similarity of words, not for emotional similarity retrieval on sentence level, which is main purpose in this paper. Taking these two sentences for example:

(1) *It is too dark outside, I am worried about my dear children.*
(2) *I like rock climbing, it is very thrilling.*

Both of the two sentences above have the emotion of *"love"* and *"anxiety"*, so they are emotionally similar, but they are not semantic or topic relevant at all.

BM25 method does not take the emotion intensity into consideration, only considers the similarity of words. For example, in BM25 method, such sentiment words like *love* and *like* have no relevant at all.

In the words based experiment, at first, we do not know the right answers, so we cannot evaluate the retrieval results. To fix the problem, we use the median of annotations of the emotional words as the right answers to evaluate our results. This is the only time in this paper we take annotations of the emotional words into consideration. But this measurement may be limited in the case of same words may express different sentiment in different sentences.

As we look thorough the dataset, the emotional logic is not a simple summation, like "*I love mom and I love dad,*" do not have the double emotional intensity of love, and its "love" intensity is not stronger than the sentence "*I love my family*". So we argue that the fuzzy logic is suitable for multi-label emotion opinion retrieval, which means it is consistent with the logic of human language when expressing emotions. In most related bibliographies with fuzzy mathematics, the introduced examples always depicted intensity analysis of human feeling, such as the oldness degree of 40 years old, or the height degree of a 180 cm man. These questions such as old and height degree are almost showed in every fuzzy theory textbook, and all got good solutions by using fuzzy set theory. And these experiments we built combine the comprehensive consideration of practicality and mathematical measurement. We use the fuzzy matrix to get a range to solve the problem.

The results has demonstrated that our model is more suitable for small text units, like sentences, especially in the response time.

6   Conclusion and Future Work

In this paper, we proposed a new way to solve the multi-label and fine-grained emotion retrieval problem. We used a fuzzy relation equation and lattice close degree methods to model and calculate the distance to the query texts. The query can be sentence level or word level. Our retrieval returns the sentences which are emotion intensity relevant but not always topic relevant with a given query.

In the future, according to linguistic logic, we will find a way to deal with some details such as adverbs and negation words. The role of them, and especially negation words should be further taken into consideration, which can much improve the performance and practicality of our multi-label and fine-grained emotion retrieval.

References

1. Atkinson, J., Salas, G., Figueroa, A.: Improving opinion retrieval in social media by combining features-based coreferencing and memory-based learning. Inf. Sci. **299**, 20–31 (2015)
2. Bisio, F., Meda, C., Gastaldo, P., Zunino, R., Cambria, E.: Sentiment-oriented information retrieval: affective analysis of documents based on the SenticNet framework. In: Sentiment Analysis and Ontology Engineering, pp. 175–197 (2016)
3. Fabbrizio, G., Aker, A., Gaizauskas, R.: Summarizing online reviews using aspect rating distributions and language modeling. IEEE Intell. Syst. **28**(3), 28–37 (2013)
4. Giachanou, A., Crestani, F.: Opinion retrieval in Twitter: is proximity effective? In: SAC 2016, pp. 1146–1151 (2016)
5. Gupta, Y., Saini, A., Saxena, A.: A new fuzzy logic based ranking function for efficient information retrieval system. Expert Syst. Appl. **42**(3), 1223–1234 (2015)
6. Järvelin, K., Kekäläinen, J.: IR evaluation methods for retrieving highly relevant documents. In: SIGIR 2000, pp. 41–48 (2000)
7. Jia, L., Yu, C., Meng, W.: The effect of negation on sentiment analysis and retrieval effectiveness. In: CIKM 2009, pp. 1827–1830 (2009)
8. Jijkoun, V., Rijke, M., Weerkamp, W.: Generating focused topic-specific sentiment lexicons. In: ACL 2010, pp. 585–594 (2010)
9. Kim, Y., Song, Y., Rim, H.: Opinion retrieval for Twitter using extrinsic information. J. UCS **22**(5), 608–629 (2016)
10. Liu, S., Liu, F., Yu, C., Meng, W.: An effective approach to document retrieval via utilizing WordNet and recognizing phrases. In: IGIR 2004, pp. 266–272 (2004)
11. Naouar, F., Hlaoua, L., Nazih Omri, M.: Collaborative information retrieval model based on fuzzy confidence network. J. Intell. Fuzzy Syst. **30**(4), 2119–2129 (2016)
12. Plutchik, R.: A psycho evolutionary theory of emotion. Soc. Sci. Inf. **21**(4–5), 529–553 (1980)
13. Quan, C., Ren, F.: A blog emotion corpus for emotional expression analysis in Chinese. Comput. Speech Lang. **24**(4), 726–749 (2010)
14. Rahman, A., Ng, V.: Narrowing the modeling gap: a cluster-ranking approach to coreference resolution. J. Artif. Intell. Res. **40**, 469–521 (2011)
15. Robertson, S., Jones, K.: Relevance weighting of search terms. JASIS **27**(3), 129–146 (1976)

16. Shankar, A.A., Kumar, K.R.: Top K-Opinion decisions retrieval in health care system. In: Computer Science & Information Technology (CS & IT), pp. 57–65 (2015)
17. Wang, C., Feng, S., Wang, D., Zhang, Y.: Fuzzy-rough set based multi-labeled emotion intensity analysis for sentence, paragraph and document. In: NLPCC 2015, pp. 444–452 (2015)
18. Wang, C., Wang, D., Feng, S., Zhang, Y.: An approach of fuzzy relation equation and fuzzy-rough set for multi-label emotion intensity analysis. In: DASFAA Workshops 2016, pp. 65–80 (2016)
19. Xia, R., Zong, C., Hu, X., Cambria, E.: Feature ensemble plus sample selection: domain adaptation for sentiment classification. IEEE Intell. Syst. **28**(3), 10–18 (2013)
20. Yu, C.: Three challenges for opinion retrieval. In: Information Studies Theory & Application (2016)
21. Zadeh, L.: Fuzzy sets. Inf. Control **8**(3), 338–353 (1965)
22. Zhang, W., Yu, C.: UIC at TREC 2006 Blog Track. In: TREC 2006

Deep Transfer Learning for Social Media Cross-Domain Sentiment Classification

Chuanjun Zhao[1], Suge Wang[1,2(✉)], and Deyu Li[1,2]

[1] Shanxi University, Taiyuan 030006, Shanxi, China
zhaochuanjun@foxmail.com, {wsg,lidy}@sxu.edu.cn
[2] Key Laboratory of Computational Intelligence,
Chinese Information Processing of Ministry of Education,
Taiyuan 030006, Shanxi, China

Abstract. Social media sentiment classification has important theoretical research value and broad application prospects. Deep neural networks have been applied into social media sentiment mining tasks successfully with excellent representation learning and high efficiency classification abilities. However, it is very difficult to collect and label large scale training data for deep learning. In this case, deep transfer learning (DTL) can transfer abundant source domain knowledge to target domain using deep neural networks. In this paper, we propose a two-stage bidirectional long short-term memory (Bi-LSTM) and parameters transfer framework for short texts cross-domain sentiment classification tasks. Firstly, Bi-LSTM networks are pre-trained on a large amount of fine-labeled source domain training data. We fine-tune the pre-trained Bi-LSTM networks and transfer the parameters using target domain training data and continuing back propagation. The fine-tuning strategy is to transfer bottom-layer (general features) and retrain top-layer (specific features) to the target domain. Extensive experiments on four Chinese social media data sets show that our method outperforms other baseline algorithms for cross-domain sentiment classification tasks.

Keywords: Transfer learning · Long short-term memory · Parameters transfer · Cross-domain sentiment classification

1 Introduction

Sentiment analysis, also known as subjectivity analysis or opinion mining, is the process of analyzing, processing, summarizing, and reasoning the subjective texts. Sentiment analysis can also be subdivided into sentiment polarity analysis, subjective and objective analysis, emotional classification, and so on [7]. Individuals can express their sentiment about emergencies, public figures, and popular products through social media directly and quickly. Being an important research direction in sentiment mining, sentiment classification for short texts, usually from social media such as online reviews and Sina Weibo, has wide application

© Springer Nature Singapore Pte Ltd. 2017
X. Cheng et al. (Eds.): SMP 2017, CCIS 774, pp. 232–243, 2017.
https://doi.org/10.1007/978-981-10-6805-8_19

prospects in the fields of public opinion analysis, consumer intention identification, and e-commerce commentary analysis. It can also provide quantitative and scientific decisions for government departments and enterprises [1].

Social media sentiment classification has always been a hotspot and difficult problem in natural language processing and artificial intelligence [19]. As we know, sentiment expression is domain-dependent, and different domains have different distributions. For example, "薄" (thin) expresses negative sentiment in hotel domain, while it expresses positive sentiment in notebook domain. Therefore, the classifier trained on the source domain may not be well adapted to the target domain. Deep neural networks (DNN) have achieved excellent results on sentiment classification tasks, but it requires massive training data, otherwise it is easy to over-fit [6]. Unfortunately, to collect and label massive domain-related samples require considerable time and efforts. Meanwhile, we have accumulated rich and fine-labeled data in traditional sentiment classification tasks, it is also extremely wasteful to discard the data completely. The goal of transfer learning is to learn the knowledge learned from the source domain to aid learning tasks about the target domain. It can take advantages of the commonality between different learning tasks to share the benefits of statistics and migration knowledge among tasks [17].

Deep transfer learning (DTL) approaches transfer deep neural networks which are trained on source domain to special target domain. It turns out to be successful in image recognition and natural language processing tasks [14]. Previous studies have proved that bottom layers can learn basic generic features, while top layers can learn data-specific and advanced features representation [4]. In other words, the features computed in higher layers of the network must depend greatly on the specific data set and tasks. In the context of deep learning, fine-tuning a deep network that pre-trained on the source domain data is a common strategy to learn task-specific features. The pre-training and fine-tuning strategies can be trained using existing data sets and adapted to target domain. In detail, it transfers bottom-layer (general) features and retrains (specific) top-layer features from the source domain to target domain.

In this paper, we propose a two-stage bidirectional long short-term memory (Bi-LSTM) and parameters transfer framework for short texts cross-domain sentiment classification tasks. There are two main advantages of our deep transfer learning framework: one is the powerful ability to capture variable length and n-gram context semantics of Bi-LSTM networks, the other is the ability to transfer knowledge from the source domain to target domain data with fine-tuning strategy. Firstly, we pre-train Bi-LSTM networks using a large number of fine-labeled source domain training samples. Then the Bi-LSTM networks are fine-tuned with limited target domain training data. In the parameters transfer process, bottom layers parameters are fine-tuned, and softmax layers parameters are retrained. Experimental results on four Chinese sentiment classification data sets show that our proposed method performs better than previous methods.

Our contributions in this paper can be summarized as follows.

- We introduce a novel Bi-LSTM and parameters transfer framework for cross-domain sentiment classification tasks. This framework can learn long-term dependence, word sequence semantic information and transfer knowledge from the source domain to target domain.
- We share bottom layers of Bi-LSTM networks and retrain top layers using a slight number of target domain training samples. This improves the effectiveness of cross-domain sentiment classification and generalization capabilities.
- Experiments demonstrate that our parameters transfer and fine-tuning schemes achieve state-of-the-art performance on Chinese short texts cross-domain classification tasks via deep transfer learning.

2   Deep Transfer Framework

In this section, we firstly introduce basic notations and problem formulation. Then we describe a deep transfer learning framework for cross-domain sentiment classification tasks in detail. Bidirectional LSTM networks are pre-trained on massive source domain training samples. Then pre-trained model parameters are transferred and fine-tuned with limited target domain data.

2.1   Notations and Problem Formulation

For a formal description of cross-domain sentiment classification tasks, $\mathcal{X} = \mathcal{R}$ denotes the instance space, $x = (x_1, x_2, \cdots, x_T)$ consists of a series of words x_i, $x \in \mathcal{X}$. $\mathcal{Y}$ is the label space for sentiment classification tasks, and $\mathcal{Y}_1 = \{very\ positive, positive, neutral, negative, very\ negative\}$ is the fine-grained sentiment classification label set, $\mathcal{Y}_2 = \{positive, negative\}$ is the binary sentiment classification label set. In this paper, x_i is a word2vec distributed representation, x_i is a d-dimensional feature vector, i.e., $x_i = (x_i^1, x_i^2, \cdots, x_i^d)$. For each instance (x, y), y is the sentiment label with x, $y \in \mathcal{Y}$.

$\mathcal{D}^S = \{(x_{s1}, y_{s1}), (x_{s2}, y_{s2}), \cdots, (x_{sm}, y_{sm})\}$ is the source domain training data set, the label space is $\{y_{s1}, y_{s2}, \cdots, y_{sm}\}$. The marginal probability distribution of source domain is $P_S(X)$. $\mathcal{D}^L = \{x_i, Y_i | 1 \leq i \leq n\}$ represents the target domain training set, $\mathcal{D}^U = \{x_i, Y_i | 1 \leq i \leq p\}$ represents the target domain testing set, $\mathcal{D}^T = \mathcal{D}^L \cup \mathcal{D}^U$ is the target domain data set. The distribution of target domain $P_T(X)$ is often different from $P_S(X)$.

There are two main transfer learning tasks: (i) transfer across domains: the data distributions between two domains are different, i.e., $P_S(X) \neq P_T(X)$, while the tasks are the same, i.e., $\mathcal{Y}^S = \mathcal{Y}^T$; (ii) transfer across tasks: both data distributions and tasks are different, i.e., $P_S(X) \neq P_T(X)$, $\mathcal{Y}^S \neq \mathcal{Y}^T$. In this paper, we verify our proposed framework on the above two tasks. The task of deep transfer learning can be formalized as follows: firstly, we learn pre-trained neural networks $f_S : \mathcal{D}^S \rightarrow \mathcal{Y}^S$, then transfer the neural networks $f_S \rightarrow f_D$ with fine-tuning the parameters weight of bottom layers and retraining the top layers on $\mathcal{D}^L$.

2.2   Bidirectional LSTM Pre-training

For sentiment classification tasks, Bi-LSTM (actually using forward and backward LSTM) can capture variable length and bidirectional n-gram context information [18]. Background topics and sentiment indicators of social media texts could be far away from the target aspect. The traditional bag-of-words based machine learning methods could not distinguish the implicit or hidden dependency in long conversations. However, the memory cell in LSTM can settle long distance dependency problems. The sequence of words in a sentence plays an important role in sentiment expression. Such as (I am very upset today.) and (I am not very happy today.) express different sentiment intensities. Compared with convolution neural networks, Bi-LSTM focuses on the reconstruction of the adjacent position, so it is more suitable for the sequence structure of language modeling.

Although Bi-LSTM model has achieved good results in sentiment classification tasks, it needs large number of related training samples, otherwise it is very prone to over-fit. However, to collect and annotate a large-scale domain-related data set require considerable time and efforts. Existing sentiment classification tasks have accumulated a large number of fine-labeled sentiment classification data [16]. An intuitive idea is to use these source domain data to assist target domain sentiment classification tasks. Bi-LSTM networks are firstly pre-trained on source domain data and then parameters are transferred into target domain.

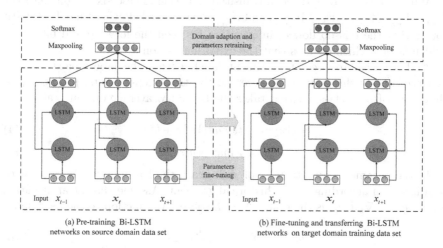

(a) Pre-training Bi-LSTM
networks on source domain data set

(b) Fine-tuning and transferring Bi-LSTM
networks on target domain training data set

Fig. 1. Flow chart of Bi-LSTM networks pre-training and fine-tuning processes. This framework can be divided into two parts: (a) Pre-training Bi-LSTM networks on source domain data set. (b) Fine-tuning Bi-LSTM networks and transferring parameters on target domain training data set. The bottom layers parameters weight are transformed to target domain, while top layers are randomized and adapted to target domain

Figure 1(a) shows the process of six layers Bi-LSTM networks pre-training on $\mathcal{D}^S$. We treat each word x_i as a time node. The input units are a sequence of words $x = (x_1, x_2, \cdots, x_T)$, $x \in \mathcal{D}^S$. Then the word sequence layer is entered into the forward hidden sequence $\overrightarrow{h}$ and the backward hidden sequence $\overleftarrow{h}$.

A LSTM memory cell consists of a memory cell c_t, an input gate i_t, a forget gate f_t, and an output gate o_t. The input gate (current cell matters) can be formalized as: $i_t = \sigma(W^{xi}x_t + W^{hi}h_{t-1} + W^{ci}c_{t-1} + b^i)$. Forget gate (gate 0, forget past): $f_t = \sigma(W^{xf}x_t + W^{hf}h_{t-1} + W^{cf}c_{t-1} + b^f)$. Output gate (how much cell is exposed): $o_t = \sigma(W^{xo}x_t + W^{ho}h_{t-1} + W^{co}c_{t-1} + b^o)$. New memory cell: $c_t = f_t \odot c_{t-1} + i_t \odot \tanh(W^{hc}x_t + W^{hc}h_{t-1} + bc)$. A d-dimensional hidden state: $h_t = o_t \odot \tanh(c_t)$. where $x = (x_1, x_2, \cdots, x_T)$ is the input feature sequence, σ is the logistic function. The symbol $\odot$ represents the element-wise operation. W is the weight matrix and the superscript indicates the matrix between two different gates.

Bi-LSTM networks compute the forward layer as the forward hidden sequence $\overrightarrow{h}$ from $t = 1$ to T, the backward layer as backward hidden sequence $\overleftarrow{h}$ by iterating from $t = T$ to 1, and the output layer y as the output sequence $y = (y_1, y_2, \cdots, y_T)$.

$$\overrightarrow{h}_t = H(W_{x\overrightarrow{h}}x_t + W_{\overrightarrow{h}\overrightarrow{h}}\overrightarrow{h}_{t-1} + b_{\overrightarrow{h}}) \tag{1}$$

$$\overleftarrow{h}_t = H(W_{x\overleftarrow{h}}x_t + W_{\overleftarrow{h}\overleftarrow{h}}\overleftarrow{h}_{t+1} + b_{\overleftarrow{h}}) \tag{2}$$

$$y_t = W_{\overrightarrow{h}y}\overrightarrow{h}_t + W_{\overleftarrow{h}y}\overleftarrow{h}_t + b_y \tag{3}$$

Where H is the LSTM block transition function. These six weight matrices $W_{x\overrightarrow{h}}, W_{x\overleftarrow{h}}, W_{\overrightarrow{h}\overrightarrow{h}}, W_{\overrightarrow{h}y}, W_{\overleftarrow{h}\overleftarrow{h}}$, and $W_{\overleftarrow{h}y}$ are repeated at every time. It is worth noting that there is no flow of information between the forward and backward hidden layers, which ensures that the expansion is non-cyclic.

And we wish to predict sentiment label y from the label space $\mathcal{Y}$. $z = (Max(y_i))_{i=1}^T$ is the maxpooling over all the time steps results. $p(y|x)$ is predicted by softmax classifier that takes the Bi-LSTM average output z as input:

$$p(y|x) = softmax(W^z z + b^z) \tag{4}$$

$$y = \arg\max p(y|x) \tag{5}$$

In the parameter update rules of **Adagrad**, the learning rate η varies with each iteration according to the historical gradient. Assume that at an iteration time t, $g_{t,i} = \nabla_\theta J(\theta_i)$ is the gradient of the objective function to the parameter.

$$\theta_{t+1,i} = \theta_{t,i} - \frac{\eta}{\sqrt{G_t + \varepsilon}} \cdot g_{t,i} \tag{6}$$

Where $G_t \in R^{d \times d}$ is a diagonal matrix, $\varepsilon = e^{-8}$ is a smoothing item to prevent G_t from being equal to 0.

We use mean squared logarithmic error (MSLE) as the loss function:

$$\varepsilon = \frac{1}{n} \sum_{i=1}^n (\log(Y + 1) - \log(y + 1))^2 \tag{7}$$

Where Y represents the true label of x, and y is the prediction label.

2.3   Fine-Tuning and Parameters Transfer

Motivation: As we all know, sentiment classification is a domain-dependent issue. The Bi-LSTM model that trained on source domain may not be necessarily well suited to target domain. The distributions between the source and target domains may be not precisely the same. In this case, to label target domain training samples is time-consuming and laborious. Besides this, the amount of training data is not normally adequate for retraining new neural networks. On account of this, the well-trained Bi-LSTM model requires a domain-adaption process. Therefore, fine-tuning and parameters transfer are just an ideal choice. Previous experiments have verified that fine-tuning performs better than the model which only trains on limited target domain samples.

Transfer Bottom Layers: Figure 1(b) shows the domain adaptation and parameters transfer processes. We pre-train Bi-LSTM networks with a low initial learning rate η and high dropout rate on $\mathcal{D}^S$. The bottom layers parameters weight of pre-trained Bi-LSTM networks W^S are $W_{x\overrightarrow{h}}$, $W_{\overrightarrow{h}\overrightarrow{h}}$, $b_{\overrightarrow{h}}$, $W_{x\overleftarrow{h}}$, $W_{\overleftarrow{h}\overleftarrow{h}}$, $b_{\overleftarrow{h}}$, $W_{\overrightarrow{h}y}$, $W_{\overleftarrow{h}y}$, and b_y. We use target domain training data $\mathcal{D}^L$ as fine-tuning source data. Then W_S is fine-tuned with a high initial learning rate η and low dropout rate from $\mathcal{D}^L$ by back propagation algorithm. We use layer-by-layer feature transference to transfer bottom layers parameters weight W_S. This is motivated by the observation that the general features of Bi-LSTM networks contain more generic features that should be useful to target domain. We do not wish to distort them too quickly or too much, so we keep learning rate low and dropout rate decay really high.

Retrain Top Layers: It is possible to fine-tune some of earlier layers fixed (due to over-fitting concerns) and retrain some higher-level portion of the networks. The later layers of the Bi-LSTM become more specific to the details of the classes contained in the target domain data set. The top-layer features depend greatly on the chosen special data set and tasks, so called as specific features. The full connection layer (softmax classifier) of transferred Bi-LSTM networks is replaced and retrained. The softmax layer parameters weight W^z and b^z are initialized randomly, and then retrained on target domain training data set $\mathcal{D}^L$. We remove the output layer, and then use the entire network as a fixed feature extractor for target domain data set. Therefore, our framework can be applied into transfer across domains and transfer across tasks problems. These pre-trained networks demonstrate a strong ability to generalize to new data set via transfer learning.

3   Experiment and Analysis

3.1   Data Sets and Experiment Setup

We use four Chinese social media sentiment classification data sets to validate our deep transfer learning framework. Hotel (H) and Notebook (N) data sets are collected from Jingdong shopping website (https://www.jd.com/). Weibo (W) data set is collected from COAE 2015 (https://www.ccir2015.com/). Fine-grained data set electronic (E) including 8000 samples is collected from COAE 2011 task 3 (https://www.ccir2011.com/). The detail of four data sets can be seen in Table 1.

Table 1. The detail of four sentiment classification data sets

Data set	Very positive	Positive	Neutral	Negative	Very negative
Hotel (H)	*	2000	*	2000	*
Notebook (N)	*	2000	*	2000	*
Weibo (W)	*	5000	*	5000	*
Electronic (E)	801	453	1139	2295	3311

We use THULAC tool (http://thulac.thunlp.org/) to get the word segmentation. After this, we use Glove vectors of 100 dimension to train the distributed word vector [12] with all source domain and target domain texts. We use 5-fold cross validation method to extract 20% target domain randomly as the target domain training data, the rest composes the target domain testing data. Back-propagation through time (BPTT) method with AdaGrad initial learning rate of 0.5 and dropout rate 0.7 on source domain, initial learning rate of 0.8 and dropout rate 0.3 on target domain, epoch number as 5, hidden layer units as 64, and mini-batch size of 20 are used to train our model. Our model is implemented by Keras deep learning library (https://keras.io/). We utilize accuracy to evaluate the baselines and our proposed framework.

3.2 Baselines and Our Framework

(1) **Active learning:** an instance-based transfer method with active learning for cross-domain sentiment classification which was proposed by Li et al. [10]. We follow original settings as bag-of-words and binary vectors representation, maximum entropy classifier, and 20% target domain data as the initial labeled data.

(2) **Multi-instance:** a hybrid strategy which combined transfer learning, deep learning and multi-instance learning which was proposed by Dimitrios et al. [9]. We use 3 epochs, mini-batch size of 50, objective function of SGD iterations with 1050 iterations and a learning rate of $\alpha = 0.0001$.

(3) **BLPT:** our proposed Bi-LSTM networks and parameters transfer method.

Three strategies are used to evaluate our framework and shown as follows:

BLPT-random: BLPT method with randomly initialized vectors;

BLPT-fixed: BLPT method with fixed word vectors which are trained by Glove method;

BLPT-tuned: BLPT method with Glove word vectors and updated in the training process.

3.3 Experimental Results

Performance with Different Parameters: We compare **BLPT-tuned** performances with different parameters, "scale of source domain training data set", "dimension of Glove word embeddings", "dropout rate", and "epoch" respectively on H→N, H→W, and H→E tasks. Figure 2a and b show the accuracy performances with different scales of source domain and word embeedings dimension under fixed dropout rate as 0.7 and epoch number as 5. We can find that more source domain training data generally performs better. The accuracy grows when the dimension of Glove word embeddings changes from 20 to 100, while the impact of the dimension on the results is not particularly obvious when dimension changes from 100 to 200. We fix the scale of source domain as 100%, word embeddings dimension as 100, and compare the impact of dropout rate and epoch of Bi-LSTM model in Fig. 2c and d. Dropout can prevent the neural network overfitting effectively, and we find that increasing dropout rate does not lead to significant improvements. A good classification performance is achieved when the epoch is 5 or 6 in Bi-LSTM networks, and our model may be over-fitting when the epoch is larger than 7.

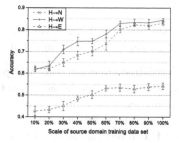

(a) Performance with different scales of source domain training data

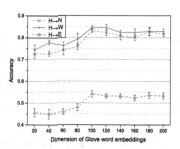

(b) Performance with different dimension of Glove word embeddings

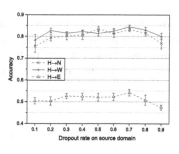

(c) Performance with different dropout rates

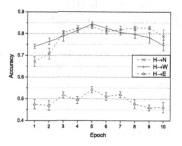

(d) Performance with different epochs

Fig. 2. The performance of transferred Bi-LSTM model trained with different source domain training scales, word embeddings dimension, dropout rates, and epoch sizes

Table 2. Mean accuracy ± standard deviation (%) results of 12 cross-domain sentiment classification tasks

Task	Active learning	Multi-instance	BLPT-random	BLPT-fixed	BLPT-tuned
H→N	80.8±0.5	82.1±0.1	79.8±0.5	82.3±0.4	**83.1±0.9**
H→W	80.6±0.8	82.8±0.5	81.1±0.7	83.0±0.5	**84.4±0.8**
H→E	51.3±0.7	53.2±0.7	51.3±1.4	53.1±1.2	**54.2±1.5**
N→H	81.7±0.6	83.1±0.6	80.1±1.2	83.7±0.3	**85.7±0.6**
N→W	80.4±0.4	82.0±0.4	81.8±0.4	82.0±0.8	**84.8±0.5**
N→E	51.0±0.9	52.1±0.8	51.8±0.4	54.8±0.4	**55.3±0.5**
W→H	82.1±0.4	83.6±0.9	81.8±1.3	82.1±0.5	**84.3±0.4**
W→N	80.9±0.6	82.3±0.5	82.0±0.6	84.7±0.5	**85.9±1.3**
W→E	56.8±0.5	55.8±0.7	54.5±0.8	57.8±1.3	**59.2±0.7**
E→H	82.3±1.2	81.8±0.8	82.1±0.7	84.4±0.4	**84.9±1.2**
E→N	82.0±0.8	82.4±1.3	82.0±0.4	83.2±1.2	**85.0±0.7**
E→W	81.5±1.1	82.1±0.4	81.1±0.8	83.0±0.4	**83.5±0.5**
Average	74.3	75.3	74.1	76.2	**77.5**

Comparing Results: Table 2 gives the mean accuracy of 12 cross-domain sentiment classification tasks on four data sets. From Table 2, we can find that:

(i) Comparing with **Active learning** and **Multi-instance** methods, our proposed framework **BLPT-tuned** generally performs better. This proves the excellent feature presentation ability and good generalization of transferred Bi-LSTM for short texts cross-domain sentiment classification tasks.

(ii) In contrast with **BLPT-random** and **BLPT-fixed** methods, our transformed Bi-LSTM networks through tuned word embeedings improve 3.4% and 1.3% respectively. The word embeddings are updated in the supervised learning process, so its semantics is more clear and the classification performance is better.

(iii) Our work can be readily adapted into transfer across domains and transfer across tasks problems. Comparing with binary sentiment classification, fine-grained sentiment classification is a more detail and difficult task. The accuracies of H→E, N→E, and W→E tasks are significantly lower than other tasks.

3.4   Discussions

(1) Deep transfer learning can obtain good performance with abundant source domain data and limited target domain data. Neural networks have achieved excellent results and need large scale training data to train the parameters weight. It is relatively rare to have a data set of sufficient size which is required for the depth of networks. In real applications, source domain

data set is relatively large in size and similar in content compared to target domain data set. Deep transfer learning depends on the scale of source domain and target domain training data, and similarity degree between source domain and target domain. It can bring feature representation of deep neural networks to a new domain. Since we have limited target domain training data, we can fine-tune the full network that trained on source domain. It is common to pre-train Bi-LSTM networks on a very large data set and then use trained parameters weights either as an initialization or a fixed feature extractor for the task of interest.

(2) Deep transfer learning including parameters transfer and fine-tuning strategies helps the training process for the target domain better. We fine-tune some of the earlier layers under lower learning rate, and retrain some higher-level portion of the networks. This is motivated by the observation that earlier features of Bi-LSTM contain more generic features, while the fully connected softmax layer becomes progressively more specific to particular data set. We can share pre-trained parameters to a new model to speed and optimize model learning to avoid learning from scratch and time-consuming training. Fine-tuning enables us to bring the power of pre-trained models to target domain with insufficient data. This can effectively exploit powerful generalization capabilities of deep neural networks, and eliminate the need to redesign complex models. Our approach solves the problem of over-fitting of deep neural model such as Bi-LSTM networks on limited samples. Besides this, our approach achieves a significant improvement of average accuracy and generalization across domains.

4   Related Work

4.1   Cross-Domain Sentiment Classification

There has been a lot of work on the issue of cross-domain sentiment classification tasks. Researchers have gradually begun to use transfer learning (TL) techniques to solve cross-domain sentiment classification tasks. The existing work can be divided into four parts: instance-based, feature-based, parameter-based, and relational-based [10]. For instance-based transfer, previous studies mainly focus on selecting valuable samples from source domain which can be used to assist the target domain sentiment classification. Feature-based transfer is to find the correlation features (shared features) between source domain and target domain, and to construct the unified feature representation space of cross-domain data. Parameter-based methods discover shared parameters or priors between the source domain and target domain models, which can benefit for transfer learning. Relational-based methods build mapping of relational knowledge from different domains. Tan et al. [15] attempted to tackle domain-transfer problem by combining source domain labeled examples with target domain unlabeled ones. The basic idea was to use source domain trained classifier to label some informative unlabeled examples in the new domain, and retrain the base classifier over these selected examples. Spectral feature alignment (SFA) was presented by

Pan et al. [13] to discover a robust representation for cross-domain data by fully exploiting the relationship between the domain-specific and domain-independent words via simultaneously co-clustering them in a common latent space. Li et al. [10] performed active learning for cross-domain sentiment classification by actively selecting a small amount of labeled data in the target domain.

4.2 Deep Transfer Learning

Deep transfer learning (DTL) focuses on adapting knowledge from an auxiliary source domain to a target domain with little or without any label information to construct neural networks model of good generalization performance [9]. It is an approach in which a deep model is trained on a source problem, and then reused to solve a target problem [5]. DTL usually trains deep neural networks in source domain, and transfer and fine-tune the parameters weight to the target domain. This strategy has been proved to improve cross-domain classification results effectively [11]. A source-target selective joint fine-tuning scheme was introduced by Ge et al. [2] for improving the performance of deep learning tasks with insufficient training data. Dimitrios et al. [9] combined transfer learning, deep learning and multi-instance learning, and reduced the need for laborious human labelling of fine-grained data when abundant labels were available at the group level. Chetak et al. [8] proposed a ensemble methodology to reduce the impact of selective layer based transference and provide optimized framework to work for three major transfer learning cases. Xavier et al. [3] studied the problem of domain adaptation for sentiment classifiers, whereby a system was trained on labeled reviews from one source domain but was meant to be deployed on another. Then a meaningful representation for each review was extracted in an unsupervised fashion.

5 Conclusions and Future Work

In this paper, we propose a deep transfer learning framework for Chinese short texts cross-domain sentiment classification tasks. Our work takes advantages of transfer learning, deep neural networks, and fine-tuning strategies. Firstly bidirectional LSTM networks are pre-trained on source domain data. Then we use transfer learning strategy to transfer and fine-tune Bi-LSTM networks on target domain training samples. We use extra massive source domain training data to enhance the performance of current learning task, including generalization accuracy, learning efficiency and comprehensibility. Experiments on four data sets show that our pre-training and fine-tuning schemes achieve better performances than previous methods. In the future, we intend to use multiple source domains training data and ensemble the final results. We will also consider attention-based RNN models for further improving the sequential representation of short texts.

Acknowledgments. This work was supported by: National Natural Science Foundation of China (61573231, 61632011, 61672331, 61432011); Shanxi Province Graduate Student Education Innovation Project (2016BY004, 2017BY004).

References

1. Zhao, C., Wang, S., Li, D.: Fuzzy sentiment membership determining for sentiment classification. In: Proceedings of ICDMW 2014, pp. 1191–1198. IEEE (2014)
2. Ge, W., Yu, Y.: Borrowing treasures from the wealthy: deep transfer learning through selective joint fine-tuning. arXiv preprint arXiv:1702.08690 (2017)
3. Glorot, X., Bordes, A., Bengio, Y.: Domain adaptation for large-scale sentiment classification: a deep learning approach. Proc. ICML **2011**, 513–520 (2011)
4. Guan, L., Zhang, Y., Zhu, J.: Segmenting and characterizing adopters of e-books and paper books based on Amazon book reviews. In: Li, Y., Xiang, G., Lin, H., Wang, M. (eds.) SMP 2016. CCIS, vol. 669, pp. 85–97. Springer, Singapore (2016). doi:10.1007/978-981-10-2993-6_7
5. Haaren, J.V., Kolobov, A., Davis, J.: Todtler: two-order-deep transfer learning. In: Proceedings of AAAI 2015 on Artificial Intelligence, pp. 3007–3015 (2015)
6. Huang, M., Cao, Y., Dong, C.: Modeling rich contexts for sentiment classification with LSTM. arXiv preprint arXiv:1605.01478 (2016)
7. Wang, S., Li, D., Zhao, L., Zhang, J.: Sample cutting method for imbalanced text sentiment classification based on BRC. Knowl. Based Syst. **37**, 451–461 (2013)
8. Kandaswamy, C., Silva, L.M., Alexandre, L.A., Santos, J.M.: Deep transfer learning ensemble for classification. In: Rojas, I., Joya, G., Catala, A. (eds.) IWANN 2015. LNCS, vol. 9094, pp. 335–348. Springer, Cham (2015). doi:10.1007/978-3-319-19258-1_29
9. Kotzias, D., Denil, M., Blunsom, P., de Freitas, N.: Deep multi-instance transfer learning. arXiv preprint arXiv:1411.3128 (2014)
10. Li, S., Xue, Y., Wang, Z., Zhou, G.: Active learning for cross-domain sentiment classification. Proc. IJCAI **2013**, 2127–2133 (2013)
11. Long, M., Cao, Y., Wang, J., Jordan, M.I.: Learning transferable features with deep adaptation networks. Proc. ICML **2015**, 97–105 (2015)
12. Mikolov, T., Sutskever, I., Chen, K., Corrado, G.S., Dean, J.: Distributed representations of words and phrases and their compositionality. Proc. NIPS **2013**, 3111–3119 (2013)
13. Pan, S.J., Ni, X., Sun, J.T., Yang, Q., Chen, Z.: Cross-domain sentiment classification via spectral feature alignment. In: Proceedings of WWW 2010, pp. 751–760. ACM (2010)
14. Papernot, N., Abadi, M., Erlingsson, U., Goodfellow, I., Talwar, K.: Semi-supervised knowledge transfer for deep learning from private training data. arXiv preprint arXiv:1610.05755 (2016)
15. Tan, S., Wu, G., Tang, H., Cheng, X.: A novel scheme for domain-transfer problem in the context of sentiment analysis. In: Proceedings of ACM 2007, pp. 979–982. ACM (2007)
16. Tang, D., Qin, B., Feng, X., Liu, T.: Target-dependent sentiment classification with long short term memory. arXiv preprint arXiv:1512.01100 (2015)
17. Tang, J., Lou, T., Kleinberg, J., Wu, S.: Transfer learning to infer social ties across heterogeneous networks. ACM Trans. Inf. Syst. (TOIS) **34**(2), 7 (2016)
18. Wang, J., Yu, L.C., Lai, K.R., Zhang, X.: Dimensional sentiment analysis using a regional CNN-LSTM model. Proc. ACL **2016**, 225–230 (2016)
19. Wang, S., Li, D., Song, X., et al.: A feature selection method based on improved fisher's discriminant ratio for text sentiment classification. Expert Syst. Appl. **38**(7), 8696–8702 (2011)

Local Contexts Are Effective for Neural Aspect Extraction

Jianhua Yuan, Yanyan Zhao, Bing Qin$^{(\boxtimes)}$, and Ting Liu

Research Center for Social Computing and Information Retrieval,
Harbin Institute of Technology, Harbin, China
{jhyuan,yyzhao,qinb,tliu}@ir.hit.edu.cn

Abstract. Recently, long short-term memory based recurrent neural network (LSTM-RNN), which is capable of capturing long dependencies over sequence, obtained state-of-the-art performance on aspect extraction. In this work, we would like to investigate to which extent could we achieve if we only take into account of the local dependencies. To this end, we develop a simple feed-forward neural network which takes a window of context words surrounding the aspect to be processed. Surprisingly, we find that a purely window-based neural network obtain comparable performance with a LSTM-RNN approach, which reveals the importance of local contexts for aspect extraction. Furthermore, we introduce a simple and natural way to leverage local contexts and global contexts together, which is not only computationally cheaper than existing LSTM-RNN approach, but also gets higher classification accuracy.

Keywords: Local contexts · Aspect extraction · Sentiment analysis

1 Introduction

Fine-grained opinion extraction involves detecting opinion holder who conveys the opinion, identifying opinion expressions and deciding their polarity and intensity, and extracting opinion aspects towards which the opinion holder expresses the opinion expression [1]. Fine-grained opinion extraction has been studied extensively recently for its benefits to a number of NLP tasks which include opinion-oriented QA and opinion summarization. In this work, we focus on fine-grained opinion aspect extraction.

The task of fine-grained aspect extraction is usually tackled as a sequence labeling problem, where each word (token) in the input sentence is assigned a label using the conventional BIO tagging schemes: B indicates the beginning of an opinion aspect, I is used for tokens inside the same opinion aspect and O stands for tokens outside any opinion-related class. Table 1 shows a sentence tagged with BIO scheme for opinion aspect extraction task.

Conditional Random Fields (CRF) and its variants have successfully applied to various opinion extraction tasks, e.g., opinion expression extraction [2]. And the state of the art models for opinion aspect extraction are also based on CRF [3, 4].

© Springer Nature Singapore Pte Ltd. 2017
X. Cheng et al. (Eds.): SMP 2017, CCIS 774, pp. 244–255, 2017.
https://doi.org/10.1007/978-981-10-6805-8_20

Table 1. A example sentence tagged with BIO labels for opinion aspect.

The	set	up	was	very	easy
O	B	I	O	O	O

However, the success of CRF models critically relies on the appropriate feature sets designed by experts which usually requires lots of feature engineering effort for each specific task.

Deep learning models that automatically learns latent features have recently achieved comparable results to CRF models even with no manual feature. For example, Irsoy and Cardie [5] apply deep recurrent neural networks (RNNs) to opinion expression extraction tasks, which then shows that deep RNNs outperform conventional CRFs. Also, Liu et al. [6] employ long short term memory networks (LSTM) to extract opinion aspects from Laptop and Restaurant reviews [7], which excels CRF baselines even when they take pre-trained word embeddings as the only feature. However, since LSTM networks are designed to model long term dependencies, it fails to leverage sufficient local context that benefits extracting opinion aspect which are usually short phrases.

Motivated by the recent success of deep learning models (LSTMs in particular) and observation that local context for neural aspect extraction, in this paper we propose new variants of LSTMs which utilize local context information. Specifically, we explore two ways of combining long term dependency and local context: one is feeding the word context window [8] directly into the input layers of LSTM which then jointly learns local context representations and long term dependency, and the other is employing a separate feed forward neural network to learn local context information which is then combined with the long term dependency learned by LSTM to form the representation of current word.

In the rest of this paper, we discuss related work in Sect. 2 and introduce our novel LSTM models in Sect. 3. We present our results and discussions in Sect. 4. Finally, we make a summary of this paper and propose some ideas for future work.

2   Related Work

This work is related to two different areas of NLP research, namely opinion mining and deep learning. And each is given a brief account of previous works due to the space constraints.

Opinion Mining. Various approaches has been applied to the recognition of fine-grained opinion elements in previous work. One branch of existing work [9–12] exploits syntactic relations in opinion extraction. For example, Qiu et al. [12] treat syntactic relationship as a crucial rule and apply double propagation method to iteratively augment the sets of opinion aspects and opinion terms. Moreover, this problem has also been tackled as a sequence labeling problem

[13–15] and many [16–18] take CRF-based methods. Li et al. [14] propose a new CRF based model to jointly extract opinion aspects and opinion words.

Deep Learning. Due to the naturally deep architecture to preserve information from the past, recurrent neural networks (RNN), long short term memory networks (LSTM) and its variants have been successfully applied to many sequential tasks, such as spoken language understanding [8], language modeling [19] and speech recognition [20]. To get information from both the past and the future, later work propose the bidirectional RNN [21] and bidirectional LSTM model. What's more, Socher et al. [22] apply recursive neural networks to hierarchically compose semantic vectors based on syntactic parsing tree and further use these vectors for sentiment classification of phrases and sentences. Recently, Yin et al. [3] and Wang et al. [4] propose joint models that first use deep learning models to learn latent syntactic features and then feed these features into a CRF labeler.

To our knowledge, the most relevant to our work are work of Mesnil et al. [8] where RNN models with a word context window are proposed to capture local context for slot filling task and of Liu et al. [6] which applies LSTM networks to extracting opinion aspects from review datasets. Our work differs from the work of Mesnil et al. [8] and Liu et al. [6] in three ways. (i) We experiment only on variants of LSTM models to avoid the vanishing gradient problems [23]. (ii) We experiment with two different ways of capturing local context within LSTM architecture. (iii) We present with a comprehensive experiment exploring the optimal settings for learning joint representations of both local and long term context in neural aspect extraction task.

3 LSTM Models

In this section, we first describe the properties those models have in common. Following subsections are dedicated to the description of common LSTM networks for modeling long term context, simple word context window for capturing local context and 2 ways to integrate these two context.

Each word is represented by a d-dimensional vector in the shared lookup table $L \in \mathbb{R}^{|V|?D}$, where $|V|$ is the total size of the vocabulary. L can be either treated as parameter randomly initialized and then jointly trained with other parameters or be initialized by pre-trained word embeddings and then be fine-tuned for specific task during model training. Given an input sentence $S = (w_1, w_2, ..., w_T)$, we first transform it into sequence of index in L for each token $w_t \in S$. Then the lookup layer create a vector $X_t \in \mathbb{R}^{m?D}$ for each token w_t in S (see Sects. 3.2 and 3.3). The generated vector is passed to the hidden layer after linear transformation. The non-linear functions in hidden layer produce a more abstract representation of the given token and feed it into $softmax$ layer that finally determines the BIO label for the input token. The probability of k-th label can be formalized as:

$$softmax \leftarrow P(y_t = k|s, \theta) = \frac{exp(w_k^T h_t)}{\sum_{k=1}^{K} exp(w_k^T h_t)} \qquad (1)$$

where, K is the class number (3 for this task), and $h_t = \phi(x_t)$ is the transformation of x_t through the non-linear function, and w_k are weights in the *softmax* layer. Models are trained by minimizing the negative log likelihood (NLL) of the training data. The NLL for an input sentence can be written as:

$$J(\theta) = \sum_{t=1}^{T} \sum_{k=1}^{K} y_{tk} log P(y_t = k|s, \theta) \qquad (2)$$

where, $y_{tk} = 1$ if $y_t = k$ and $y_{tk} = 0$ if $y_t \neq k$. The loss function minimizing the cross-entropy between the predicted labels and gold labels.

All the following models first obtain high level abstractions of input tokens in a bottom up fashion and then feed them into succeeding *softmax* layer for classification. The main difference between those models is how they construct their input and the way they compute the abstract representations.

3.1   Global Context

Neural networks with recurrent connections have been widely used to compute high level representations for sentence with variant length. In a standard Elman-type RNN, the output from the hidden layer at time step t is computed from a nonlinear transformation of the current input vector x_t and the previous hidden state h_{t-1}. The mathematical form of h_t can be written as:

$$h_t = f(Ux_t + Vh_{t-1} + b) \qquad (3)$$

where f is a nonlinear function (e.g. sigmoid function), U and V are the weight matrices between the input and hidden layer and between two consecutive hidden layers respectively, and b is the bias vector connected the hidden units. h_0 is randomized for the base case.

However, vanishing gradient problem [23] has limited the ability RNN has for capturing long term dependencies. To address this problem, a Long Short-Term Memory(LSTM) architecture with purpose-designed hidden units called memory blocks has been proposed to model long range context. In a standard LSTM, a memory block is made up of four components: (1) a memory cell conveying the state (2) an input gate i to control the values to update (3) a forget gate f to decide the portion of the current state to be forget (4) an output gate o to produce a filtered output for other neurons. For each time step t, a layer of a memory block can be described by the following sequence of equations:

$$i_t = \sigma(U_i x_t + V_i h_{t-1} + C_i c_{t-1} + b_i) \qquad (4)$$

$$f_t = \sigma(U_f x_t + V_f h_{t-1} + C_f c_{t-1} + b_f) \qquad (5)$$

$$c_t = i_t \odot g(U_c c_t + V_c h_{t-1} + b_c) + f_t \odot c_{t-1} \qquad (6)$$

$$o_t = \sigma(U_o x_t + V_o h_{t-1} + C_o C_t + b_o) \qquad (7)$$

$$h_t = o_t \odot h(c_t) \qquad (8)$$

where U, V, C are weight matrices between the input and hidden layers, between two consecutive hidden layers, and between two consecutive cell activations respectively, and b is the corresponding bias vector. Moreover, σ is the *sigmoid* function, g and h are the cell input and cell output activations, usually a *tanh* function, and $\odot$ denotes the element-wise product (i.e. Hadamard Product) of two vectors. All of our LSTM models are trained with full BPTT.

Yet uni-directional LSTM models can only acquire history information while neglecting the future clues. Given the example sentence: *The set up is very easy*, to correctly assign *set* with a **B**, the next word *up* turns out to be very crucial. And bi-directional LSTMs with forward and backward links in hidden layers have been proposed to solve this problem. The backward pass has a counterpart for each of Eqs. from (4) to (8).

Note that the forward and backward pass are done independently until their computation results are concatenated in the output layer. This means, during training, after backpropagating error from output layer to forward hidden layer and backward hidden layer separately, two independent BPTT can be applied to two directions respectively.

3.2   Local Context

Since opinion aspects are usually short phrases, not taking local context into consideration may hinder the performance of sequence labeler. A straightforward way to get local context information of the current word is constructing a word context window [8] made up of current word and w words from its left side and right side respectively, where $(2 * w + 1)$ is the window size. The ordered concatenation of corresponding embeddings of words in the context window can be used as local context of current word. We give an illustration of constructing a word context window of size 3 ($w = 1$):

$$w(t) = [the, set, up]$$

$$set \rightarrow x_{set} \in \mathbb{R}^d$$

$$w(t) \rightarrow x(t) = [x_{the}, x_{set}, x_{up}] \in \mathbb{R}^{3d}$$

where $w(t)$ is a 3-word context window surrounding center word *set*, x_{set} is the word embedding of the word *set* in the lookup table, and d is the dimension of embedding vector. $x(t)$ is the corresponding concatenation of embedding vectors for the words in $w(t)$.

A feed-forward neural network [24] is a simple network without recurrent connections between neutrons where information only flows in single direction. In this kind of network, the raw input vector x is first fed into the hidden layer for non-linear transformation. Then the abstract representation generated by hidden layer is fed into output(usually a softmax) layer to yield final output y. Usually:

$$h_t^{lc} = \tanh\left(W_1 x(t) + b_1\right) \tag{9}$$

$$f^{lc}(t) = softmax(W_2 h_t^{lc} + b_2) \tag{10}$$

where $x(t)$ is the input context window, h_t^{lc} is the hidden layer output and $f^{lc}(t)$ is the classification result for current word. W_1, b_1 and W_2, b_2 are weight matrix and bias between input layer and hidden layer, and between hidden layer and output layer respectively. We refer to this model as LC.

We conduct contrast experiments between the basic LSTM and LC model in Sect. 4 to see whether local context captured by the context window can contribute to the extraction of opinion aspect.

3.3   Hybrid of Global and Local Context

Since opinion aspects are usually very short phrases, both long term dependency and short term context shall be crucial to neural aspect extraction. So we combine these two kind of information to further improve the performance of basic LSTM models and LC model. One approach (we refer to this series of models as LC1) taken by many works [6,8] is first feeding the raw word context window instead of only the current word into hidden layer of LSTM and then learning these local context representation together with long term dependencies through non-linear transformations in memory blocks. Figure 1(a) gives an illustration of LSTM with this kind of context window. Formally:

$$x(t) \leftarrow x_{t-w}...x_t...x_{t+w}$$

where x_t in Eqs. (4) to (8) is replaced by $x(t)$, a concatenation of vectors in context window.

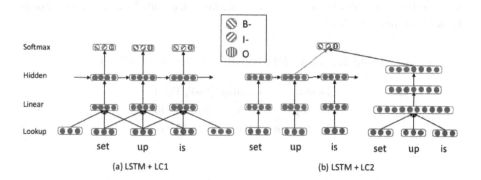

Fig. 1. Two ways of integrating local context for LSTM

Different from previous work, we proposed a novel model to incorporate local context and long term dependency, where local context is first generated by a separate LC model and then combined with the long term dependency computed by hidden layers of LSTM network. We refer to this series of models as LC2 and an illustration is presented in Fig. 1(b). Since the non-linear transformation in

LSTM is computationally expensive, in our model the local context is generated by a separate network instead of being mixed with long term dependency in the previous case, which not only is computationally cheaper but also yields more *pure* local context. Our propose model can be described with following equations:

$$h(t) = [h_t^{lstm}, h_t^{lc}] \tag{11}$$

$$f(t) = softmax(W_j h(t) + b_j) \tag{12}$$

where $h(t)$ is the concatenation of hidden layer output from (8) and (9). W_j and b_j are weight matrix and bias between hidden layer and output layer.

4   Experiments

In this section, we present our experimental setting, results and analysis for extractions of opinion aspects in SemEval2014 datasets.

4.1   Datasets and Settings

We evaluate our models on the SemEval 2014 datasets including data from laptop and restaurant domains. The detail of these datasets are given in Table 1. Standard precision, recall, and F1 score are applied in evaluation. We use two pre-trained word embeddings, namely Senna Embeddings (50 dimension) and Google Embeddings (300 dimension) for model input. Each dataset was preprocessed by lowercasing all characters and replacing each digit number with Digit. The lookup table was built from training set by marking rare word with no more than one occurrence as UNK. Paddings are added for boundary words to make context windows.

Table 2. SemEval2014 Laptop and Restaurant dataset

Domain	Training	Test	Total
Laptop	3,045	800	3,845
Restaurant	3,041	800	3,841

We spare 10% of the training data as the development set and the remaining 90% training set as training set in order to implement *early stoping* in SGD. The weights and biases of the models are initialized by randomly sampling from a small uniform distribution $\mu(-0.2, 0.2)$. The learning rate is fixed at 0.01. We run SGD for 30 epochs. We experiment with window size $\in \{1, 3, 5, 7, 9, 11, 13, 15\}$ and with hidden layer size $\in \{50, 100, 150, 200\}$. We report the best results together with their window size and hidden layer size.

4.2   Experimental Results

In Table 3, we give two examples labeled by LSTM and LC model to show the complementarity of local context and global context. Table 4 shows our results of aspect extraction on the standard test set in F1 scores. We report the best results of each model at each window size in Fig. 2. In the following, we highlight our main findings.

Local Context vs. Global Context. From Table 4, we can see that LC model achieve comparable results with the LSTM model, which indicates the importance of local context. Then we further analyze the extraction results of these two models to see whether local context captures information ignored by the LSTM network. From examples in Table 3, we can see that LC model can find some missing part of opinion aspects left out by LSTM model.

Table 3. Two examples showing complementarity of LSTM and LC

GOLD	The [set up] was easy
LSTM	The [set] up was easy
LC	The [set up] was easy
GOLD	I am please with the products ease of [use] out of the box ready [appearance] and [functionality]
LSTM	I am please with the products ease of [use] out of the box ready appearance and [functionality]
LC	I am please with the products ease of use out of the box ready [appearance] and [functionality]

Window Size. From Fig. 2, we can see that most of models reach the peak of performance at window size 3. If not, the F1 score at size 3 is still comparable to the peak one. Smaller window size can't capture enough local context while larger window size usually introduces noisy information which harms the performance.

Effectiveness of Local Context for LSTMs. Notice that LSTM models with LC1 are standard LSTMs when window size is 1. We can see that most joint models achieve best results at window size 3, which shows the importance of proper local context. Though we achieve comparable results with SemEval2014 best systems and outperform similar models of Liu et al. [6], our results are still behind the state of art results of Yin et al. [3] and Wang et al. [4] which explicitly encodes syntactic information as features. Still, our model provides a good way for better capture of local context which can also be applied to these models.

Table 4. F1-score for LSTM models

Model	Dim.	win_l	hid_l	Laptop	win_r	hid_r	Restaurant
LC	Senna (50)	3	50	72.26	3	50	79.3
LC	Google (300)	3	50	71.33	3	50	77.7
LSTM	Senna (50)	1	100	71.19	1	150	79.09
LSTM	Google (300)	1	200	71.22	1	200	77.47
LSTM+LC1	Senna (50)	3	50	71.92	3	150	79.72
	Google (300)	3	50	71.66	3	200	78.39
BLSTM+LC1	Senna (50)	1	100	73.04	3	150	80.53
	Google (300)	3	50	72.12	9	50	78.47
LSTM+LC2	Senna (50)	3	200	**74.78**	3	100	80.53
	Google (300)	3	200	74.54	3	100	78.47
BLSTM+LC2	Senna (50)	3	100	73.37	3	200	**80.62**
	Google (300)	3	100	72.27	3	200	78.5
LSTM (Liu)	-	-	-	73.40	-	-	79.89
IHS_RD	-	-	-	**74.55**	-	-	79.62
DLIREC	-	-	-	73.78	-	-	**84.01**
WDEmb+B+CRF	-	-	-	75.16	-	-	**84.97**
RNCRF+F	-	-	-	**78.42**	-	-	84.93

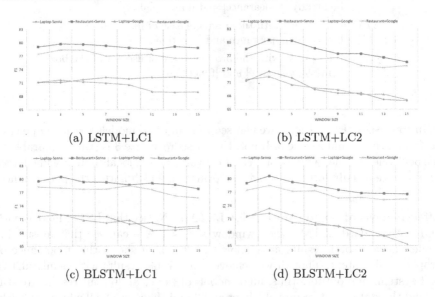

(a) LSTM+LC1

(b) LSTM+LC2

(c) BLSTM+LC1

(d) BLSTM+LC2

Fig. 2. F1 score for different window size

Comparison Among LC1 and LC2. A comparison in Table 4 tells that LSTM models with LC2 generally outperforms LSTM models with LC1 with maximum 2.86% gains, which confirms that separate feed forward neural network can keep more local context than the mixed fashion.

Training Time. Furthermore, we evaluate the training time for these models. Table 5 shows the training time for different models with window size 3 on the same dataset. The LC2 models runs faster than corresponding LC1 models for that the LSTM structure is more computationally expensive and the parameters of separate network can be parallel trained using GPUs. Our proposed LC2 models are more time-efficient.

Table 5. Training time of each model (window size = 3) on Laptop dataset using Senna Word Embeddings per epoch(s)

Models	LC	LSTM+LC1	LSTM+LC2	BLSTM+LC1	BLSTM+LC2
Time	9.11	32.99	26.06	42.81	38.2

5   Conclusions and Future Work

We propose two ways of integrate local context and long term dependency: one is changing the input from single word to a window of words and the other is concatenating the hidden layer output of LSTM with that of a window-based feed forward neural network.

We experiment with different window sizes ranging from 1 to 15 on SemEval2014 datasets and find that 3 is optimal for most cases. Comprehensive experiments show that incorporating long term history with local context can further boost the performance of opinion aspect extraction. Results also show that the using a separate network to train local context representations is slightly superior to the mixed one.

The local context we use here is acquired by a hard "selected" window. One potential future direction is to apply the attention mechanism to automatically choosing the appropriate context. Additionally, memory network which has multi-hop structure for learning good context can be investigated for this task as an alternative to LSTM.

Acknowledgements. Thanks to the help of my mentor and senior.

References

1. Wiebe, J., Wilson, T., Cardie, C.: Annotating expressions of opinions and emotions in language. Lang. Resour. Eval. **39**(2–3), 165–210 (2005)
2. Yang, B., Cardie, C.: Extracting opinion expressions with semi-Markov conditional random fields. In: Proceedings of the 2012 Joint Conference on Empirical Methods in Natural Language Processing and Computational Natural Language Learning, pp. 1335–1345. Association for Computational Linguistics (2012)

3. Yin, Y., Wei, F., Dong, L., et al.: Unsupervised word and dependency path embeddings for aspect term extraction. arXiv preprint arXiv:1605.07843 (2016)
4. Wang, W., Pan, S.J., Dahlmeier, D., et al.: Recursive neural conditional random fields for aspect-based sentiment analysis. arXiv preprint arXiv:1603.06679 (2016)
5. Irsoy, O., Cardie, C.: Opinion mining with deep recurrent neural networks. In: EMNLP (2014)
6. Liu, P., Joty, S., Meng, H.: Fine-grained opinion mining with recurrent neural networks and word embeddings. In: Conference on Empirical Methods in Natural Language Processing (EMNLP 2015) (2015)
7. Pontiki, M., Galanis, D., Pavlopoulos, J., et al.: Semeval-2014 task 4: aspect based sentiment analysis. In: Proceedings of the 8th International Workshop on Semantic Evaluation (SemEval 2014), pp. 27–35 (2014)
8. Mesnil, G., He, X., Deng, L., et al.: Investigation of recurrent-neural-network architectures and learning methods for spoken language understanding. In: INTER-SPEECH, pp. 3771–3775 (2013)
9. Hu, M., Liu, B.: Mining opinion features in customer reviews. In: AAAI, vol. 4, no. 4, pp. 755–760 (2004)
10. Popescu, A.M., Etzioni, O.: Extracting Product Features and Opinions from Reviews. Natural Language Processing and Text Mining. Springer, London (2007)
11. Wu, Y., Zhang, Q., Huang, X., et al.: Phrase dependency parsing for opinion mining. In: Proceedings of the 2009 Conference on Empirical Methods in Natural Language Processing: Volume 3, vol. 3, pp. 1533–1541. Association for Computational Linguistics (2009)
12. Qiu, G., Liu, B., Bu, J., et al.: Opinion word expansion and aspect extraction through double propagation. Comput. Linguist. 37(1), 9–27 (2011)
13. Jin, W., Ho, H.H., Srihari, R.K.: A novel lexicalized HMM-based learning framework for web opinion mining. In: Proceedings of the 26th Annual International Conference on Machine Learning, pp. 465–472 (2009)
14. Li, F., Han, C., Huang, M., et al.: Structure-aware review mining and summarization. In: Proceedings of the 23rd International Conference on Computational Linguistics, pp. 653–661. Association for Computational Linguistics (2010)
15. Jakob, N., Gurevych, I.: Extracting opinion aspects in a single-and cross-domain setting with conditional random fields. In: Proceedings of the 2010 Conference on Empirical Methods in Natural Language Processing, pp. 1035–1045. Association for Computational Linguistics (2010)
16. Choi, Y., Cardie, C., Riloff, E., et al.: Identifying sources of opinions with conditional random fields and extraction patterns. In: Proceedings of the Conference on Human Language Technology and Empirical Methods in Natural Language Processing, pp. 355–362. Association for Computational Linguistics (2005)
17. Breck, E., Choi, Y., Cardie, C.: Identifying Expressions of Opinion in Context. In: IJCAI 2007, vol. 7, pp. 2683–2688 (2007)
18. Johansson, R., Moschitti, A.: Extracting opinion expressions and their polarities: exploration of pipelines and joint models. In: Proceedings of the 49th Annual Meeting of the Association for Computational Linguistics: Human Language Technologies: Short Papers-Volume 2, pp. 101–106. Association for Computational Linguistics (2011)
19. Mikolov, T., Karafit, M., Burget, L., et al.: Recurrent neural network based language model. In: Interspeech 2010, vol. 2, no. 3 (2010)
20. Graves, A., Jaitly, N.: Towards end-to-end speech recognition with recurrent neural networks. In: ICML 2014, vol. 14, pp. 1764–1772 (2014)

21. Schuster, M., Paliwal, K.K.: Bidirectional recurrent neural networks. IEEE Trans. Signal Process. **45**(11), 2673–2681 (1997)
22. Socher, R., Perelygin, A., Wu, J.Y., et al.: Recursive deep models for semantic compositionality over a sentiment treebank. In: Proceedings of the Conference on Empirical Methods in Natural Language Processing (EMNLP), pp. 1631–1642 (2013)
23. Bengio, Y., Simard, P., Frasconi, P.: Learning long-term dependencies with gradient descent is difficult. IEEE Trans. Neural Netw. **5**(2), 157–166 (1994)
24. Hinton, G., Deng, L., Yu, D., et al.: Deep neural networks for acoustic modeling in speech recognition: the shared views of four research groups. IEEE Signal Process. Mag. **29**(6), 82–97 (2012)

Context Enhanced Word Vectors
for Sentiment Analysis

Zhe Ye$^{(\boxtimes)}$ and Fang Li

Department of Computer Science and Engineering,
Shanghai Jiao Tong University, Shanghai 200240, China
{yezhejack,fli}@sjtu.edu.cn

Abstract. Word vectors have become very important features for sentiment analysis. The aim of this paper is to encode sentimental context into pre-trained word vectors for sentiment analysis. The negation and intensity words in a context, as well as the sentimental words are combined to form context enhanced word vectors. Experiments on the datasets of SemEval show that the method of using intensity words has improved the result comparing with the baseline. The context enhanced words vectors from a distant supervision data can increase the similarity of the same polarity and decrease the similarity of the different polarity.

Keywords: Sentiment classification · Linguistic resources · Word vectors

1 Introduction

Sentiment analysis has played an important role in many real-world applications. The objective of sentiment classification is to classify a message, sentence or document as positive, neural or negative [1]. In recent years, a variety of neural networks have been proposed for sentence-level classification tasks such as convolutional neural network (CNN) and so on. CNN has become one of the most attractive neural network model in sentiment analysis [2,3]. Deriu [4] leveraged large amounts of data with distant supervision to train an ensemble of 2-layer CNN and achieved the best results for Message Polarity Classification task in SemEval2016 [5]. Among the 10 top-ranked teams, 7 teams used either general-purpose or task-specific word vectors generated via word2vec [6] or GloVe [7]. Although those word embeddings work well for many tasks, it is not good enough for sentiment classification. The most obvious problem of word vectors for general purpose is that they only model the syntactic context of words but ignore the sentiment information of messages. Therefore, words with opposite polarity such as *good* and *bad*, are clustered [1].

In this paper, we propose a simple approach to enhance the sentimental context for sentimental word vectors. The approach will encode linguistics features into word vectors without training a model. The intuition of our approach is to "pull" the words away from its opposite polarity by leveraging sentimental,

X. Cheng et al. (Eds.): SMP 2017, CCIS 774, pp. 256–267, 2017.
https://doi.org/10.1007/978-981-10-6805-8_21

negation and intensity words in the training data or distant supervision data. For example, *good* is a word with positive polarity. It may be used to express a negative sentiment like *the movie is not good*. The negation word *not* changes the sentiment polarity of *good* in this local context. Another example, *the movie is bad*, expresses a negative sentiment but the local context of *bad* is different from the *good* in the previous example because of *not*. In order to use those sentimental contexts, we propose a method of context enhancement for word vectors.

The major contributions of our work in this paper are as follows: (1) the semantics of sentimental words are enhanced with their contexts in the supervised corpus; (2) negation and intensity words are used in a context to improve sentimental word vectors.

2 Related Work

In recent years, deep learning models have achieved remarkable results in natural language processing especially the works involved neural language models [6–8]. In 2003, Bengio [8] proposed a neural network model for learning distributed representation for words in the unannotated corpus. Following his work, Mikolov developed two simple approaches, Skipgram and CBOW, to learn the distributed representation for words [6]. The word vectors have become the general features for many tasks including sentiment analysis.

The word vectors for general purpose are not good enough for specific-tasks. The unsupervised methods which use co-occurrence information in corpora to learn distributed representation of words would coalesce the notations of semantic similarity and relatedness [9]. In general, there are three kinds of methods to adapt word vectors for specific-tasks.

The first kind of the methods is to change the context that used by the neural language model. Schwartz [10] used symmetric pattern contexts instead of bag-of-words contexts to improve the similarity performance of verb on word2vec by up to 15%. Ling [11] adapts word2vec models so that they are sensitive to the positioning of the words. For the Skipgram of word2vec, the number of output prediction matrices has increased so that every position of output has its own matrix. For the CBOW of word2vec, the size of output prediction matrix is extended so that every position of the input can affect the output through its own parameters. These adapted word2vec models have the sparsity issue because of too many parameters.

The second kind of the methods is to encode linguistics knowledge or other supervised information into word vectors during the procedure of learning word vectors. Tang [1] proposed a neural network for learning sentiment-specific distributed representation of words by using distant supervision method and C&W model [12]. The language model learned by the neural network can distinguish positive and negative sentimental words. But it also introduces noise sentiment information into the word vectors because of using distant supervision method. For sentiment analysis task, Rouvier [13] used three kinds of word embeddings:

lexical embeddings, part-of-speech embeddings and sentiment embeddings. The latter two embeddings are respectively encoded with POS and sentiment information during the procedure of learning word vectors.

The third kind of the methods is to encode linguistics knowledge or other supervised information into pre-trained word vectors. This kind of method is efficient compared to the second kind of the methods because it does not need to train a language model on large corpora. Faruqui [14] proposed a graph-based learning technique called retrofitting to leverage semantic lexicons to obtain higher quality word vectors based on pre-trained word vectors. It is reported in [14] that the retrofitting process is fast. It takes about 5 s to process 100,000 word vectors of 300 dimensions. In the spirit of retrofitting, Mrkšić proposed counter-fitting method to fine-tune the pre-trained word vectors. It introduces three constraints into the post-processing procedure [9]. They are antonym repel, synonym attract and vector space preservation. The first two constraints are used to distinguish between synonyms and antonyms. The last constraint is used to preserve the semantic information contained in the original vectors. The objective function for the training procedure is the weighted sum of the three constraints. It has been reported that an end-to-end run of counter-fitting takes less than two minutes on a laptop with four CPUs.

Our method is based on the third kind of the above. Yancheva [15] purposed a method to classify a person to be a dementia patient based on their description text about a picture. They have investigated the contexts of the same talking topics from two groups people. One group is dementia patients. The other one is healthy people. To this end, they augment the word vectors with local context windows from DementiaBank [16]. Every word vector is constructed by a linear combination of its global vector from the trained GloVe model and the vectors of the surrounding context words where each context word is weighted inversely to its distance from the central word. Inspired by their idea, we propose the method to enhance the sentimental context of sentimental words.

3   Method

It has been proved that sentimental words are very important for sentiment analysis [1,2,17,18]. Negation words will inverse the polarity of a sentence, while those intensity words will enhance the polarity of a sentence. If the context of a sentimental word includes a negation word or an intensity word, it is an important feature with which the sentimental word vectors should be combined. In the following, a simple method to enhance contextual sentiments of sentimental word vectors will be introduced and a CNN based model for sentiment analysis will be discussed.

3.1   Context Enhanced Word Vectors for Sentimental Words

Figure 1 shows the basic idea of our method. The **good** appears in many different contexts, the **good** word vector will be combined with the context of words **not** or *very*. There are two linguistics lexicons (1) negation lexicon, (2) intensity lexicon with positive meaning such as *very*, *extremely* and so on. We propose three ways of producing context enhanced word vectors based on (1) negation lexicon (NEG), (2) intensity lexicon (INT), and (3) negation and intensity lexicon (NEG+INT).

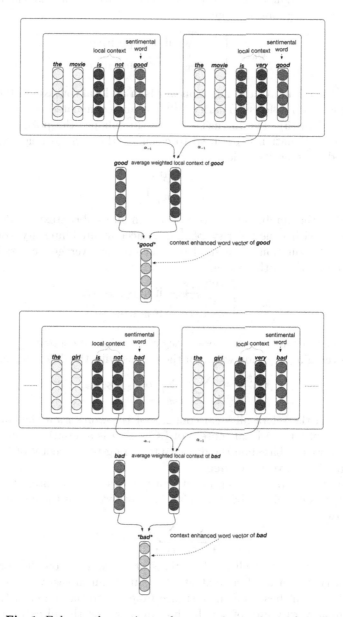

Fig. 1. Enhance the sentimental context for sentimental words.

Equation (1) is to calculate context enhanced word vectors of the word for three methods:

$$v'_w = v_w + \phi_w \tag{1}$$

where w is the original word vectors, v'_w is the new word vector, and ϕ_w is the average weighted context vector. Negation words will reverse the polarity of a sentimental word. Equation (2) is used to calculate the average weighted context vector based on negation lexicon:

$$\phi_w = \begin{cases} \frac{\phi_w^{\text{NEG}}}{n_w^{\text{NEG}}} & \text{if } w \text{ is positive} \\ -\frac{\phi_w^{\text{NEG}}}{n_w^{\text{NEG}}} & \text{if } w \text{ is negative} \end{cases} \tag{2}$$

where n_w^{NEG} is the number of negation words in all local contexts of the word w on the corpus with different polarity. ϕ_w^{NEG} is an accumulated negation context vector defined in (5). The intensity words can enhance the polarity of a sentimental word. Equation (3) is used to calculate the average weighted context vector based on intensity lexicon:

$$\phi_w = \frac{\phi_w^{\text{INT}}}{n_w^{\text{INT}}} \tag{3}$$

where n_w^{INT} is the number of intensity words in all local contexts of the word w on the corpus with same polarity. ϕ_w^{INT} is an accumulated intensity word vector defined in (6). Equation (4) is used to calculate the average weighted context vector based on the both lexicon:

$$\phi_w = \begin{cases} \frac{\phi_w^{\text{INT}}+\phi_w^{\text{NEG}}}{n_w^{\text{INT}}+n_w^{\text{NEG}}} & \text{if } w \text{ is positive} \\ \frac{\phi_w^{\text{INT}}-\phi_w^{\text{NEG}}}{n_w^{\text{INT}}+n_w^{\text{NEG}}} & \text{if } w \text{ is negative} \end{cases} \tag{4}$$

For a sentimental word w, its average negation weighted context ϕ_w^{NEG} is calculated by leveraging negation words in the corpus with the different polarity:

$$\phi_w^{\text{NEG}} = \sum_{t \in \mathbf{LC}_w} \sum_{\text{neg} \in t} \alpha_{\text{neg},t} \times v_{\text{neg}} \tag{5}$$

where $\mathbf{LC}_w$ is the set of all local contexts of the word w on the corpus with different polarity, t is a local context of $\mathbf{LC}_w$, $\mathbf{neg}$ is a negation word of t, and $\alpha_{\text{neg},t}$ is the weight based on the distance from $\mathbf{neg}$ to the center of t, and v_{neg} is the original word vector of $\mathbf{neg}$.

In a similar way, for a sentiment word w, its average intensity weighted context ϕ_w^{INT} is calculated by leveraging intensity words in the corpus with the same polarity:

$$\phi_w^{\text{INT}} = \sum_{t \in \mathbf{LC}_w} \sum_{\text{int} \in t} \alpha_{\text{int},t} \times v_{\text{int}} \tag{6}$$

where $\mathbf{LC}_w$ is the set of all local contexts of the word w on the corpus with same polarity, t is a local context of $\mathbf{LC}_w$, $\mathbf{int}$ is an intensity word of t, and $\alpha_{\text{int},t}$ is the weight based on the distance from $\mathbf{int}$ to the center of t, and v_{int} is the original word vector of $\mathbf{int}$. In this paper, we set the all $\alpha_{\text{int},t}$ and $\alpha_{\text{neg},t}$ equal to 1.

3.2 The Framework of Sentiment Analysis

We use the CNN model (shown in Fig. 2) for sentiment analysis. The CNN model is on the top of an embedding layer. The bottom of Fig. 2 is our work to use supervised corpus to enhance the context sentimental information for word vectors. A message is transformed to a matrix by concatenating the context enhanced word vectors of the word sequence in the message. The matrix is feed to the CNN model. The CNN outputs the probability distribution of three classes (positive, neutral and negative).

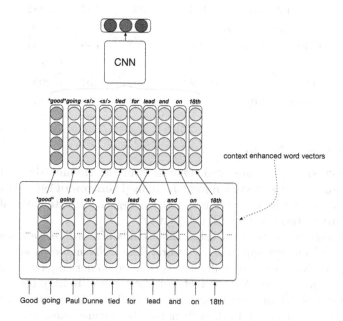

Fig. 2. The framework of sentiment analysis.

4 Experiments

4.1 Datasets

We conduct experiments on the Message Polarity Classification dataset of SemEval 2016 [5]. It is labeled in three categories (positive, neutral and negative). Table 1 shows statistics for SemEval dataset. The training data and development data of Twitter2013 and Twitter2016 are used as the training data for the experiment. The development-test data of Twitter2016 is used as the development data for the experiment. The test data of Twitter2016, Twitter2015, Twitter2014, Twitter2014Scarasm, SMS2013 and LiveJournal2014 are used for our model test.

Table 1. SemEval dataset statistics

Dataset	Positive	Negative	Neutral	Total
Twitter2013-train	3662	1466	4600	9728
Twitter2013-dev	575	340	739	1654
Twitter2016-train	3094	863	2043	6000
Twitter2016-dev	844	765	391	2000
Twitter2016-devtest	994	681	325	2000
Twitter2016-test	7059	10342	3231	20632
Twitter2015-test	1040	365	987	2392
Twitter2014-sarcasm	33	44	13	86
Twitter2014-test	982	202	669	1853
Twitter2013-test	1572	601	1640	3813
SMS2013-test	492	394	1207	2093
LiveJournal2014-test	427	304	411	1142

4.2   Experiment Setup

Two supervised corpora are used. One is from the training data (TD). The other is using distant supervision method, named **sentiment140**[1] (DSD) which contains 800,000 positive tweets and 800,000 negative tweets with labels. The original word vectors are from Google News.[2]

Experiments are conducted using three kinds of word vectors described in Sect. 3.1 based on two supervised corpora. The CNN model is similar to the model used by Kim [19]. The difference is that we do not employ regularization on the penultimate layer. We use rectified linear units (ReLU), dropout rate of 0.5, constraint of 0.4, and mini-batch size of 32. Three filter windows with sizes of 3, 4, 5 are experimented. Each of them has 100 feature maps. These values are not carefully tuned for higher accuracy result or macro F1 score. Our baseline CNN model is the model initialized with the original word vectors trained on Google News. Macro F1 score is used to measure the performance on SemEval dataset.

4.3   The Result on SemEval Dataset

Table 2 shows the result. The context enhanced word vectors based on intensity lexicon have achieved the best results, comparing with the other two ways. The context enhanced word vectors from the training data is better than from the distant supervision data. The human labeled training data is much more accurate than the way of distant supervision. Comparing with the baseline, our context enhanced word vectors can improve the performance of sentiment analysis in most test corpora of SemEval.

---

[1] http://help.sentiment140.com/for-students/.

[2] https://code.google.com/archive/p/word2vec/.

Table 2. Macro F1 on SemEval dataset.

	Twitter2016	Twitter2015	Twitter2014Sarcasm	Twitter2014	Twitter2013	SMS2013	LiveJournal2014
baseline	0.600	0.597	0.434	0.621	0.644	0.561	0.700
NEG+TD	**0.604**	**0.598**	0.418	**0.624**	**0.651**	0.558	0.675
NEG+DSD	**0.604**	0.587	0.433	0.617	0.625	0.541	0.664
INT+TD	**0.607**	<u>**0.614**</u>	0.442	0.639	<u>**0.663**</u>	<u>**0.583**</u>	<u>**0.710**</u>
INT+DSD	**0.606**	0.612	<u>**0.472**</u>	0.630	0.648	0.562	0.696
NEG+INT+TD	0.599	**0.606**	**0.455**	**0.636**	0.631	0.534	0.679
NEG+INT+DSD	<u>**0.613**</u>	0.611	0.464	<u>**0.643**</u>	0.648	0.556	0.692

4.4 The Similarity Between Sentimental Word Vectors

In order to evaluate our context enhanced ways for sentimental word vectors, the average cosine similarity of opposite polarity words and same polarity words are presented in the Table 3. Three enhancement methods from distant supervision data can decrease the similarities of different polarities and increase the similarities of the same polarity.

Table 3. Average of cosine similarity of opposite polarity words and same polarity words.

Method	Opposite polarity	Same polarity
Baseline	0.0934870	0.1271270
NEG+TD	**0.092970**	0.1251890
NEG+DSD	**0.064399**	**0.129778**
INT+TD	0.0954210	**0.128010**
INT+DSD	**0.060834**	**0.132146**
NEG+INT+TD	0.0942730	0.1264420
NEG+INT+DSD	**0.054429**	**0.133068**

Table 4 shows the most similar words of *good*. The baseline includes *bad*, *lousy*. Our context enhancement methods have excluded *bad* and *lousy*.

Table 4. The most similar words of *good*.

Method	Top-10 most similar words of *good*
Baseline	great **bad** terrific decent nice excellent fantastic better solid **lousy**
NEG+TD	great nice fantastic better decent terrific happy definitely okay perfect
NEG+DSD	great decent nice better excellent fantastic terrific solid okay definitely
INT+TD	great nice decent excellent terrific pretty really better fantastic okay
INT+DSD	great nice terrific decent excellent fantastic better pretty really wonderful
NEG+INT+TD	great nice decent fantastic excellent terrific better pretty really okay
NEG+INT+DSD	great nice terrific decent fantastic excellent better really pretty wonderful

Table 5 shows the most similar words of *bad*. Our method has decreased the order of *good* to appear in the list. That is the aim to "pull" the *good* away from *bad* and vice versa.

Table 5. The most similar words of *bad*.

Method	Top-10 most similar words of *bad*
Baseline	**good** terrible horrible lousy crummy horrid awful dreadful horrendous nasty
NEG+TD	terrible horrible lousy **good** horrid crummy dreadful nasty horrendous crappy
NEG+DSD	terrible horrible **good** lousy crummy horrid dreadful crappy awful nasty
INT+TD	terrible horrible lousy **good** horrid dreadful horrendous nasty crummy rotten
INT+DSD	terrible horrible lousy **good** horrid crummy dreadful nasty rotten crappy
NEG+INT+TD	terrible horrible lousy **good** horrid dreadful horrendous nasty crummy rotten
NEG+INT+DSD	terrible horrible lousy **good** horrid dreadful rotten nasty crummy poor

4.5 The Result on SimLex-999

There are seven kinds of word embeddings in this work. We conduct an intrinsic experiment on SimLex-999 [20]. SimLex-999 is a gold standard resource for evaluating distributional semantic models. The result is presented in Table 6 measured by Spearman's ρ correlation. Three enhancement methods based on the distant supervision data and the training data have all improved the performance on the SimLex-999. The result of this experiment shows that our methods do not decrease the semantic information of the word embeddings but even increase the semantic information.

Table 6. The result on SimLex-999.

Method	ρ
Baseline	0.446
NEG+TD	**0.453**
NEG+DSD	**0.473**
INT+TD	**0.453**
INT+DSD	**0.471**
NEG+INT+TD	**0.457**
NEG+INT+DSD	**0.469**

5 Conclusion

In this paper, we propose a simple method to enhance the sentimental context for word vectors. The method leverages some linguistics information, such as intensity and negation words in a context. The experiments on SemEval data show that the context enhanced word vectors based on intensity words have achieved the best results comparing with the context enhanced word vectors based on negations. Comparing with the baseline using original word vectors, three ways of context enhancements are better in most of the test corpora. Our method only focuses on pre-trained word vectors. We do not use very complicated models like [4] for sentiment analysis.

Using a large distant supervision data can increase the similarity between the same polarities, decrease the similarity of different polarities, and thus enhance the sentimental information. In the experiments of sentiment analysis, the performance of context vectors from a distant supervision data does not show better than those from the training corpus. The reason is that the distant supervision data may introduce noise as well.

Using a large distant supervision data can also increase the semantic information of the word vectors. In the experiment on SimLex-999, the result of our methods is better than the baseline, however has not outperformed significantly. There are two reasons. One is that the lexicons we use are not large enough,

the other reason is that there are no constraints to keep the semantic information of the pre-trained word embeddings during the procedure of "pulling" the
sentiment words away.

In our future work, more experiments need to be conducted. How to improve
word vectors for sentiment analysis is still our main focus.

Acknowledgments. We would like to thank the National Natural Science Foundation
of China (Grant No. 61673266) for the financial support of this research.

References

1. Tang, D., Wei, F., Yang, N., Zhou, M.: Learning sentiment-specific word embedding
   for twitter sentiment classification. In: ACL, pp. 1555–1565 (2014)
2. Miura, Y., Sakaki, S., Hattori, K., Ohkuma, T.: TeamX: a sentiment analyzer
   with enhanced lexicon mapping and weighting scheme for unbalanced data. In:
   Proceedings of SemEval, pp. 628–632 (2014)
3. Hagen, M., Potthast, M., Büchner, M., Stein, B.: Webis: an ensemble for twitter
   sentiment detection. In: Proceedings of SemEval, pp. 582–589 (2015)
4. Deriu, J., Gonzenbach, M., Uzdilli, F., Lucchi, A., De Luca, V., Jaggi, M.: Swiss-
   Cheese at SemEval-2016 task 4: sentiment classification using an ensemble of convolutional neural networks with distant supervision. In: Proceedings of SemEval,
   pp. 1124–1128 (2016)
5. Nakov, P., Ritter, A., Rosenthal, S., Sebastiani, F., Stoyanov, V.: SemEval-2016
   task 4: sentiment analysis in Twitter. In: Proceedings of SemEval, pp. 1–18 (2016)
6. Mikolov, T., Chen, K., Corrado, G., Dean, J.: Efficient estimation of word representations in vector space. arXiv preprint arXiv:1301.3781 (2013)
7. Pennington, J., Socher, R., Manning, C.: GloVe: global vectors for word representation. In: EMNLP, pp. 1532–1543 (2014)
8. Bengio, Y., Ducharme, R., Vincent, P., Janvin, C.: A neural probabilistic language
   model. J. Mach. Learn. Res. **3**, 1137–1155 (2003)
9. Mrkšić, N., Séaghdha, D.O., Thomson, B., Gašić, M., Rojas-Barahona, L., Su, P.H.,
   Vandyke, D., Wen, T.H. and Young, S.: Counter-fitting word vectors to linguistic
   constraints. arXiv preprint arXiv:1603.00892 (2016)
10. Schwartz, R., Reichart, R., Rappoport, A.: Symmetric patterns and coordinations:
    fast and enhanced representations of verbs and adjectives. In: HLT-NAACL, pp.
    499–505 (2016)
11. Ling, W., Dyer, C., Black, A., Trancoso, I.: Two/Too simple adaptations of
    Word2Vec for syntax problems. In: NAACL, pp. 1299–1304 (2015)
12. Collobert, R., Weston, J., Bottou, L., Karlen, M., Kavukcuoglu, K.: Natural lan-
    guage processing (almost) from scratch. J. Mach. Learn. Res. **12**, 2493–2537 (2011)
13. Rouvier, M., Favre, B.: SENSEI-LIF at SemEval-2016 task 4: polarity embed-
    ding fusion for robust sentiment analysis. In: Proceedings of SemEval, pp. 202–208
    (2016)
14. Faruqui, M., Dodge, J., Jauhar, S., Dyer, C., Hovy, E., Smith, N.: Retrofitting
    word vectors to semantic lexicons. In: NAACL, pp. 1606–1615 (2015)
15. Yancheva, M., Rudzicz, F.: Vector-space topic models for detecting Alzheimer's
    disease. In: ACL, pp. 2337–2346 (2016)
16. Becker, J.T., Boller, F., Lopez, O.L., Saxton, J., McGonigle, K.L.: The natural
    history of Alzheimer's disease. Arch. Neurol. **51**, 585–594 (1994)

17. Wilson, T., Wiebe, J., Hoffmann, P.: Recognizing contextual polarity in phrase-level sentiment analysis. In: EMNLP, pp. 347–354 (2005)
18. Mohammad, S., Kiritchenko, S., Zhu, X.: NRC-Canada: building the state-of-the-art in sentiment analysis of Tweets. In: Proceedings of SemEval, pp. 321–327 (2013)
19. Kim, Y.: Convolutional neural networks for sentence classification. arXiv preprint arXiv:1408.5882 (2014)
20. Hill, F., Reichart, R., Korhonen, A.: SimLex-999: evaluating semantic models with genuine similarity estimation. Comput. Linguist. **41**(4), 665–695 (2015)

Social Network Analysis and Social Computing

Divergence or Convergence: Interaction Between News Media Frames and Public Frames in Online Discussion Forum in China

Lun Zhang[(✉)]

Department of Digital Media, School of Arts and Communication,
Beijing Normal University, Beijing 100875, China
zhanglun@bnu.edu.cn

Abstract. When information provided by news media is incomplete, biased, and untrustworthy, how will social media interact with news media to construct public discourse? With a focus on the issue of violent resistance in Xinjiang Uyghur Autonomous Region in China, this study identified and compared the frames employed by news and online discussion forum and further expanded the research horizon to explore the frame interactions in these two media platforms. Adopting an automatic content analysis approach, this study identified four news media frames and four public frames in online discussion forum. Frames in news and online discussion forum differed in terms of thematic structure and weight of frames. News media frames were found to well represent government interests and the political ideology of the state. Public frames in online discussion forum were more deliberate, as evidenced by the concentration on the responsibility frame that attributes the violent resistance issue to the failure of the government's ethnic policy, the long-term conflict between ethnicities, and the intervention of foreign political and religious forces. By employing Granger causality analysis, this study further found that news and online discussion forum influence each other in placing the issue in their respective agenda to build and strengthen their existing frames instead of presenting a dominating relationship.

Keywords: Framing · Online discussion forum · Terrorism · Topic modeling

In China, two distinct universes of discourses form and distribute public opinions about political issues: the official discourse universe and the nonofficial discourse universe (He 2008). The official discourse universe comprises government discursive activities, as exemplified by the news media and other official outlets. The nonofficial discourse universe conventionally refers to the oral sphere in which information is communicated by personal networks. Increasing access to the Internet and prevalence of social media, however, have provided new channels for information distribution and started a universe of nonofficial discourse (He 2008).

Although social media has been regarded as having the potential to act as a liberating force in the civil society in contemporary China (Tai 2013), the central government in China continues to have a strong control on public opinion concerning politically sensitive issues to maintain government legitimacy. How can online discussions on politically sensitive issues resonate when such issues are severely

© Springer Nature Singapore Pte Ltd. 2017
X. Cheng et al. (Eds.): SMP 2017, CCIS 774, pp. 271–282, 2017.
https://doi.org/10.1007/978-981-10-6805-8_22

controlled and censored by the authoritarian government in China? This study aims to examine the role of news media and social media in constructing public opinion on a politically sensitive issue—the violent resistance in XUAR. This issue is highly politically sensitive. It has the potential to cause damage to the legitimacy of the government in XUAR. The Chinese government has to establish normative justifiability by exhibiting its dominance during a crisis. Any discourse deemed detrimental to the legitimacy of the government would be strictly banned in news media. Therefore, when the official discourse is ambiguous, indoctrinated, biased, and untrustworthy, will the nonofficial discourse in social media comply with news media or play a complementary role in producing an alternative or opposing discourse on political issues in news media? By using the present case, this study seeks to contribute to the growing body of knowledge concerning China's evolving media system, which has undergone considerable transition, and further understand the media dynamics in China.

Furthermore, this study adopts framing theory to compare discourses and examine the interplay of framing processes in news media and online discussion forum on violent resistance in XUAR. This study would contribute to framing theory by empirically investigating the interaction of frames of politically sensitive issue in two media platforms.

1   Frame Building in News Media and Online Discussion Forum

According to Entman (1993), the frame-building process aims "to select some aspects of a perceived reality and make them more salient in a communicating text" (p. 52), which allows people to identify, label, understand, remember, and evaluate social realities (Goffman 1974). Frames in news media are embedded in the discourse authored by journalists who are instituted in the news agency. News frames are built by selecting, emphasizing, and excluding certain aspects of social issues. News reporting frequently uses four frames, namely, conflict frame, human interest frame, morality frame, and responsibility frame (Luther and Zhou 2005). Conflict frame highlights conflicts between evolved individuals or groups. Human interest frame personalizes news stories by focusing more on the emotional reactions of people and using individual examples to illustrate stories. For news media, the adoption of human interest frame, such as the description of the seriousness of the terrorist attack and the illustration of individual death because of the terrorist attack, would increase the newsworthiness of a terrorism event (Chermak and Gruenewald 2006). Morality frame places the news story in the context of moral prescriptions or tenets. Responsibility frame attributes responsibility for a cause or solution to specific individuals, social groups, or institutions (Entman 1993).

According to the "five-tier hierarchical model" influencing media content (Shoemaker and Reese 2011), frames in news media are formed based on ideological and professional routines. Compared with news media, public frame in online discussion forum is the discourse authored by ordinary users in the online discussion forum. Online discussion forum may shift the traditional framing of news media to an interactive, socially constructive, and bottom-up process. Therefore, given the different antecedents influencing the mechanisms of frame building in news and social media, the present study first obtains an answer to the following question.

Research Question 1: *What conflict, human interest, morality, and responsibility frames on the issue of Xinjiang violent resistance are present in news media and online discussion forum?*

Furthermore, unpacking and comparing the dynamics of frame interaction between these two media platforms in China merit further attention. Therefore, the second objective of this study aims to investigate

Research Question 2: *What is the dynamic pattern of public frames in online discussion forum and news media frames in describing the violent resistance in XUAR? Specifically, this study hypothesizes that*

H1: News media will positively influence online discussion forum in forming the frame of the violent resistance in XUAR.

H2: Online discussion forum will negatively influence news media in forming the frame of the violent resistance in XUAR.

2 Method

2.1 Data Collection

Data were collected from Baidu News and a Chinese online discussion forum, Tianya. cn (known as "Tianya"), from May 1, 2014 to May 30, 2014. During the said period, on May 22, 2014, a violent resistance occurred in a busy street market in Urumqi, XUAR. The incident, which targeted civilians, resulted in the deaths of 29 civilians and 4 perpetrators and the injury of more than 140 others.

The two sources of data collection, Tianya and Baidu News, albeit not exclusive, can validly represent the population of online public opinion and news media outlets respectively. The validity and popularity of Tianya in representing online public opinion have been demonstrated in a series of studies. Tianya is one of the most popular and influential bulletin board systems in China. Baidu News is a news archive that collects news from traditional news media (e.g., People's Daily) and news portals (e.g., www.yangtse.com). This database was compiled using automatic news-gathering robot programs that collect approximately 120,000 to 130,000 articles per day from more than 1,000 most-visited Chinese websites.

News or posts regarding the incident had the keyword "Xinjiang" combined with "attack", "resistance", or "terrorist/terrorism." Titles or the main body of the posts having these terms were considered as qualified entries. Irrelevant news and posts were excluded via manual selection. A total of 1,836 news stories were considered valid, of which the number of posts was 7,440.

2.2 Frame Identification

For each headline or original forum post, the present study adopted an automatic content analysis method to identify the frames. This method is the Latent Dirichlet

Allocation (LDA) topic modeling, which has been widely used in the field of natural language processing in computer sciences (Blei et al. 2003; DiMaggio et al. 2013). The algorithm of LDA topic modeling is an unsupervised machine-learning method for identifying hidden thematic structures based on words and their co-occurrence in a collection of documents (DiMaggio et al. 2013). This approach generates groups of terms (i.e., words used in documents) that are associated by themes, and the strength of the documents exhibiting such themes are assessed (DiMaggio et al. 2013).

This study limits the number of topics from 2 to 15. That is, the program was requested to generate 2 to 15 clusters of themes. By manually checking the classification result, for both news media and online discussion forum, this study found that the number of topics (i.e., groups of words associated under a frame) assigned less than 7 and more than 3 have higher face validity than do other topics.

The final step of LDA topic modeling is validation. Previous studies have suggested the validity evaluation by checking a sample of automatically coded articles manually (DiMaggio et al. 2013). This study randomly selected 10% of the documents (i.e., 184 news articles and 744 discussion forum posts) to manually calculate the accuracy rate. The classification result shows that the four-frame clusters in both news and online discussion forum has the highest accuracy scores of 80% and 74% respectively.

2.3   Data Analysis

Granger causality test was conducted to identify the interplay of frames between news media and online discussion forum. If B can be predicted using the past value (i.e., the time lags) of A while controlling the past value of B, then A is said to Granger cause B. In the present study, Granger causality occurred when the distribution of frames in one media platform explains a significant amount of variance of frame distribution in another media platform. This variance exceeds that explained by endogenous frame distribution. In other words, this technique can statistically determine whether frames in one media precede and predict frames in another media. Considering that news production is performed daily, the present study selects one day as an appropriate time lag. In this study, the model is specified as

$$Y_{i.t} = \alpha + \sum_{i=1}^{4} \phi_i Y_{i.t-1} + \sum_{j=1}^{4} \beta_j X_{j.t-1} + e_t$$

$Y_{i.t}$ is the frequency of the focal frame at time t in one media platform. β_j is a measure of the influence of frequency of other frames in this media platform on Y_i at time $t - 1$. Θ_i is the regression coefficient of the effects of frames in another media platform at time $t - 1$.

3   Findings

3.1   Frames of XUAR Violent Resistance in News Media and Online Discussion Forum

This study identified four frames in news media and another four in online discussion forum. Tables 1a and 1b present the classification results of the frames by presenting the first 12 key words with the most differentiation scores in the topic modeling analysis. The third row of Tables 1a and 1b is the name of issue-specific frames.

The first news media frame labeled "crackdown" highlights the instrumental actions of the state in response to the violent resistance (e.g., the government crackdown on terrorism, policies implemented, and actions taken). Key words representing this frame include "President Xi," "pledge to," "request," "strike-hard operation," and "anti-terrorism." The second news media frame, labeled as "fact depiction," describes facts of the violent resistance, particularly the suffering and damages caused by the event. Key words in this frame include "Urumchi," "5.22 explosion," "train," "train station," "attack," and "explosion," all of which are key elements of the violent resistance. The third frame, labeled as "condemn & mourn," contains public condemnation and mourning on the loss of the victims during the violent resistance. The key words are mainly about people expressing their condolences, such as "condemn," "joint letter," "death," "5.22 explosion," "mourn," and "strongly." The fourth media frame, labeled as "ethnic harmony," indicates the solution to the violent resistance as the maintenance of ethnic unity and harmony. Key words in this frame include "people," "all ethnicities in China," "unity," "maintenance," and "anti-terrorism."

Four frames in online discussion forum were identified. The "ethnic conflict" frame attributes violent resistance to the conflict between ethnicities, especially between the Han Chinese and Uyghur. Key words in this frame include "Xinjiang," "Uyghur," "they," "problem," "Han," "we," and "poverty." Posts belonging to the "ethnic policy" frame attributes the violent resistance issue to the failure of the ethnic policy implemented by the central government for years. This policy prohibits local people from expressing their religious beliefs and revealing their cultural identity. The key words include "ethnicity," "policy," "they," "Xinjiang minority," "religion," and "state cadre." The third frame is labeled as "foreign intervention." Users of online discussion forum attribute the event to the intervention by foreign religious and political organizations. For example, some posts mentioned the financial support of the U.S. to Rebiya Kadeer, a political activist advocating for the independence of XUAR. Key words in this frame include "the United States," "Rebiya Kadeer," "terrorism," "support," "Eastern Turkistan," "Islam," and "organize." Similar to news media, online discussion forum also has a "crackdown" frame, which focuses on the "crackdown" of violent resistance. Key words for this frame include "state," "terrorist," "determined," "anti-terrorism," "attack," "safety," and "crackdown."

These frames both in news media and online discussion forum describe the uniqueness of discourse of this politically sensitive issue. For example, the "ethnic harmony" frame, which contains the key words like "all ethnicities in China", "unity maintenance", is a frame that is unique in Chinese context but invisible in Western media. The "crackdown" frames both in news media and in online discussion forum represent riot conflicts and terrorist attacks in particular but are invisible in political news in general. The "ethnic conflict" frame and "ethnic policy" frame are specific in online discussion forum whereas invisible shown in Chinese news media, concerning the severe censorship on Chinese news media.

Table 1a. Key words extracted to represent news media frames

General	Conflict	Human interest	Morality	Responsibility
Specific	Crackdown	Fact depiction	Condemn mourn	Ethnic harmony
Key words	President Xi	Urumchi	Condemn	People
	Request	5.22 explosion	Joint letter	All ethnicities in China
	Strike-hard operation	Train	Death	Firmly
	Crackdown	South station	Mourn	Unity
	Antiterrorism	Attack	5.22 explosion	Maintenance
	Activity	Explosion	University	Anti-terrorism
	Punish severely	Train station	Beijing	Ethnicity
	Pledge to	Policy	Strongly	We
	Separatist	Event	Urumchi	5.22 explosion
	Sentence	Detail	Youth	People
	Court	Death	College students	Stability
	Announcement	Implement	Denounce	Society

Table 1b. Key words extracted to represent public frames

General	Conflict	Responsibility frame	Responsibility frame	Responsibility frame
Specific	Crackdown	Ethnic conflict	Ethnic policy	Foreign intervention
Key words	State	Xinjiang	Ethnicity	United States
	Terrorist	Uyghur	Policy	Rebiya Kadeer
	Determined	They	They	Terrorism
	Organization	Problem	Xinjiang	China
	Anti-terrorism	Han	Minority	Xinjiang
	Case	We	We	Support
	Terrorism	Poverty	Religion	Terrorist
	Attack	Ethnicity	Problem	Eastern Turkistan
	Safety	Han people	Han	Islam
	Crack down	Fundamental	Region	Organize
	People	Hope	Country	Government
	Activity	Han	State cadre	Event

In addition to thematic differences, this study further compared the salience of news media frames and public frames in online discussion forum. All eight frames were categorized into generic frames according to the classic frame typology, as shown in the second row of Tables 1a and 1b. Entman (1993) noted that the "ethnic conflict" and "crackdown" frames belong to the conflict frame, "fact depiction" to the human interest frame, "condemn and mourn" to the morality frame, and "ethnic harmony", "ethnic policy", and "foreign intervention" to the responsibility frame.

As shown in Table 2, the discourses in online discussion forum and news media shared a common interest on the crackdown of the violent resistance (for news media frame, N = 1,011, 55%, whereas for public frame in online discussion forum, N = 1,657, 22%). However, the crackdown was mentioned much less often in online discussion forum than in news media, where the "crackdown" frame was the single prominent frame. As shown in Table 2, opinions expressed in news media were also different compared with those in online discussion forum. First, the public frames in online discussion forum were less emotional, as evidenced by the absence of human interest and morality frames. By contrast, about 20% (N = 399) of the stories in news media were denouncement of terrorists and mourning as a result of loss. Second, news coverage was fact-centered, whereas posts in online discussion forum were opinion-oriented. A total 6% (N = 104) of the stories in news media provided basic facts of the violent resistance, whereas the discourse in online discussion forum was more subjective, centering on the issue (i.e., the conflict between ethnicities, N = 1,246, 17%), emphasizing the intervention by foreign organizations and the U.S. (N = 2,203, 29%), and criticizing the failure of the ethnic policy of the central government (N = 1,315, 17%). Meanwhile, news media made no reference to either ethnic conflict or failure of the ethnic policy. Third, the public frame in online discussion forum did not indicate appropriate solutions to address the problem, whereas in news media, a sizable proportion of stories (N = 157, 9%) called for continued ethnic unity as a primary solution to the crisis.

Table 2. Distribution of news frames & public frames on social media

Frame categories	Frames	News frames (N = 1,836)	Public frames (N = 7,560)
Conflict frame	Government crackdown on terrorism	1011 (55%)	1657 (22%)
Human interest frame	Fact	104 (6%)	–
Morality	Condemn & Mourn	399 (20%)	–
Responsibility frame	Solution: ethnic harmony	157 (9%)	–
	Ethnicity tension	–	1246 (17%)
	Ethnic policy	0	1315 (17%)
	Foreign intervention	0	2203 (29%)
Undefined		165 (9%)	1116 (15%)

3.2 Interplay Between News Media Frames and Public Frames in Online Discussion Forum

Table 3 presents the results of the interplay between news media frames and public frames in online discussion forum. For each frame on either online discussion forum or news media, this study examined the Granger causality relationship between the frequency of the focal frame and that (i.e., the value of one-day time lag) of other frames in another media platform while controlling for the possible auto-correlation.

We identified the intertwined relationship between news media frames and public frames in online discussion forum, unlike previous studies on frame setting that identified the transmission of attribute salience of political events reported in news media to those in the public, or vice versa (Scheufele 1999). The results in Table 3 indicate that news media and online discussion forum do not set frames for each other, but influence each other in strengthening their own existing frames. Therefore, Hypothesis 1 and Hypothesis 2 are not supported.

Particularly, the portrayal of news media on official actions to the violent resistance (i.e., the "crackdown" frame) strengthened the salience of the opinion frames in online discussion forum (i.e., "ethnic policy,", "ethnic conflict," and "foreign intervention" frames), whereas public frames in online discussion forum did not influence the frames that describe fact and official government actions in news media. This result suggests that users of online discussion forum refer to news media as a primary official source for constructing their own discourse. The emphasis of government actions in news media will stimulate the questions of online discussion forum users on government ethnic policy and the connection of the violent attack with foreign intervention and ethnic conflict.

In the same vein, online discussion forum does not set their frames in news media. Instead, online discussion forum influences the salience of existing frames (i.e., moral judgment and responsibility) in news media. The responsibility and morality frames in news media (i.e., "ethnic harmony" and "condemn and mourn" frames) negatively interacted with the responsibility frame in online discussion forum (i.e., "ethnic policy" frame). The increase in questions on government ethnic policy over the past decades in online discussion forum will decrease the call for ethnic harmony and condemn terrorists in news media. This result suggests a so-called adjustment effect. Journalists will face and respond to the discourse challenge in online discussion forum by monitoring the discourse and adjusting the salience of their own issue framing. In addition, the conflict frame (i.e., "ethnic conflict" frame) in online discussion forum positively influences the responsibility and morality frames in news media (i.e., "ethnic harmony" frame and "condemn and mourn" frames). The emphasis on ethnic conflict in online discussion forum will introduce the focus on the legitimacy of ethnic harmony and on condemning violent activities in news media. The co-evolution dynamics between news media frames and public frames in online discussion forum shows that journalists attempted to clearly distinguish between "amateur netizens" and "professional journalists (Tong 2015). News media denies the definition of a violent resistance by online discussion forum as "ethnic conflict." On behalf of the party-state, news media sustains the legitimacy of the government in enforcing the law against the violent attack by emphasizing the moral judgment and highlighting the maintenance of ethnic harmony.

Interestingly, this study also found divided opinions within online discussion forum, as evidenced by the negative correlation between the "crackdown" and "ethnic policy" frames at the succeeding time point. The increase in the salience of enforcing regulation against terrorism will trigger the decrease in doubts on the government ethnic policy at the next time point, and vice versa. Opinions in online discussion forum compete with one another; thus, opinions in online discussion forum are more divergent than the discourse in news media.

Table 3. Interplay between news frame and public frame on social media.

		Public frame on social media				News frame			
		Ethnic policy	Ethnic conflict	Foreign intervention	Crack down	Fact	Ethnic harmony	Condemn and mourn	Crack down
Public frame on social media	Lag.Ethnic Policy	.31 (.21)	.29 (.22)	.37 (.31)	−.35 (.28)	−0.02 (.02)	−0.07 (.02)*	−0.28 (.06)***	−0.14 (.13)
	Lag. EthnicConflict	−.21(.49)	.38 (.51)	.56 (.71)	0.93 (.64)	0.04 (.05)	0.13 (.05)*	0.46 (.13)**	0.08 (.29)
	Lag.Foreign Intervention	.35 (.37)	−.34 (.39)	−.84 (.54)	−0.76 (.49)	−0.03 (.04)	−0.06 (.04)	−0.15 (.1)	0.0001 (.22)
	Lag. CrackDown	−1.82 (.38)***	−.62 (.39)	.48 (.55)	1.03 (.49)*	0.03 (.04)	0.05 (.04)	0.13 (.1)	0.44 (.22)
News frame	Lag.Fact	0.35 (1.91)	−.05 (1.99)	.14 (2.78)	−1.09 (2.51)	0.33 (.2)	−0.11 (.21)	0.34 (.51)	−0.9 1.12)
	Lag.Ethnic Harmony	3.1 (2.43)	−0.25 (2.54)	−1.63 (3.55)	−1.36 (3.2)	−0.27 (.26)	0.07 (.26)	−0.48 (.65)	−0.38 (1,43)
	Lag.condemn & Mourn	.92 (.73)	−.31 (.76)	−0.43 (1.06)	0.25 (.95)	0.03 (.08)	−0.02 (.08)	−0.1 (.19)	−0.05 (.43)
	Lag.crack down	3.77 (.52)***	2.5 (.54)***	1.92 (.76)*	0.51 (.69)	−0.03 (.06)	0.04 (.06)	0.2 (.14)	0.02 − (.31)
R2 (%)		77.2	64.6	62.6	58.9	21.4	37.3	66.4	64.4
F-Value		13.27***	7.61***	7.06***	6.2***	0.716	3.16*	8.164***	7.516***

Notes: * $p = 0.05$; ** $p = 0.01$; ***$p = 0.001$

4 Discussion

This study, which focuses on the violent resistance issue in XUAR, examines the role of online discussion forum and news media in shaping political discourse in China. News media frames and public frames in online discussion forum were found to differ in terms of thematic structure and weight of frames (i.e., the proportion of number of articles/posts in a frame to the total number of articles/posts in a given media platform).

Moreover, the current study further examines the interplay between news media frames and public frames in online discussion forum. In contrast to research findings in other political contexts, this study found that news media and online discussion forum did not set frames for each other. Instead, they influenced each other by strengthening their own existing frames. The theoretical and methodological implications of this study will be discussed in subsequent sections.

4.1 Frame Interaction Between News Media and Online Discussion Forum in China

This study contributes to addressing a gap in the literature on framing by investigating the interplay of the framing dynamics between online discussion forum and news media using longitudinal data. Previous research on frame setting effects has maintained that news media obtains facts to make some aspects of political issues more salient in stimulating individuals to activate their prior knowledge and efficiently process information conveyed in news coverage; these facts further influence the recipients' cognition, transmit the news frames to the public, and then shape public dialogues about political issues (Scheufele 1999).

In contrast to the general notion of frame-setting effects, this study found that news media failed to set frames for the public in online discussion forum, and the reverse frame setting from social media to news media did not raised on the issue which is politically sensitive in China. The two media platforms had mutual influences, that is, news and online discussion forum interacted to place the issue on their own agenda of building and strengthening their existing frames instead of presenting a dominating relationship. Specifically, news media has been attentive to the challenge of online discussion forum. However, with the pressure of ideology considerations and media routines, news media did not correspond to the challenge by actively reframing the issues or adopting frames that had been presented in online discussion forum. Chinese news media still plays the role of "mouthpiece," as assigned by the Party-run state. Public opinion in online discussion forum hardly penetrates mainstream news media. On the other hand, online discussion forum is not slavishly dependent on the voice in news media. The publics in online discussion forum persist in their alternative perspectives, despite the enforcement of elite frames in news media.

The interplay further encourages the separation between official and nonofficial discourse spaces, as noted by He (2008). Although the emergence of social media has generated renewed attention to frame setting in other social contexts (Russell Neuman et al. 2014), we should not be overly optimistic about the disappearance or convergence of dualistic discourse spaces in China on politically sensitive issues. Among the challenges raised in social media, state ideology and media professional routines are still the two dominant factors influencing frame building in news media. These two factors prevent the frame transition and convergence of separate discourse spaces between social and news media.

4.2   Topic Modeling as an Alternative Approach for Frame Identification

Previous framing studies adopted manual content analysis and discourse analysis to categorize news stories into several predefined frames. The manual content analysis assumed that researchers have knowledge before the analysis, knowledge that they apply in the analysis (DiMaggio et al. 2013). The application of prior knowledge is very impractical in most framing analysis. Many scholars have pointed to threats to reliability and validity in frame analysis because of the subjective judgment and bias of the researcher (D'Angelo 2002). To reflect these concerns, this investigation hopes to make a methodological contribution by introducing an alternative approach for the content analysis of media frames.

Frames, which serve as the thematic unit of social issues, are produced through the emphasis, repetition, and structure of narratives (Entman 1993). (Pan and Kosicki 1993) and Scheufele (1999) argued that the arrangement of words and phrases is one of the most important framing elements that can be empirically operationalized. These frame elements group together in a systematic way, thereby forming frames.

Topic modeling extracts and classifies frames by calculating the frequency of words and their co-occurrence relationship in a collection of documents. This new data-driven approach is inherently consistent with the general principle of frame identification. In fact, topic modeling is not a brand new approach to detect frames. Based upon Entman's notion that frames are manifested in the use of specific words (Entman 1993), Miller (1997) has suggested a similar computer-assisted approach, known as "frame mapping" to detect frames twenty years ago. Frame mapping is a method of finding particular words that occur together in some text while not tend to occur together in other texts (Miller 1997). Words that tend to occur together in texts are identified with the help of clustering algorithms (Matthes 2008). In comparison with manual-holistic approach of frame detection, this computer-assisted approach is believed more objective and reliable (Matthes 2008; Miller 1997). In this regard, topic modeling is a technical extension of Miller's computer-assisted approach of frame detection.

Topic modeling avoids subjectivity and bias of manual coding. This approach helps uncover frames that a researcher might not otherwise be aware of by manual coding and therefore could identify frames that are substantively plausible and statistically validated (DiMaggio et al. 2013). In addition, as Golder and Macy (2014) mentioned, computational approaches address fundamental puzzles of social science. The availability of large-scale online data has equipped social science scholars with a deep understanding of the frame-building process, while more powerful approaches are being required. LDA topic modeling, as one of the new textual analysis methods, allows large quantities of textual data to be analyzed. Thus, future research can automatically reduce text complexity, extract latent frames, and monitor the dynamics of public opinion.

References

Blei, D.M., Ng, A.Y., Jordan, M.I.: Latent Dirichlet allocation. J. Mach. Learn. Res. **3**, 993–1022 (2003)

Chermak, S.M., Gruenewald, J.: The media's coverage of domestic terrorism. Justice Q. **23**(4), 428–461 (2006)

D'Angelo, P.: News framing as a multiparadigmatic research program: a response to Entman. J. Commun. **52**(4), 870–888 (2002)

DiMaggio, P., Nag, M., Blei, D.: Exploiting affinities between topic modeling and the sociological perspective on culture: application to newspaper coverage of US government arts funding. Poetics **41**(6), 570–606 (2013)

Entman, R.M.: Framing: toward clarification of a fractured paradigm. J. Commun. **43**(4), 51–58 (1993)

Goffman, E.: Frame Analysis: An Essay on the Organization of Experience. Harvard University Press, Cambridge (1974)

Golder, S.A., Macy, M.W.: digital footprints: opportunities and challenges for online social research. Annu. Rev. Sociol. **40**(1), 129–152 (2014)

He, Z.: SMS in China: a major carrier of the nonofficial discourse universe. Inf. Soc. **24**(3), 182–190 (2008)

Luther, C.A., Zhou, X.: Within the boundaries of politics: news framing of SARS in China and the United States. Journal. Mass Commun. Q. **82**(4), 857–872 (2005)

Matthes, J., Kohring, M.: The content analysis of media frames: Toward improving reliability and validity. J. Commun. **58**(2), 258–279 (2008)

Miller, M.: Frame mapping and analysis of news coverage of contentious issues. Soc. Sci. Comput. Rev. **15**, 367–378 (1997)

Pan, Z., Kosicki, G.M.: Framing analysis: an approach to news discourse. Polit. Commun. **10**(1), 55–75 (1993)

Russell Neuman, W., Guggenheim, L., Mo Jang, S., Bae, S.Y.: The dynamics of public attention: agenda-setting theory meets Big Data. J. Commun. **64**(2), 193–214 (2014)

Scheufele, D.A.: Framing as a theory of media effects. J. Commun. **49**(1), 103–122 (1999)

Shoemaker, P., Reese, S.D.: Mediating the message. Routledge, New York (2011)

Tai, Z.: The Internet in China: Cyberspace and Civil Society. Routledge, New York (2013)

Tong, J.: The defence of journalistic legitimacy in media discourse in China: an analysis of the case of Deng Yujiao. Journalism **16**(3), 429–446 (2015)

A Unified Framework of Lightweight Local Community Detection for Different Node Similarity Measurement

Jinglian Liu[1,2], Daling Wang[1,3(✉)], Weiji Zhao[2,4], Shi Feng[1,3], and Yifei Zhang[1,3]

[1] School of Computer Science and Engineering, Northeastern University, Shenyang, People's Republic of China
datamining@163.com,
{wangdaling,fengshi,zhangyifei}@cse.neu.edu.cn
[2] School of Information Engineering, Suihua University, Suihua, People's Republic of China
sdzhaoweiji@163.com
[3] Key Laboratory of Medical Image Computing of Ministry of Education, Northeastern University, Shenyang, People's Republic of China
[4] School of Computer Science and Technology, Harbin University of Science and Technology, Harbin, People's Republic of China

Abstract. Local community detection aims at finding a local community from a start node in a network without requiring the global network structure. Similarity-based local community detection algorithms have achieved promising performance, but still suffer from high computational complexity. In this paper, we first design a unified local community detection framework for fusing different node similarity measurement. Based on this framework, we implement eight local community detection algorithms by utilizing different node similarity measurements. We test these algorithms on both synthetic and real-world network datasets. The experimental results show that the local community detection algorithms implemented in our framework are better at detecting local community compared with related algorithms. That means the performance of discovering local community would be largely improved by using good node similarity measurements. This work provides a novel view to evaluate similarity measurements, which can be further applied to link prediction, recommendation system and so on.

Keywords: Local community detection · Node similarity measurement · Community structure · Lightweight · Unified framework

1 Introduction

Detecting community structure in complex networks such as social networks [6, 8, 21], collaboration networks [14], the Internet [4], and E-mail networks [22] has attracted lots of attention in recent years. These complex networks are usually modeled as graphs which are composed of nodes representing entities and edges representing relationships

© Springer Nature Singapore Pte Ltd. 2017
X. Cheng et al. (Eds.): SMP 2017, CCIS 774, pp. 283–295, 2017.
https://doi.org/10.1007/978-981-10-6805-8_23

between entities. A community refers to a group of nodes that are more densely connected to each other than with the rest of the network [5, 6, 18, 19].

Traditional community detection methods are based on global network structure [3, 6, 15, 17, 20, 22]. However, for some real-world networks, such as Web network and some online social networks, they are usually too huge to get their global network structure [7]. Therefore, the global network structure based methods have difficulty in handling these huge networks. For solving this problem, local community detection was first proposed in [2] and has recently attracted intensive research interests.

Local community detection aims at finding a community from a start node without requiring the global network structure, and various algorithms have been proposed in recent years [1, 2, 7, 12, 13, 24]. The existing local community detection methods can be classified into two categories: degree-based methods and similarity-based methods [13]. Degree-based methods discover local community by using the nodes' degree, such as l-shell expansion algorithm [1], greedy optimization local modularity R algorithm [2], and greedy optimization local modularity M algorithm [12]. Similarity-based methods discover local community by using similarities between pairs of nodes, such as LTE [7] and GMAC algorithm [13]. In previous studies, the similarity-based local community detection algorithms have achieved promising performance.

In this paper, we focus on similarity-based local community detection methods. The previous LTE and GMAC which need to calculate the similarities between pair of nodes whose shortest path length is within a predefined threshold, and the calculation suffers from high computational complexity. Inspired by LTE [7] and GMAC [13] algorithms, we propose a unified local community detection framework. Moreover, eight local community detection lightweight algorithms are implemented in our framework. Here "lightweight" measurement means only calculating the similarity between pairs of connected nodes. Our main contributions are summarized as follows:

- We propose a unified framework of local community detection for node similarity measurement. New local community detection algorithms are implemented by embedding different node similarity measurements into this framework.
- We implement eight local community detection lightweight algorithms compared with LET [7] and GMAC [13] algorithms.
- We test the algorithms on both synthetic and real-world network datasets. The experimental results show that the algorithms implemented by our framework are better at local community detection compared with related algorithms.

The rest of the paper is organized as follows. Section 2 gives the problem definition of local community detection and some related definitions, and introduces some related work. We describe our framework and algorithm in Sect. 3 and report experimental results in Sect. 4, and followed by conclusions in Sect. 5.

2   Preliminaries

In this section, we first give the definition of local community detection problem in network, and then introduce some related work in local community detection based on node similarity measurement and necessary definitions.

2.1 Problem Description

Definition 1 (Network). Let $G = (V, E)$ be an undirected graph, where V is the set of nodes and E is the set of edges in G, and $n = |V|$ is the number of nodes in G. For two nodes, $x, y \in V$, $(x, y) \in E$ indicates that there is an edge between nodes x and y. Moreover, $m = |E|$ is the number of edges in G. The set of nodes adjacent to node x is denoted by $\Gamma(x)$, $\Gamma(x) = \{y \mid y \in V, (x, y) \in E\}$. The degree of node x is the number of nodes in $\Gamma(x)$, denoted by k_x.

Local community detection focuses on finding local community D in an undirected graph $G = (V, E)$ starting from a node $s \in V$. The networks we focus on in this paper are of undirected and unweighted type.

Note that the entire network structure is unknown at the beginning of the community detection. As shown in Fig. 1, we can dynamically divide the entire network nodes into three parts: interior node set D, D's shell node set N, and unknown node set U, $U = V - D - N$. The interior node set D and the edges among them constitute a local community because of higher similarity among the nodes in D (We call D as local community without ambiguity). Every node in N has at least one adjacent node in D. During the process of detecting local community, only partial network information, i.e., nodes in $D \cup N$ and their linkage information are available. Similar definitions of local community detection can be found in [2, 11, 23].

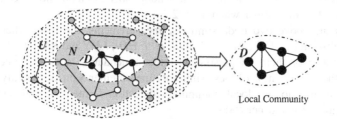

Fig. 1. An illustration of division of a network nodes into interior node set D (block nodes), D's shell node set N (white nodes), and unknown node set U (grey nodes).

2.2 Related Work

Similarity-based local community detection methods evaluate the local community quality by using the similarities between nodes. These quality metrics focus on the internal similarity and external similarity of the community [13]. For a local community D, the internal similarity of D is the sum of similarities between any two adjacent nodes both inside the community D, while the external similarity of D is the sum of similarities between any two adjacent nodes with one node in D and the other outside of D.

LTE [7] and GMAC [13] algorithms are representatives of this kind of methods. LTE introduced a local community quality metric called *tightness* which is defined as internal similarity divided by the sum of internal and external similarity of the local community. GMAC introduced a local community quality metric called *Compactness-Isolation* which is defined as the ratio of internal similarity to external similarity of the local community. These metrics are based on node similarity measurements which play key role in this kind of local community detection methods.

Moreover, [10, 26] surveyed eight node similarity measurements, they are *Common Neighbor* (*CN* for short), *Salton Index* (*SAL* for short), *Jaccard Index* (*JAC* for short), *Sorenson Index* (*SOR* for short), *Hub Promote Index* (*HPI* for short), *Hub Depressed Index* (*HDI* for short), *Leicht-Holme-Newman Index* (*LHN* for short), and *Resource Allocation index* (*RA* for short). In our unified framework, we implemented eight algorithms with these node similarity measurements.

2.3 Node Similarity

For measuring the similarities between pairs of nodes, LTE algorithm [7] adopted a structure similarity between two adjacent nodes, while GMAC algorithm [13] adopted d-neighbors based similarity measurement which takes into account non-adjacent nodes within a distance away. Suppose the nodes' average degree in G is k, the above two methods' computational complexities are $O(k)$ and $O(k^d)$. The latter method uses a relatively larger level nodes to represent a node, so its computational complexity is much higher than the former when $d \geq 2$.

Besides the above two node similarity measurements, there are other similarity measurements, such as *CN*, *SAL*, *JAC*, *SOR*, *HPI*, *HDI*, *LHN* and *RA*. [10, 26] surveyed the performance of them on personalized recommendation and link precision. For evaluating the effectiveness of these similarity measurements on community detection, we embed them into our local community detection framework and test them on synthetic and real-world networks.

2.4 Local Community Quality

Inspired by [7, 12, 13], we adopt a similarity-based metric Compactness-Isolation (*CI* for short) to quantify the local community quality based on node similarity.

Definition 2 (Compactness-Isolation Metric). For a network $G = (V, E)$ and any two nodes $x, y \in V$, the similarity between x and y is denoted by s_{xy}. For a local community D with shell node set N, the Compactness-Isolation Metric of D, denoted by $CI(D)$, is defined as

$$CI(D) = \frac{\sum\limits_{x,y \in D, (x,y) \in E} s_{xy}}{1 + \sum\limits_{u \in D, v \in N, (u,v) \in E} s_{uv}} \tag{1}$$

where the numerator is the internal similarity of D, and the denominator is one plus the external similarity of D. Such a denominator is to avoid dividing by zero [13]. Different node similarity measurements can lead to different local community quality metrics.

A good local community should have high internal similarity and low external similarity, which will lead to a high CI value.

3   Lightweight Local Community Detection Framework

In this section, we first give a local community detection framework based on node similarity, and then give an explanation of why the algorithm implemented in our framework is lightweight.

3.1   Local Community Detection Framework Based on Node Similarity

We give a local community detection framework in Fig. 2, and a lot of local community detection algorithms can be implemented by utilizing different node similarity measurements in this framework.

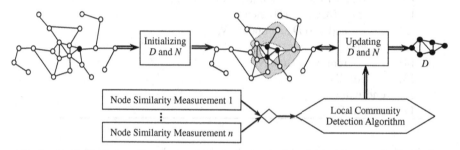

Fig. 2. Local community detection framework embedded multi algorithms with different similarity measurement between nodes

According to Fig. 2, the process of our local community detection is further described. Firstly, initialize $D = \{s\}$ and $N = \Gamma(s)$. At each step, the node in N which has maximum similarity with nodes in D is selected as candidate node. If agglomerating the candidate node into D will cause an increase in CI, then add it to D, otherwise, remove it from N. Repeat this step until N is empty. Finally, D is the local community discovered from node s. The pseudo code of above process is shown as follows.

Algorithm: Lightweight Local Community Detection Framework

Input: a start node s, a network $G = (V, E)$;
Output: local community D;
Describe:
   1)   initialize $D=\{s\}$, $N=\Gamma(s)$;
   2)   **while** $N \neq \phi$ **do**
   3)   create a new dictionary variable *dic_sim* to store the similarities
         of nodes belonging to N with D;
   4)   **for each** $i \in N$ **do**
         //compute similarities between node i and its neighbors in D
   5) $dic\_sim[i] = 0$;
   6)   **for each** $j \in \Gamma(i) \cap D$ **do**
            // different algorithms can be implemented by
            // embedding different node similarity measurements
   7) $dic\_sim[i]+=s_{ij}$; //s_{ij} is the similarity between nodes i and j
   8)   **end for**
   9)   **end for**
  10)   find a whose $dic\_sim[a]$ is maximum;
  11)   **if** $CI(D \cup a) > CI(D)$ **then**
  12)   add a to D and update N;
  13)   **else**
  14)   remove a from N;
  15)   **end if**
  16) **end while**
  17) **return** D;

A lot of local community detection algorithms can be implemented by calculating s_{ij} via different node similarity measurements. They are described in line $(6) \sim (8)$ of the algorithm and shown in "Node Similarity Measurement $1 \sim n$" of Fig. 2. For evaluating the performance of these algorithms, we compare eight node similarity measurements on both synthetic and real-world network datasets in next section.

3.2   Lightweight Local Community Detection Algorithm

Compared with traditional similarity-based algorithms, the local community detection algorithms implemented in our framework are lightweight.

Traditional similarity-based local community detection algorithms, such as LTE [7] and GMAC [13], choose node i in N as candidate node by calculating the sum of similarities between i and all the nodes in D. In fact, the similarities between i and those nodes in D that are not adjacent to i have no effect on CI except for more calculations. This restricts these algorithms from huge networks. For solving this problem, our algorithms only calculate the similarities between i and the nodes in D that are adjacent to i as its similarity with D. This means that only for $(i, j) \in E$, the similarity between

i and j will be calculated by one of the existing similarity measurements. This operation reduces computational complexity on the premise of optimizing CI.

For example, in Fig. 3, for node 7, when computing similarity between 7 with local community D, our algorithm only considers node 2, but ignores node 1, 3, 4, 5, 6. This is main difference between our lightweight algorithm with LTE and GMAC.

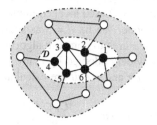

Fig. 3. An example of computing similarity between nodes

3.3   Computational Complexity Analysis

In our algorithm, we store network as a hash table of nodes in this network. Each node is denoted by a unique integer and associated with a vector of its adjacent nodes. The values in vectors are sorted for faster access.

Let t denote the size of $D \cup N$, E_{in} denote the number of edges with two nodes in D, E_{out} denote the number of edges with one node in D and the other in N, and k denote the mean node degree of nodes in $D \cup N$. The computational cost of our algorithm mainly consists of two parts: calculating the similarities of $(E_{in} + E_{out})$ pairs of connected nodes and choosing a node in N as candidate node.

Before calculating the similarities of $(E_{in} + E_{out})$ pairs of connected nodes, we need to compute t nodes' neighbor nodes firstly, and then compute $(E_{in} + E_{out})$ pairs of connected nodes. Their time complexity is $O(k \cdot t)$ and $O(k \cdot (E_{in} + E_{out}))$ respectively. Adding these together, the time complexity is $O(k \cdot (E_{in} + E_{out} + t))$. Another most computational expensive steps is in lines $(4) \sim (10)$, which is the time to find $a \in N$ having the maximal similarity with the current local community D. Except the above computations, in each while-loop, the time complexity is $O(|N|)$. So, the running time of our algorithm depends on the size of the union of local community and its shell node set rather than that of the entire graph.

4   Experiments

By embedding *CN, SAL, JAC, SOR, HPI, HDI, LHN* and *RA index* [10, 26], we implement eight local community detection algorithms, denoted by LLCDF-CN, LLCDF-SAL, LLCDF-JAC, LLCDF-SOR, LLCDF-HPI, LLCDF-HDI, LLCDF-LHN and LLCDF-RA. In this section, we evaluate the performance of these algorithms on synthetic as well as real-world networks.

4.1   Related Methods and Evaluation Criteria

We compare our algorithms with three representative local community detection algorithms: (1) Clauset's algorithm [2] is the first algorithm to find local community by maximizing metric R. Note that the same as [13, 23], we improve its stopping criteria by detecting changes in R. (2) Luo et al.'s algorithm [12] (LWP for short) is a well-known algorithm to find the sub-graph with maximum metric M. (3) GMAC algorithm [13] is a popular similarity-based local community detection algorithm which uses d-neighbors to represent node. We fix $d = 3$ as suggested by authors.

We use three evaluation measures to compare algorithmic performance: *precision*, *recall* and *F-score*, which are widely adopted by many community detection methods [7, 11, 13, 23, 24]. The *precision* and *recall* are calculated as follows.

$$precision = \frac{|C_F \cap C_R|}{|C_F|} \tag{2}$$

$$recall = \frac{|C_F \cap C_R|}{|C_R|} \tag{3}$$

where C_R is the set of nodes in real local community which contains the given node and C_F is the set of nodes discovered by local community detection algorithm which starts from the given node.

F-score is the harmonic mean of *precision* and *recall*. Its formula is as follows.

$$F - score = 2 \times \frac{precision \times recall}{precision + recall} \tag{4}$$

4.2   Evaluation on Synthetic Networks

For comparing the performance of various local community detection algorithms, we first generate 10 LFR benchmark networks [9] with ground-truth community structure. There are 500 nodes in each network.

LFR benchmark networks, introduced by Lancichinetti et al. [9], are widely used to test algorithms identifying community structure in networks [7, 13]. The important properties of this network generating model are defined as follows: the number of nodes is denoted by n, the average degree of nodes is denoted by k, the maximum degree is denoted by k_{max}, mixing parameter is denoted by μ, minus exponent for the degree sequence is denoted by $t1$, minus exponent for the community size distribution is denoted by $t2$, number of overlapping nodes is denoted by on, number of memberships of the overlapping nodes is denoted by om, minimum for the community sizes is denoted by $minc$, maximum for the community sizes is denoted by $maxc$. These parameters are set as follows: $n = 500$, $k = 10$, $k_{max} = 50$, others except μ use default values. Mixing parameter u is the fraction of edges of each node outside its community, which is used to control the difficulty of community detection [19]. So we generate 10 networks with different mixing parameter μ ranging from 0.05 to 0.5 with a span of 0.05. These networks are generated with ground-truth communities.

For each network in our experiments, we use every node in this network as a start node once, and repeat the local community detection experiments for 500 times which start from different node every time, then report algorithmic average *precision, recall* and *F-Score* on this network. Figure 4 shows the comparison results of *precision, recall, F-score* for eleven algorithms on these networks, respectively.

We discuss the experimental results in detail. Firstly, with the increase of μ, all the eleven algorithms suffer varying degree of performance degradation. This is because the higher the mixing parameter μ of a network, the weaker community structure it has. Secondly, with the increase of μ, the performance of LWP, Clauset and GMAC drops rapidly, meanwhile our algorithms drop slowly. This is because both LWP and Clauset simply depend on the number of edges incident to the node, and neglect the fact that the similarity between nodes of external edge is smaller than that of internal edge. GMAC algorithm returns too many nodes that are not relevant and the low *precision* leads to bad performance.

The *precision, recall,* and *F-score* of the LWP algorithm is zero or nearly zero when $\mu \geq 0.35$. This is because all the local communities discovered by LWP algorithm satisfy $M > 1$, which means the number of internal edges should be more than the number of external edges. However, almost no local community can satisfy $M > 1$ when $\mu \geq 0.35$, so LWP algorithm performs badly in this case. This conclusion is in accordance with the results reported in Ref. [7].

In general, the algorithms implemented in our framework outperform the other three algorithms. Among these eight similarity measurements, LLCDF-CN has the best overall performance, and LLCDF-RA performs the second best.

4.3   Evaluation on Real-World Networks

In this subsection, we use additional three real-world networks to evaluate the performance of our algorithms. (1) The first network is Zachary Karate Club Network (Karate for short) [25], in which $n = 34$ and $m = 78$. (2) The second is NCAA football network (Football for short) [6], in which $n = 115$ and $m = 613$. (3) The third is Books about US politics (Polbooks for short) [16], in which $n = 105$ and $m = 441$.

In our experiments, we use every node in these network as a start node once, and repeat the local community detection experiments for n times which start from different node every time (n is the number of nodes in this network), and report algorithmic average *precision, recall* and *F-Score* on this network. The comparison results on these real-world networks are reported in Fig. 5. The algorithms implemented in our framework usually outperform the other three algorithms. Among these similarity measures, LLCDF-CN has the best overall performance, and LLCDF-RA performs the second best.

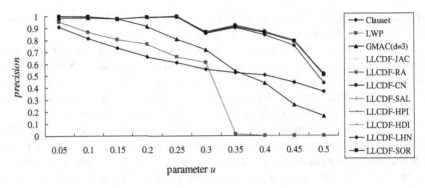

(a) Comparison result of Precision

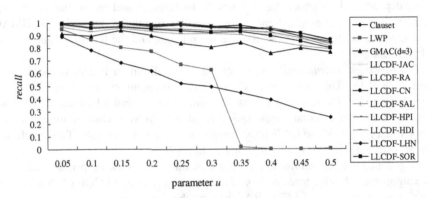

(b) Comparison result of Recall

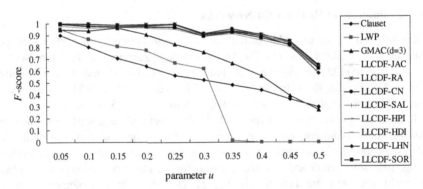

(c) Comparison result of F-score

Fig. 4. Comparison results on LFR benchmark networks

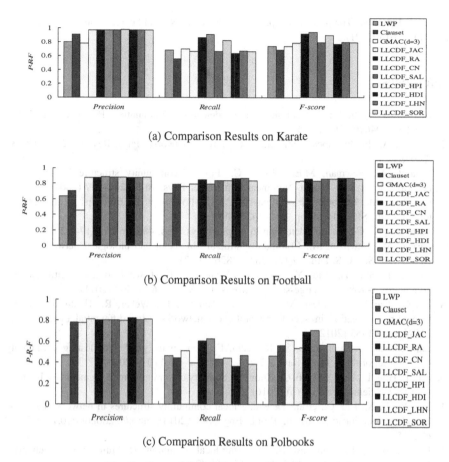

(a) Comparison Results on Karate

(b) Comparison Results on Football

(c) Comparison Results on Polbooks

Fig. 5. Comparison results on real-world networks

5   Conclusion and Future Work

For fusing different node similarity measurement algorithms, we design a unified local community detection framework based on node similarity. Our framework opens a rich space for researches, new local community detection algorithms can be implemented by utilizing different node similarity measures in this framework. We present an evaluation of eight widely used similarity measurements on both synthetic and real-world networks. Compared with other related algorithms, the local community detection algorithms implemented in our framework don't need any manual parameters, and achieve good performance on both synthetic and real-world networks, especially *CN* and *RA* have outstanding performance.

In the future, we will consider the community detection in heterogeneous social media networks, and study on the node similarity measurement between heterogeneous nodes and between multimodal nodes.

Acknowledgments. The project is supported by National Natural Science Foundation of China (61370074, 61402091).

References

1. Bagrow, J., Bolt, E.: A local method for detecting communities. Phys. Rev. E **72**(4), 046108-1–046108-10 (2005)
2. Clauset, A.: Finding local community structure in networks. Phys. Rev. E **72**(2), 026132 (2005)
3. Clauset, A., Newman, M.E., Moore, C.: Finding community structure in very large networks. Phys. Rev. E Stat. Nonlinear Soft Matter Phys. **70**(6), 264–277 (2004)
4. Faloutsos, M., Faloutsos, P., Faloutsos, C.: On Power-law relationships of the internet topology. In: SIGCOMM, pp. 251–262 (1999)
5. Fortunato, S.: Community detection in graphs. Phys. Rep. **486**(3/5), 75–174 (2010)
6. Girvan, M., Newman, M.: Community structure in social and biological networks. Proc. Natl. Acad. Sci. U.S. Am. **99**(12), 7821–7826 (2002)
7. Huang, J., Sun, H., Liu, Y., Song, Q., Weninger, T.: Towards online multiresolution community detection in large-scale networks. PLoS ONE **6**(8), 492 (2011)
8. Jia, G., Cai, Z., Musolesi, M., Wang, Y., Tennant, D., Weber, R., Heath, J., He, S.: Community detection in social and biological networks using differential evolution. In: LION, pp. 71–85 (2012)
9. Lancichinetti, A., Fortunato, S., Radicchi, F.: Benchmark graphs for testing community detection algorithms. Phys. Rev. E **78**(4), 046110-1–046110-5 (2008)
10. Liu, J., Hou, L., Pan, X., Guo, Q., Zhou, T.: Stability of similarity measurements for bipartite networks. Scientific reports (6) (2016)
11. Liu, Y., Ji, X., Liu, C., et al.: Detecting local community structures in networks based on boundary identification. Math. Probl. Eng. 1–8 (2014). http://dx.doi.org/10.1155/2014/682015
12. Luo, F., Wang, J., Promislow, E.: Exploring local community structures in large networks. Web Intell. Agent Syst. (WIAS) **6**(4), 387–400 (2008)
13. Ma, L., Huang, H., He, Q., Chiew, K., Wu, J., Che, Y.: GMAC: A Seed-Insensitive Approach to Local Community Detection. In: Bellatreche, L., Mohania, Mukesh K. (eds.) DaWaK 2013. LNCS, vol. 8057, pp. 297–308. Springer, Heidelberg (2013). doi:10.1007/978-3-642-40131-2_26
14. Newman, M.: The structure of scientific collaboration networks. Working Pap. **98**(2), 404–409 (2000)
15. Newman, M.: Fast algorithm for detecting community structure in networks. Phys. Rev. E Stat. Nonlinear Soft Matter Phys. **69**(6), 066133-1–066133-5 (2004)
16. Newman, M.: Modularity and community structure in networks. Proc. Natl. Acad. Sci. **103** (23), 8577–8582 (2006). http://www-personal.umich.edu/~mejn/netdata/
17. Newman, M., Girvan, M.: Finding and evaluating community structure in networks. Phys. Rev. E Stat. Nonlinear Soft Matter Phys. **69**(2), 026113-1–026113-15 (2004)
18. Radicchi, F., Castellano, C., Cecconi, F., et al.: Defining and identifying communities in networks. Proc. Natl. Acad. Sci. U.S. Am. **101**(9), 2658–2663 (2004)
19. Schaeffer, S.: Graph clustering. Comput. Sci. Rev. (CSR) **1**(1), 27–64 (2007)
20. Shao, J., Han, Z., Yang, Q., Zhou, T.: Community detection based on distance dynamics. In: Proceedings of the 21th ACM SIGKDD International Conference on Knowledge Discovery and Data Mining, pp. 1075–1084 (2015)

21. Takaffoli, M.: Community evolution in dynamic social networks - challenges and problems. In: ICDM Workshops, pp. 1211–1214 (2011)
22. Tyler, J.R., Wilkinson, D.M., Huberman, B.A.: Email as spectroscopy: automated discovery of community structure within organizations. Inf. Soc. **21**(2), 143–153 (2005)
23. Wu, Y., Huang, H., Hao, Z., Chen, F.: Local community detection using link similarity. J. Comput. Sci. Technol. (JCST) **27**(6), 1261–1268 (2012)
24. Wu, Y., Jin, R., Li, J., Zhang, X.: Robust local community detection: on free rider effect and its elimination. In: VLDB, pp. 798–809 (2015)
25. Zachary, W.: An information flow model for conflict and fission in small groups. J. Anthropol. Res. **33**(4), 452–473 (1977)
26. Zhou, T., Lü, L., Zhang, Y.: Predicting missing links via local information. Eur. Phys. J. B **71**(4), 623–630 (2009)

Estimating the Origin of Diffusion in Complex Networks with Limited Observations

Shuaishuai Xu, Yinzuo Zhou$^{(\boxtimes)}$, and Zike Zhang

Alibaba Research Center for Complexity Sciences, Hangzhou Normal University,
Hangzhou 311121, People's Republic of China
zhouyinzuo@163.com

Abstract. A disease propagating in a community or a rumor spreading in a social network can be described by a contact network whose nodes are persons or centers of contagion and links heterogeneous relations among them. Suppose that a disease or a rumor originating from a single source among a set of suspects spreads in a network, how to locate this disease/rumor source based on a limited set of observations? We study the problem of estimating the origin of a disease/rumor outbreak: given a contact network and a snapshot of epidemic spread at a certain time, root out the infection source. Assuming that the epidemic spread follows the usual susceptible-infected (SI) model, we introduce an inference algorithm based on sparsely placed observers. We present an algorithm which utilizes the correlated information between the network structure (shortest paths) and the diffusion dynamics (time sequence of infection). The numerical results of artificial and empirical networks show that it leads to significant improvement of performance compared to existing approaches. Our analysis sheds insight into the behavior of the disease/rumor spreading process not only in the local particular regime but also for the whole general network.

Keywords: Locating source · Observer nodes · Correlation · Complex network

1 Introduction

Spreading of epidemics and information cascades through social networks is ubiquitous in the modern world [1]. Examples include the propagation of infectious diseases, information diffusion in the Internet. In general, any of these situations can be modeled as an epidemic-like rumor spreading in a network [2–8]. As a result, a piece of information or rumor posted by one individual in a network can

This work is supported by the Natural Science Foundation of Zhejiang Province LQ16F030006, National Natural Science Foundation (NNSF) of China under Grant 61503110, 11405059, the General Science Foundation of the Education Department of Zhejiang Province Y201431653, the Startup Foundation of Hangzhou Normal University PF15002004010.

X. Cheng et al. (Eds.): SMP 2017, CCIS 774, pp. 296–307, 2017.
https://doi.org/10.1007/978-981-10-6805-8_24

be propagated to a large number of people in a relatively short time. In some locales like China [9], publication of such rumors and opinions are illegal, and law enforcement agencies need to act to identify the rumor and opinion sources. This may be difficult if the source is anonymous and significant time and effort is expended to trace the IP addresses and identities of the individual profiles carrying or linking to the opinion piece. In another example, it is important to identify the index case of a disease spreading in a community in order to determine the epidemiology of the disease.

Understanding and controlling the spread of epidemics on networks of contacts is an important task of today's science. It has far-reaching applications in mitigating the results of epidemics caused by infectious diseases, computer viruses, rumor spreading in social media, and others. Identifying the source of a rumor or a virus by leveraging the network topology suspect characteristics and the observation of infected nodes is an extremely desirable but challenging task. This is the case, for example, when an infectious disease spreads through human populations across a large region, as observed with the worldwide H1N1 virus pandemic in 2009. Here, the system is more conveniently modelled as a network of interconnected people, and source localization reduces to identifying which person in the network was first infected. Rooting out the disease or rumor source has practical applications and also allows us to better understand the amplification role of the network in information cascades.

The stochastic nature of infection propagation makes the estimation of the epidemic origin intrinsically hard: indeed, different initial conditions can lead to the same configuration at the observation time. Finding an estimator that locates the most probable origin, given the observed configuration, is, in general, computationally intractable, except in very special cases such as the case where the contact network is a line or a regular tree [9–11]. The methods that have been studied in the existing works are mostly based on various kinds of graph-centrality measures. Examples include the distance centrality or the Jordan center of a graph [9–12]. The problem was generalized to estimating a set of epidemic origins using spectral methods in [13,14]. Another line of approach uses more detailed information about the epidemic than just a snapshot at a given time [15–17].

In this paper, we assume that the epidemic spread follows the widely used susceptible-infected (SI) model [18]. Here, we introduce an algorithm for the estimation of the origin of an SI epidemic from the knowledge of the network and the snapshot of some nodes–we call them observation nodes–at a certain time. Our algorithm estimates the probability that the observed information resulted from a given patient zero in a way which is crucially different from existing approaches. For every possible origin of the epidemic, we use a fast dynamic method to estimate the probability that a given node in the network was in the observed state (I–state). We then use a shortest paths and Pearson relation coefficient to compute the probability of each node to be the origin of the epidemic or rumor.

Our goal is to locate the source of diffusion under the practical constraint that only a small fraction of nodes can be observed. This is the case, for example, when locating a spammer who is sending undesired emails over the internet, where it is clearly impossible to monitor all the nodes. Thus, the main difficulty is to develop tractable estimators that can be efficiently implemented (i.e., with subexponential complexity), and that perform well on multiple topologies. We test our algorithm on synthetic spreading data and show that it performs better than existing approaches. The algorithm is very robust; for instance, it remains efficient even in the case where the states of only a fraction of nodes in the network are observed. From our tests, we also identify a range of parameters for which the estimation of the origin of epidemic or rumor spreading is relatively easy, and a region where this problem is hard.

The rest of this paper is organized as follows. We first give our model in Sect. 2. In Sect. 3, we give the detail of the algorithm. The results were given in Sect. 4. At last, we summarized the paper and prospected the future work in Sect. 5.

2   Model

We introduce our model for identifying the source nodes. The underlying network on which diffusion takes place is a finite, undirected graph $G = \{V, E\}$, where $|V|$ and $|E|$ are the number of nodes and edges in the network, respectively. The topology of the network G is assumed to be known, at least approximately, as is often verified in practice–e.g., rumors spreading in a social network, or electrical perturbations propagating on the electrical grid. The information source, $s* \in G$, is the vertex that originates the information and initiates the diffusion. We model $s*$ as a random variable (RV) whose prior distribution is uniform over the set V; i.e., any node in the network is equally likely to be the source a priori.

The diffusion process is under SI model. In the SI model, all the nodes are divided into two categories: the susceptible population (S) and the infected ones (I). The infected nodes are the epidemic sources, they have a certain probability of infection to infect the susceptible nodes. Once the susceptible nodes are infected, they have become new sources of infection. In the SI model, the disease can not be cured, or for being lack of the effective control to the sudden outbreak of the epidemic. The diffusion process is modeled as follows. At time t, each vertex $u \in G$ has one of two possible states: (i) infected, if it has already been infected by one of his neighbors or received the information from any neighbor; or (ii) susceptible, if it has not been infected or informed so far. This diffusion model is general enough to accommodate various scenarios encountered in practice.

How is the source location recovered from the measurements taken at the observers? Our method based on two hypotheses:

– The distance between the neighbor node who transmit the epidemic or information to observation node and the source node should be closer to the distance between the observation node and the source node.

- The order of the observation nodes getting infected or informed should be consistent with the order of the distance between the observation nodes to the source nodes.

In this paper, we use the Pearson correlation coefficient to quantitatively describe the correlation we proposed between nodes infected order and network structure. In order to utilize this correlation to locate the source of diffusion, we measure $O \doteq \{o_k\}_{k=1}^K \subset G$ denote the set of K observers, whose location on G is chosen or known. Each observer measures from which neighbor and at what time it received the information. Specifically, O_n nodes in the network record the time of infection moments about these nodes as vector $T = (t_1, t_2, \cdots, t_{O_n})$, and we can calculate shortest path length from any node s' to O_n nodes respectively, denoted by vector $D_{s'} = (d_{1s'}, d_{2s'}, \cdots, d_{O_n s'})$. What needs to be noted is the component in this the vector corresponds to the component in the time vector T. Then the correlation between vector T which represents the infected time of O_n nodes and the vector $D_{s'}$ which represents the respective distance between s' and O_n nodes can be calculated using the following equation:

$$r_{s'} = \frac{\sum_{i=1}^{O_n}(d_{is'} - \bar{d})(t_i - \bar{t})}{\sqrt{\sum_{i=1}^{O_n}(d_{is'} - \bar{d})^2 \sum_{i=1}^{O_n}(t_i - \bar{t})^2}}, \tag{1}$$

where $\bar{d}$ is the average of s' to all O_n nodes, and $\bar{t}$ is also the average of infected time of all O_n nodes, and $d_{is'}$ is the length of shortest path from i to s'. According to the structure-temporal correlation we believe that if s' is the source of diffusion, the value of $r_{s'}$ should be higher. Conversely, if the value of $r_{s'}$ is higher, the possibility that s' is the source of diffusion is greater.

Base on the above discussion, the main process of our algorithm is as follows:

- In a connected network, we randomly assign a node as virus source and the top O_n nodes who have the highest degree centrality score are chosen as observers. Then the disease starts to diffuse under SI model in the network. In the process of diffusion, each observer measure the absolute time at which the observer receives the information from which of its neighbor.
- In general, if node j is infected by node i, node i is closer to the source s than node j. This situation is more intuitive when the rates of infection is greater, or the network is relatively sparse who has few loops. This knowledge can be formalized as the following inequality $T_{sj} : d_{sj} > d_{si}$. So after each node are recorded for the first time and direction to get infected or informed, select those nodes among un-observer ones whose shortest paths to the neighbor node–transmit the epidemic or information to observation nodes are less than to the observation nodes, marked as set C. Here, each node c in C needs to have no less than half of observer nodes (we call one of them node o) to satisfy the inequality $T_{co} : d_{co} > d_{co'}$, where o' is the direct source node of infection of node o, i.e. each node c and at least $1/2O_n$ observers satisfy the distance from c to this kind of observer node o larger than the distance form c to node o's direct source of infection.

– Calculate the order of the shortest distance between C and each observation node. Here, we defined set $T_c = \{T_{c1}, T_{c2}, \cdots, T_{cO_n}\}$, and $N(T_c)$ is the number of conditions established in the T_c. So the condition that any node c' becomes an element in C need to be satisfied can be mathematically represented as $N(T_{c'}) \geq 1/2O_n$. Through the experimental analysis, we find it is difficult to form the set C if the coefficient is set to 1, and the coefficient $1/2$ can make our algorithm have better localization accuracy. After building the set C, we

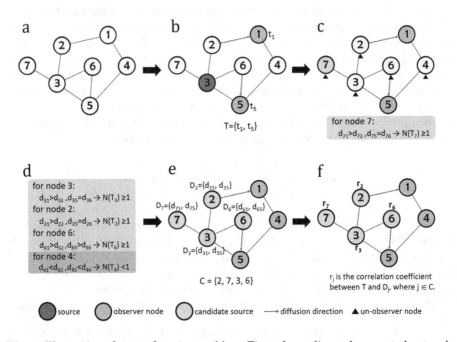

Fig. 1. Illustration of source location problem. First, the undirected connected network has been generated, as shown in (a). The source node of diffusion and observer nodes are selected in (b), and marked in red and blue, respectively. Then the diffusion process is completed under the SI model. During this process each observer measure from which neighbor and at what time it is infected the disease. The former measurement is represented by the red arrow, and we learn from it that observer node 1 is infected by 2, observer node 5 is infected by 6. The later measurement can form the vector T which represents the infected time of all observers. Next we build the candidate source nodes set C from un-observer nodes, which marked by black triangle in (c). For example, the node 7 is not an observer node and $T_7 = \{T_{71}, T_{75}\}$, where $T_{71} : d_{71} > d_{72}$ is established but $T_{75} : d_{75} > d_{76}$ is not. So we obtain $N(T_7) = 1$, which is greater than or equals to $1/2O_n$, in this case, node 7 becomes a member of set C. (d) shows the process of determining whether the remaining un-observer node are the members of C. We get $C = \{2, 7, 3, 6\}$ by using the same method as node 7 does, in particular, node 4 can not become a member of C. And then, for each node c in set C we get the distance vector $D_c = \{d_{c1}, d_{c5}\}$ from c to all observers. Finally, according to formula 1 to calculate r_c, where $c \in C$. The node with the largest r_c value is the predicted source based on our algorithm given. (Color figure online)

calculate and the time series correlation between the shortest distance order and the real time series of observation nodes getting infected or informed based on Eq. 1 for each node in C. The correlation value is higher, the node is more likely the source.

Figure 1 shows an example of the algorithm process.

3   Detail of Algorithm

3.1   The Set of Observation Nodes

In this section, we exam our algorithm where the observer nodes are chosen with different measures. The first measure is the degree centrality, that who have the largest degree are chosen as the observers. The second measure we adopt is the closeness centrality, which is defined in the following way. For a node i, suppose the shortest distance form i to j is d_{ij}, then the average distance of all nodes in the network to node i could be expressed as

$$d_i = \frac{1}{N} \sum_{j=1}^{N} d_{ij}$$

then the closeness centrality of node i is the reciprocal value of d_i, i.e. $1/d_i$. Thus, higher closeness centrality score means a smaller distance from a node to the rest of the network in average. We then choose top O_n nodes in the both centrality score lists as the observers.

To manifest the impact of different set of observer nodes, we choose four different set of nodes as observers, which are (i) the top O_n nodes have the highest degree centrality score, (ii) the top O_n nodes have the lowest degree centrality score, (iii) the top O_n nodes have the highest closeness centrality score, (iv) the top O_n nodes have the lowest closeness centrality score.

The experimental results show that our algorithm achieve the best performance when the nodes with highest degree centrality are chosen as observers, that is, in this case, our algorithm can locate the source of diffusion with higher accuracy. Figure 2 shows the effect of the results when the observers is selected in a different way. In order to quantify the results of algorithm, we define two indices "*hitting rate*" and "*getting rate*" to describe the accuracy of source locating. The index *hitting rate* is the ratio of the times that the highest score in the sequence of the predict results given by our algorithm is exactly the true source of diffusion to the total number of times we carry out our algorithm. The index *getting rate* is the ratio of the times that source of diffusion appears in the top 10% of the sequence of the predict results given by our algorithm.

3.2   Ratio of Observer Nodes

Figure 2 further shows the effect of the number of the observers. One may expect that assigning more observers (or monitor) into the networks may reward with

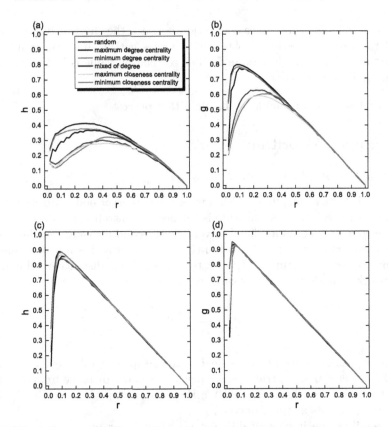

Fig. 2. The influence of different observers set on algorithm accuracy. r in the figure represents the ratio of the observer's proportion, and the value is the ratio of the observers' number to the size of the network. Where (a), (b) and (c), (d) are the predict results of algorithm when infection rate is 0.5 and 1.0, respectively. The size of network is 100. Each cure in the figure is obtained by average of 5000 networks, and for each network, the source of diffusion has been assigned randomly, then running our algorithm to predict the location of source diffusion.

a higher accuracy in source location. However, our results in Fig. 2 show that the performance of the algorithm does not increase with the increasing of the observers, there exists an optimal range of the fraction of observers where the location of the source could be correctly identified with an accuracy up to 95%. Specifically, as shown in Fig. 2(a) and (b) when infection rate $\beta = 0.5$ the algorithm reaches its optimal performance when $r \approx 0.3$ for *hitting rate h* and $r \approx 0.2$ for *getting rate g*, here r is the ratio of the observer nodes chosen from the whole network. While when $\beta = 1$, the optimal range of r is only around 0.1. These results suggests that (i) our algorithm perform better for the case of high infection rate, and (ii) more importantly, only a small fraction of observers, such as the case in Fig. 2(c) and (d), may guarantee a high accuracy in the source identification.

3.3   Method to Calculate Correlation

We also consider measuring the correlation between temporal information and network structure by Cosine similarity. Figure 3 shows the location results of the algorithm adopt different correlation indices for the case of $\beta = 0.1, 0.2, \cdots, 1.0$. One may observe that with the increasing of β, the performance of the algorithm also improves which is accordant with the previous result shown in Fig. 2 and we can obtain that our algorithm achieved almost identical result regardless of which index has been used to measure the temporal-structure correlation.

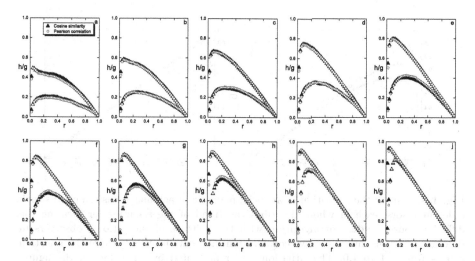

Fig. 3. The impact of Pearson correlation coefficient and Cosine similarity are used to calculate the correlation between D_i and T of our algorithm. r in the graph is still the ratio of the observers. From a to j in the figure above, corresponding to the infectious rate from 0.1 to 1.0 with interval 0.1, respectively. The circle marked with red in each picture is the result of using the Cosine similarity, and the triangle marked with black is the result of using the Pearson correlation coefficient, each result is represented by the index of *hitting rate h* and *getting rate g*, respectively. In the figure h/g represents the success rate of *hitting rate* and *getting rate*. From the diagram, we find the two methods have almost identical results. (Color figure online)

4   Results

4.1   Artificial Networks

Our previous results are obtained from artificial networks such as ER model and BA model. Now, we systematically presented them in Fig. 4 for easy comparison. One may clearly observe that our algorithm performs better for higher infection rate and has similar efficacy for both ER model and BA model.

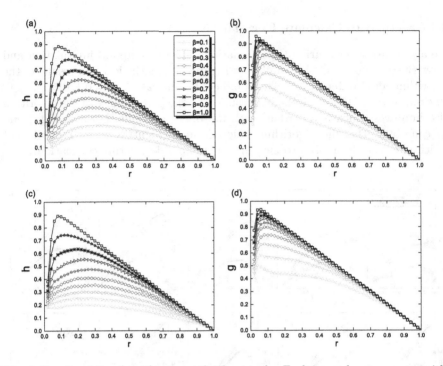

Fig. 4. Results of the algorithm in artificial networks. Each image has ten curves with different colors, each of whom describes the trend of the accuracy for locating the diffusion source under our algorithm with the increase of the ratio of observers in different infection rate. Here, we consider ten different situations where the infection rate is from 0.1 to 1.0. The situations are represented by ten colors in the figure. Among them, (a) and (b) are results of undirected ER network. (c) and (d) are results of undirected SF networks. (Color figure online)

4.2   Empirical Networks

We further apply the algorithm to four different empirical networks. (i) FWFW: The food web in Florida Bay during wet season. (ii) C.elegans: The neural network of C.elegans. (iii) USAir: The network of the US air transportation system. (iv) Metabolic: The metabolic network of the nematode worm C.elegans. The basic topological features of such network are summarized in Table 1.

In Fig. 5, we observe that our algorithm could predict the location of the source in high accuracy with a small fraction of observers, especially when the infection rate β is large.

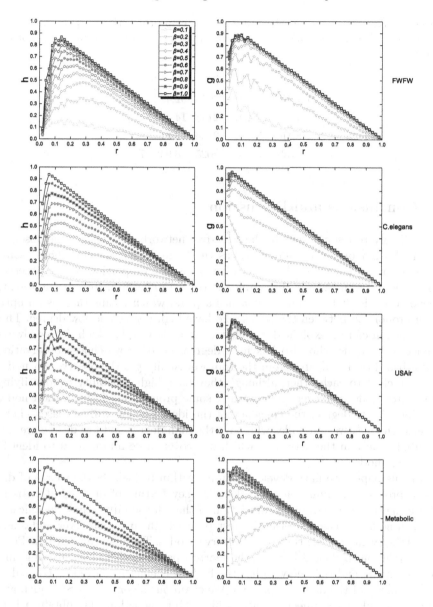

Fig. 5. The influence of observer ratio r on accuracy of the algorithm under different infection rates β in the empirical networks. Four different networks from various research fields are tested. Assume that the network size is N. Under a specific infection rate, for a particular observer ratio, we set each node in the network in turn to be the source of diffusion, and each node spread the disease m times. So we need to run $N * m$ times to identify the location of the source at the specific infection rate and observer ratio. Suppose there are n_0 times the algorithm observer ratio can be calculated by $n_0/(N * m)$. And there are n_1 times the algorithm rooting out the source of the diffusion, then the value of *getting rate* g at the specific infection rate and observer ratio can be calculate by $n_1/(N * n)$.

Table 1. The basic topological feature of four empirical networks. |V| and |E| are the number of nodes and links. $\langle k \rangle$ is the average degree, $\langle d \rangle$ is the average shortest distance.

	| V |	|E|	$\langle k \rangle$	$\langle d \rangle$
FWFW	128	2075	32.422	3
C.elegans	297	2148	14.47	5
USAir	332	2126	12.81	6
Metabolic	453	2025	8.940	7

5   Conclusions and Discussion

Specially, we test some cases in the scale-free networks. We choose the observation nodes not only the nodes who has the large degree, we also choose some nodes who has the small degree. We set n as the proportion of the observation nodes who are chosen based on the degree index from top. A interesting phenomenon is that, we find that n has a peak, which means there is an optimum proportion between the nodes who have high degree and low degree. This phenomenon can be explained as follows: the node who has high degree always connects the node who also has high degree. So once we choose observation nodes based on the degree from the top, we usually get the observation nodes who are close to each other, although they have high capability to identifying the source node. However, once we mix some proportion of observation nodes who have lower degree, the efficiency of the identification will increase. That's because the observation nodes who have the high degree and low degree are distributed widely in the networks, which can cover more information to identify the source node.

In this paper, we have developed an algorithm to identify the source of diffusion processes running on networks. The key feature of our algorithm is to utilize the structure-temporal correlation of the diffusion dynamics. Specifically, for each node, we calculate the shortest path length to the observer nodes and record the event arriving time of the observer nodes. Then, we calculate the Pearson correlation coefficient between the shortest path length and the arriving time for each node, and the one who has the highest coefficient score is predicted to be the source of the diffusion process. We exam our algorithm of SI model, and found that when the infection rate is high. Our algorithm may obtain a high accuracy in locating the source of diffusion, while when the infection rate is low, the results may suffer from a larger fluctuation. We further study the performance of the algorithm based on different observer sets collected under different topology measures. Considering the computational complexity and accuracy of the results, the results suggest adopting those with the highest degree centrality as observers. Since our results may have larger fluctuation for lower infection rate, improving the proposed algorithm with a more stable performance under such case, and the comparison between our algorithm and existing similar algorithms could be a further study.

References

1. Barabasi, A.L., Albert, R.: Emergence of scaling in random networks. Science **286**(5439), 509 (1999)
2. Zhou, Y.Z., Liu, Z.H., Zhou, J.: Periodic wave of epidemic spreading in community networks. Chin. Phys. Lett. **24**(2), 581–584 (2007)
3. Zhou, J., Liu, Z.H.: Epidemic spreading in communities with mobile agents. Phys. A **388**(7), 1228–1236 (2009)
4. Zhou, J., Xiao, G., Cheong, S.A., Fu, X., Wong, L., Ma, S., et al.: Epidemic reemergence in adaptive complex networks. Phys. Rev. E. **85**(3 Pt 2) (2012)
5. Zhou, J., Chung, N.N., Chew, L.Y., Lai, C.H.: Epidemic spreading induced by diversity of agents' mobility. Phys. Rev. E **86**(2), 026115 (2012)
6. Zhou, J., Xiao, G., Chen, G.: Link-based formalism for time evolution of adaptive networks. Phys. Rev. E **88**(3), 032808 (2013)
7. Zhou, Y., Xia, Y.: Epidemic spreading on weighted adaptive networks. Phys. A **399**(4), 16–23 (2014)
8. Yao, Y., Zhou, Y.: Epidemic spreading on dual-structure networks with mobile agents. Phys. A **467**, 218–225 (2017)
9. http://www.techinasia.com/china-tweeting-rumors-land-years-jailor-worse/
10. http://investor.fb.com/releasedetail.cfm?ReleaseID=780093
11. Kossinets, G., Watts, D.J.: Empirical analysis of an evolving social network. Science **311**(5757), 88–90 (2006)
12. Alcarria, R., Robles, T., Camarillo, G.: Towards the convergence between IMS and social networks. In: International Conference on Wireless and Mobile Communications, pp. 196–201. IEEE Computer Society (2010)
13. Kempe, D., Kleinberg, J., Tardos, E.: Maximizing the spread of influence through a social network. In: ACM SIGKDD International Conference on Knowledge Discovery and Data Mining, pp. 137–146. ACM (2003)
14. Leskovec, J., Adamic, L.A., Huberman, B.A.: The dynamics of viral marketing. In: ACM Conference on Electronic Commerce, vol. 1, pp. 228–237. ACM (2006)
15. Chen, W., Wang, Y., Yang, S.: Efficient influence maximization in social networks. In: ACM SIGKDD International Conference on Knowledge Discovery and Data Mining, Paris, France, 28 June–1 July, pp. 199–208. DBLP (2009)
16. Pinto, P.C., Thiran, P., Vetterli, M.: Locating the source of diffusion in large-scale networks. Phys. Rev. Lett. **109**(6), 068702 (2012)
17. Zhesi, S., Shinan, C., Wen-Xu, W., Zengru, D., Eugene, S.H.: Locating the source of diffusion in complex networks by time-reversal backward spreading. Phys. Rev. E **93**(3), 032301 (2016)
18. Shah, D., Zaman, T.: Rumors in a network: who's the culprit? IEEE Trans. Inf. Theory **57**(8), 5163–5181 (2011)

Exploring the Country Co-occurrence Network in the Twittersphere at an International Economic Event

Xinzhi Zhang[(✉)]

Department of Journalism, Hong Kong Baptist University, Hong Kong Baptist
University Road, Kowloon Tong, Hong Kong SAR
xzzhang2@gmail.com

Abstract. This paper explores how international relations are represented on social media in the context of an international economic event, specifically the "Belt and Road Initiative" proposed by the government of mainland China. The present study focuses on the country co-occurrence network represented in the Twittersphere, such that a link is established between two countries if they appear in the same tweet. The study also investigates how the formation of such a network can be explained by geographical, political, and economic factors. An application programming interface (API) harvested all relevant public tweets (n = 26,515) in a one-month time span (2 June–28 June 2017). The names of the countries or regions were extracted to establish the network, with 52 nodes (countries or regions) and 86 edges. Social network analysis revealed that mainland China, Hong Kong, Pakistan, Greece, Kenya, and Iran were in the network's important positions, as indicated by their high betweenness centrality. Exponential random graph modeling (ERGM) results suggested that West Asian countries engaging heavily in international polities, countries with lower levels of press freedom, and those receiving less direct investment from mainland China, were more likely to be tweeted together.

Keywords: Social network analysis · Twittersphere · International relations · Text mining · Country co-occurrence network · Exponential random graph models

1 Background and Objectives

1.1 Media and International Relations

Contents appearing in media represent and reconstruct the relationships among nation-states. Such international relations are observable when crucial international events are disseminated among the global public and generate discussions via global communication channels, both conventional media [1–3] and online media [4, 5]. While most studies have focused on the construction of international politics in the mass media, there is an emerging trend to focus on social media, where not only political institutions such as media agencies or governmental branches but also ordinary online citizens known as netizens can contribute to the discussion. The rapid

© Springer Nature Singapore Pte Ltd. 2017
X. Cheng et al. (Eds.): SMP 2017, CCIS 774, pp. 308–318, 2017.
https://doi.org/10.1007/978-981-10-6805-8_25

development of social media has enabled people to disseminate, communicate, discuss, and reconstruct developments in international politics in the sphere formed via social media [6], and may have an impact on diplomacy in an unprecedented way [7]. A sizable literature has documented that the media's representation of international relations are influenced by real-world factors, such as political factors like national interests and ideologies [8–10] and economic factors like bilateral trade and economic development [3, 4]. Research on factors contributing to the representation of politics, especially international relations on social media, is scarce [3, 11].

1.2   Country Co-occurrence Network on the Twittersphere

The present study explores international relations as represented on social media by focusing on the country co-occurrence network represented on Twitter, the largest public social media platform. The theoretical and empirical foundations of the present study are threefold. First, a co-occurrence network is defined as a network in which nodes represent entities such as persons, companies, countries, etc., and links represent observations of these entities' existing together [12, 13]. In the present study, a link is established between two countries if they appear in the same Twitter post. The link is unidirectional because it denotes the co-presence of the two countries. As reported by Barnett et al. [11, p. 38], the country co-occurrence network appearing on social media demonstrates how international relations are perceived semantically in the "global social media sphere" [14, p. 77]. When such a network is constructed, it can be investigated with social network analysis, an approach that examines the formation and characteristics of the network and thereby offers a new perspective for examining international relations [11]. Second, the present paper focuses on Twitter as the research context. Twitter is an online venue for people to discuss, share, contextualize, and make sense of news [15]; it is known as the Twittersphere [16, 17]. The Twittersphere is an ideal conduit for exploring the extent to which reality is constructed and disseminated by the "transnational electronic public sphere" [11, p. 38]. Third, by applying social network analysis, this study explores the structural features of this network and determines which countries occupy crucial positions within it. The study explicates the possible mechanism of the formation of this country co-occurrence network. Informed by the theoretical framework of network science, the study examines the role of real-world political, geographical, and economic factors.

The empirical case for the present study was derived from the Twitter representations of the "Belt and Road Initiative" proposed by mainland China, along with its economic and political considerations. The initiative has been widely covered by global media. Anecdotal as it may be, the case offers an optimal opportunity for observing the dynamics between the Twittersphere and real-world factors in an international context, especially given that multiple countries are involved in the initiative. The objectives of the present study are:

- to establish the country co-occurrence network from the tweets harvested from Twitter;
- to describe the structural features of the country co-occurrence network;

- to explicate the factors contributing to the formation of such country co-occurrence network, and
- to provide a better understanding on the construction of reality on social media.

2 Theoretical Frameworks

Prior studies have offered three theoretical frameworks to describe the formation of inter-connections among social entities. The present study applies these frameworks to explicate the formation of interrelations among countries. First, the homophily hypothesis argues that individuals or social entities are more likely to associate and interact with similar others [18]. This hypothesis has been applied in numerous studies on the formation of interpersonal relationships in both offline [19] and online environments [20]. To explain the formation of relationships among countries, prior studies have found that the development of international relationships was based on the homophily process [21], such as shared political interests [22] and political and values similarities [23]. Geographical factors also contribute to the formation of international relations in the field of media and communication. For example, international Facebook friendships tend to exist between people in countries sharing geographical borders, language, culture, and migration [24]; in addition, the greater the geographical proximity, the greater the likelihood for media in one country to cover news originating from another country [9]. Therefore, homophily between two countries in terms of institutions and geographical proximity facilitates interaction between two countries in both real-world settings and in socially constructed venues like news media and social media. It is reasonable to propose that countries sharing similar political ideologies and geographical locations are more likely to be discussed together and thus more likely to co-occur in the representation of social media. Hence, it is hypothesized that:

H1: Based on the homophily hypothesis, countries with similar political contexts or who share geographical proximity are more likely to co-occur in one tweet.

Second, beyond political and geographical factors, the world polity theory [25] argues that if a nation-state constructs itself as an institution within international society, rather than an entity fulfilling only domestic needs, then this state is a product of global culture. Operationally, countries situating themselves into this global society will have more connections with international organizations. Countries within this world polity, despite possible differences in political context, economic development, culture, or religion, are more likely to interact and share similar public concern [26]. It is therefore proposed that countries actively engaging in the global polity are more likely to co-occur in tweets than those which are not. Hence, it is proposed that:

H2: Based on the world polity theory, the more international polity in which any two countries are involved, the more likely those two countries will co-occur in one tweet.

Thirdly, while the homophily hypothesis and the world polity theory both articulate a bilateral relationship between two countries, the relations between any country pair should also take into consideration the relations between each of the two countries and other countries. As Hoff and Ward [25, p. 160] explicated, this kind of investigation involves "third-order dependencies such as transitivity, clustering, and balance." The transitivity hypothesis derived from network science offers another mechanism for two entities to link with each other, such that if node X is linked to node Y and node Y is linked to node Z, then it is reasonable to infer that node X is also linked to node Z [25–27]. This implies the common wisdom that the friend of friend is a friend. Moving beyond interpersonal relationships, studies on international relations have suggested that the bilateral relationships between two countries depend largely on the nature of how those two countries are related to a third country, such that countries with common allies or enemies are more likely to be allies or enemies, respectively [25, 26]. Examining transitivity can reveal the balance of relationships among several countries [28]. In the present study, in which mainland China is the driver of the Belt and Road initiative, it is proposed that:

H3: Based on the transitivity hypothesis, countries having close relations with mainland China will be more likely to co-occur in one tweet.

3   Method

3.1   Data Collection

The present project used the Twitter search Application Program Interface (API) to harvest relevant public tweets on the "Belt and Road Initiative" theme with the "*twitteR*" package in R. After consulting the expressions used by official media in mainland China (such as *China Daily*) and news reports from leading international media such as the *New York Times* and Hong Kong's *South China Morning Post*, a set of search enquiries and hashtags were adopted. The complete list of search enquiries is as follows (as appearing in the *twitteR* search syntax): "\"one belt one road\"", "\"one belt, one road\"", "\"one belt and one road\"", "\"belt and road\"", "#onebeltoneroad", "#onebeltandoneroad", "#beltandroad", "#OBOR." Using the backslash symbol ("\") and quotation marks ensured that the search included exact phrasings.

The queries harvested 26,515 pieces of tweets created from 2 June 2017 to 28 June 2017, approximately a one-month time span. Due to API's limitation, each searching action could retrieve Tweets dating back seven to ten days, so the data were collected in two instalments (15 June and 28 June). All harvested tweets were cleaned by converting lower case letters and removing numbers, punctuation, URLs, extra white space, and non-English text, before analysis was performed.

3.2   Measurements

The Freedom Index proposed by the Freedom House was used to measure the political environment [29]. To measure geographical location, the continent on which a country was located was documented. To measure involvement within the global polity, the number of international environmental non-governmental organizations (INGOs) registered in the country was documented. This measurement is a typical proxy variable used in studies employing the world polity theory to indicate the extent to which a country is involved in that world polity [30, 31]. The number of INGOs were summed from 2001 to 2005. Finally, to measure the relationship with mainland China, its outbound foreign direct investment (FDI) was used as a proxy indicator. The data for mainland China's FDI in each country was obtained from the China Global Investment Tracker published by the American Enterprise Institute and The Heritage Foundation [32]. The volume of investment from mainland China consisted of summed values from the latest report issued in January 2017, with records dating from January 2005 to December 2016, in millions of US dollars.

4 Results

A country co-occurrence network was established such that two countries were related to each other if those two countries appeared in the same portion of a tweet, as shown in the example below:

"China Sambut Bergabungnya Mongolia Dalam Program Belt and Road https://t.co/6kiVwWyPvJ.*"* (posted by @ jurnascom on 5/12/2017 7:32)

First, using the *"stringr"* package in R, all country names were extracted, then combined into an edge list. In the above example, an undirected link between mainland China and Mongolia was established. After manually combining entities referring to the same country such as "Kyrgyzstan" and "Kyrgyz" and noun and adjective forms like "China" and "Chinese" or "France" and "French" and removing self-loop and redundant edges, the present study only focused on the largest connected component of the network; and this connected component was regarded the focal network to be further analyzed. It contained 52 nodes (countries/regions) and 86 edges. The network density was 0.065. The network's transitivity, also known as the global clustering coefficient, was 0.104, indicating that the network was sparse and not closely connected.

The network is presented in Fig. 1 using the *"igraph"* package in R and employing the Fruchterman-Reingold layout. By simply eyeballing the network, most relationships were bilateral between mainland China and another country, whereas there were limited triad relationships involving countries other than mainland China, with exceptions such as a closure formed among Kenya, Philippines, and Indonesia and one formed among Hong Kong, Netherlands, and Canada.

Meanwhile, Table 1 reports the node-level (country/region) statistics. All the node-level centrality statistics were standardized.

Countries or regions with the largest numbers of connections were mainland China, India, the US, Pakistan, Iran, Sri Lanka, the Hong Kong SAR, and Russia. In terms of betweenness centrality—the extent to which a node connects different clusters of the network as bridges—the most important countries and regions were: mainland China (betweenness centrality = 0.090), Hong Kong (betweenness centrality = 0.078), Pakistan (betweenness centrality = 0.063), Kenya (betweenness centrality = 0.039), and Greece (betweenness centrality = 0.039).

To explain the tie-formation mechanism among countries and regions, an exponential random graph model (ERGM) was estimated by the *"statnet"* package in R.

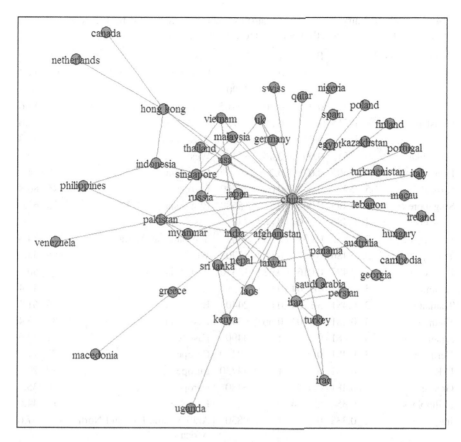

Fig. 1. The country co-occurrence network in the Twittersphere discussing the "One Belt One Road." The network was plotted by the *igraph* package in R.

The ERGM compares the network statistics from an empirically observed network to the distribution of network statistics generated from randomly simulated networks. The predictors of the present study were geographical location (each geographical area was dummy coded), freedom of the press, the number of INGOs registered in the country, and the amount of FDI from mainland China, using Monte Carlo MLE estimation methods. The results are presented in Table 2.

The results indicated that the log-odds for any two countries to co-occur in one tweet were -4.40, and the probability was 1.17% ($=\exp(-4.44)/(1 + \exp(-4.44))$). To interpret the GEWSP (geometrically-weighted edgewise shared partners) coefficient [33], if any pair of two countries had co-occurred countries in common, and each of these countries was in at least one triadic relation with each of those countries, then the log-odds of two countries to co-occur increased to -3.71 ($= -4.44 + 0.73$). Compared with countries located in East Asia and the Pacific Rim, countries located outside these regions were more likely to co-occur (coefficient $= -1.16$, s.e. $= 0.35$, $p < .01$). Countries located in West Asia were more likely to co-occur (coefficient $= 1.33$, s.e. $= 0.50$, $p < .01$). A country's freedom of the press was negatively related to

Table 1. Node (country/region) level statistics, countries are ranked by the degree centrality (i.e., the number of link directly linked to the country/region)

Country/region	D	C	B	T	I	Geo	O	F
China	44	0.879	0.897	0.035	.	East Asia	33	84.2
India	10	0.531	0.017	0.311	7190	West Asia	39	36.4
The U.S.	9	0.52	0.015	0.306	1E+05	America	80	18.6
Pakistan	8	0.515	0.063	0.25	10720	West Asia	19	62.6
Iran	6	0.5	0.024	0.333	4720	West Asia	9	90.8
Sri Lanka	6	0.51	0.007	0.467	3630	West Asia	9	73
Hong Kong	5	0.505	0.078	0.2	.	East Asia	.	34
Russia	5	0.5	0.002	0.6	28090	Europe	46	80.8
Singapore	4	0.495	0.002	0.5	15580	East Asia	13	67.4
Taiwan	4	0.486	8E−04	0.667	740	East Asia	.	25.2
Kenya	3	0.486	0.039	0.333	360	Sub-Saharan Africa	20	54.6
Turkey	3	0.486	0.017	0.667	4290	West Asia	18	55.6
Indonesia	3	0.495	0.016	0.333	13370	East Asia	18	50.4
Panama	3	0.481	4E−04	0.667	.	America	5	46.4
Thailand	3	0.481	3E−04	0.667	2440	East Asia	19	61.2
Vietnam	3	0.481	3E−04	0.667	6240	East Asia	8	83.4
Japan	3	0.481	0	1	3490	East Asia	51	22.6
Germany	3	0.481	0	1	19470	Europe	77	17
U.K.	3	0.481	0	1	44730	Europe	80	20.6
Greece	2	0.481	0.039	0	6480	Europe	46	35.2
Philippines	2	0.354	8E−04	0	.	East Asia	12	44.6
Iraq	2	0.338	0	1	9830	Arab Middle East and North Africa	0	67.6
Persian	2	0.481	0	1	.	Arab Middle East and North Africa	.	.
Saudi Arabia	2	0.481	0	1	4910	Arab Middle East and North Africa	5	83.4
Australia	2	0.477	0	1	84610	Australia	47	21.4
Laos	2	0.477	0	1	4390	East Asia	0	84.2
Malaysia	2	0.477	0	1	17230	East Asia	17	63.8
Myanmar	2	0.486	0	1	5510	East Asia	0	83.2
Afghanistan	2	0.477	0	1	3270	West Asia	0	71.6
Nepal	2	0.477	0	1	1840	West Asia	10	57.2
Egypt	1	0.472	0	.	5200	Arab Middle East and North Africa	29	62.4
Lebanon	1	0.472	0	.	.	Arab Middle East and North Africa	0	53
Qatar	1	0.472	0	.	100	Arab Middle East and North Africa	5	66.6
Cambodia	1	0.472	0	.	3540	East Asia	0	63.8

(*continued*)

Table 1. (*continued*)

Country/region	D	C	B	T	I	Geo	O	F
Macau	1	0.472	0	.	.	East Asia	.	.
Finland	1	0.472	0	.	9890	Europe	67	10.4
Hungary	1	0.472	0	.	3610	Europe	48	32
Ireland	1	0.472	0	.	6800	Europe	57	15.8
Italy	1	0.472	0	.	19820	Europe	64	32.8
Macedonia	1	0.327	0	.	.	Europe	0	52.2
Netherlands	1	0.338	0	.	11180	Europe	69	12.2
Poland	1	0.472	0	.	440	Europe	58	25.4
Portugal	1	0.472	0	.	7190	Europe	47	17
Spain	1	0.472	0	.	5010	Europe	62	25.2
Swiss	1	0.472	0	.	14640	Europe	65	12.4
Canada	1	0.338	0	.	45980	North America	53	19.2
Venezuela	1	0.342	0	.	4370	South America	19	76.2
Nigeria	1	0.472	0	.	7550	Sub-Saharan Africa	16	51.6
Uganda	1	0.329	0	.	4670	Sub-Saharan Africa	6	55.6
Georgia	1	0.472	0	.	370	West Asia	10	52.4
Kazakhstan	1	0.472	0	.	18060	West Asia	15	81.6
Turkmenistan	1	0.472	0	.	400	West Asia	10	95.6

Notes for column names: D = Degree; C = closeness centrality; B = betweenness centrality; T = Transitivity; I = the amount of Foreign Direct Investment from China; Geo = Geographical location; O = the number of International NGOs registered within the county; F = the country/region level of the freedom of the press. A "." in the cell denotes the missing or not applicable values.

Table 2. Summary of ERGM results

Model terms	Estimate	Std. error	p-value
Edges	−4.43908***	0.566225	<1e−04
GWESP	0.734168***	0.183776	<1e−04
East Asia and Pacific (=0)	−1.15837***	0.346603	0.000855
East Asia and Pacific (=1)	0.516684	0.352903	0.143407
West Asia (=0)	−0.01212	0.278562	0.965295
West Asia (=1)	1.334991**	0.504182	0.008198
Europe (=0)	0.103064	0.298437	0.729889
Europe (=1)	0.507127	0.79375	0.523
Press freedom index (reversed)	−0.01856**	0.006754	0.006079
Number of INGOs	0.032204***	0.008703	0.000224
Investment form China (logged)	−0.01539*	0.00628	0.014379

Model AIC = 546.3; BIC = 6031.4; Residual Deviance = 1838.2 (df = 1326); Null Deviance = 524.3 (df = 1315)

Note: GWESP refers to geometrically-weighted edgewise shared partners, which is to estimate the triadic relationships of nodes and being more robust to degeneracy [33].

***p < .001; **p < .01; *p < .05

tie-formation (coefficient = −0.02, s.e. = 0.006, $p < .01$). Countries more deeply involved in the global polity, as measured by the number of INGOs registered within the country, were more likely to co-occur (coefficient = 0.03, s.e. = 0.009, $p < .001$). Finally, FDI from mainland China was negatively related to co-occurrence (Coefficient = −0.02, s.e. = 0.006, $p < .05$). To sum up, both H1 (homophily) and H2 (world polity) were supported, but H3 was not supported (transitivity).

5 Discussion

First, the country co-occurrence network presented in Fig. 1 reveals that the number of bilateral linkages between mainland China and other countries vastly outnumbered the linkages or triads that involved countries other than mainland China. This pattern is in line with the empirical implications of the Belt and Road Initiative proposed by the mainland China government, which is to facilitate bilateral collaboration between mainland China and other countries. Secondly, in the Twittersphere, the most frequently connected and important countries—as reflected by both the number of ties linked to the country and the betweenness centrality of the country—had close collaboration with mainland China (such as Hong Kong and Pakistan) or are geographically proximate to mainland China (i.e., in the Eastern and Western Asian regions). This result partly echoes previous explanations of international relations from the perspective of geographical locations [23, 24]. Thirdly, the ERGM results confirmed the homophily hypothesis: countries appeared together in the Twittersphere because of similar political environments (press freedom) and geographical proximity. The study also supports the world polity theory, by countries with closer connections to international organizations were more likely to be clustered together. However, the greater the FDI from mainland China, the less likely it was for two countries to form a virtual relationship in the Twittersphere. One possible explanation is the nature of the Belt and Road Initiative, which strives to promote collaboration between mainland China and other countries. The nature of the initiative, however, triggers adversarial relationships among country pairs that are involved in this collaboration event. To maximize national interest, a particular country might adopt the strategy of strengthening its linkage with mainland China while weakening the linkage between mainland China and third countries. Nevertheless, such triadic and multilateral relationships among the countries call for further research, which is beyond the current study's scope.

This study has certain limitations. The Twittersphere cannot represent the entire online environment, which in turn cannot represent the entire public, either in a given country or globally. The study used English search keywords and focused only on English tweets; it did not examine the multilingual scenario of the social media sphere. In addition, the paper focuses only on country names, but organizations, communities, or individual political figures may also indicate the international relations of any two countries. Future studies might analyze the contents of Twitter texts in greater depth. The present study focuses only on a single breaking event over a roughly one-month period, which is anecdotal. Future studies can lengthen the time span of data collection and consider the longitudinal variation of the focal research question. However, that event is treated as an empirical case. The patterns revealed by the dynamics between online and offline scenarios are likely to be replicated in other contexts. Finally, the present study fails to identify the authors of the tweets in the sample, and the statements regarding and reactions to this global event differ across government agencies, media agencies, opinion leaders, and ordinary users.

Acknowledgments. The study was partly funded by the Start-up Grant for New Academics (no. RC-1617-1-A2) by Hong Kong Baptist University. The author would like to thank Professor David John Frank for generously sharing the data of international NGOs, together with Dr. Li Chen (the Department of Computer Science at Hong Kong Baptist University), Dr. Lun Zhang (Beijing Normal University), Ms. Mengyi Zhang (Hong Kong Baptist University), and the three anonymous reviewers of the *2017 National Conference of Social Media Processing*.

References

1. Gans, H.J.: Deciding What's News: A Study of CBS Evening News, NBC Nightly News, Newsweek, and Time. Northwestern University Press, Evanston (1979)
2. Golan, G.J.: Where in the world is Africa? Predicting coverage of Africa by US television networks. Int. Commun. Gazette **70**(1), 41–57 (2008)
3. Zhu, J.H.: Between the prescriptive and descriptive roles: a comparison of international trade news in China and Taiwan. Asian J. Commun. **2**(1), 31–50 (1991)
4. Zhu, Q.: Citizen-driven international networks and globalization of social movements on Twitter. Soc. Sci. Comput. Rev. (2015). doi:10.1177/0894439315617263
5. Wang, C.J., Wang, P.P., Zhu, J.J.: Discussing occupy wall street on Twitter: longitudinal network analysis of equality, emotion, and stability of public discussion. Cyberpsychol. Behav. Soc. Networking **16**(9), 679–685 (2013)
6. Qin, J.: Hero on Twitter, traitor on news: how social media and legacy news frame Snowden. Int. J. Press/Polit. **20**(2), 166–184 (2015)
7. Duncombe, C.: Twitter and transformative diplomacy: social media and Iran–US relations. Int. Aff. **93**(3), 545–562 (2017)
8. Chang, T.K., Lau, T.Y., Hao, X.: From the United States with news and more: international flow, television coverage and the world system. Gazette (Leiden, Netherlands) **62**(6), 505–522 (2000)
9. Chang, T.K., Shoemaker, P.J., Brendlinger, N.: Determinants of international news coverage in the US media. Commun. Res. **14**(4), 396–414 (1987)
10. Wu, H.D.: Systemic determinants of international news coverage: a comparison of 38 countries. J. Commun. **50**(2), 110–130 (2000)
11. Barnett, G.A., Xu, W.W., Chu, J., Jiang, K., Huh, C., Park, J.Y., Park, H.W.: Measuring international relations in social media conversations. Gov. Inf. Q. **34**(1), 37–44 (2017)
12. Sluban, B., Grčar, M., Mozetič, I.: Temporal multi-layer network construction from major news events. In: Cherifi, H., Gonçalves, B., Menezes, R., Sinatra, R. (eds.) Complex Networks VII. SCI, vol. 644, pp. 29–41. Springer, Cham (2016). doi:10.1007/978-3-319-30569-1_3
13. Popović, M., Štefančić, H., Sluban, B., Novak, P.K., Grčar, M., Mozetič, I., Puliga, M., Zlatić, V.: Extraction of temporal networks from term co-occurrences in online textual sources. PloS One **9**(12), e99515 (2014)
14. Castells, M.: The new public sphere: global civil society, communication networks, and global governance. Ann. AAPSS **616**, 77–93 (2008)

15. Holton, A.E., Baek, K., Coddington, M., Yaschur, C.: Seeking and sharing: motivations for linking on Twitter. Commun. Res. Rep. **31**(1), 33–40 (2014)
16. Marchetti, R., Ceccobelli, D.: Twitter and television in a hybrid media system: the 2013 Italian election campaign. Journalism Pract. **10**(5), 626–644 (2016)
17. Zhang, X.: Visualization, technologies, or the public? Exploring the articulation of data-driven journalism in the Twittersphere. Digit. Journalism (2017). doi:10.1080/21670811.2017.1340094
18. McPherson, M., Smith-Lovin, L., Cook, J.M.: Birds of a feather: homophily in social networks. Annu. Rev. Sociol. **27**, 415–444 (2001)
19. Kandel, D.B.: Homophily, selection, and socialization in adolescent friendships. Am. J. Sociol. **84**(2), 427–436 (1978). Homophily in interpersonal
20. Eklund, L., Roman, S.: Do adolescent gamers make friends offline? Identity and friendship formation in school. Comput. Hum. Behav. **73**, 284–289 (2017)
21. Maoz, Z.: Preferential attachment, homophily, and the structure of international networks, 1816–2003. Conflict Manag. Peace Sci. **29**(3), 341–369 (2012)
22. Johnson, J.D., Tims, A.R.: Communication factors related to closer international ties. Hum. Commun. Res. **12**(2), 259–273 (1985)
23. Sheafer, T., Shenhav, S.R., Takens, J., van Atteveldt, W.: Relative political and value proximity in mediated public diplomacy: the effect of state-level homophily on international frame building. Polit. Commun. **31**(1), 149–167 (2014)
24. Barnett, G.A., Benefield, G.A.: Predicting international Facebook ties through cultural homophily and other factors. New Media Soc. **19**(2), 217–239 (2017)
25. Hoff, P.D., Ward, M.D.: Modeling dependencies in international relations networks. Polit. Anal. **12**(2), 160–175 (2004)
26. Louch, H.: Personal network integration: transitivity and homophily in strong-tie relations. Soc. Networks **22**(1), 45–64 (2000)
27. Wasserman, S., Faust, K.: Social Network Analysis: Methods and Applications, vol. 8. Cambridge University Press, Cambridge (1994)
28. Maoz, Z., Terris, L.G., Kuperman, R.D., Talmud, I.: What is the enemy of my enemy? Causes and consequences of imbalanced international relations, 1816–2001. J. Polit. **69**(1), 100–115 (2007)
29. Freedom House Index Report. https://freedomhouse.org
30. Meyer, J.W., Boli, J., Thomas, G.M., Ramirez, F.O.: World society and the nation-state. Am. J. Sociol. **103**(1), 144–181 (1997)
31. Givens, J.E., Jorgenson, A.K.: Individual environmental concern in the world polity: a multilevel analysis. Soc. Sci. Res. **42**(2), 418–431 (2013)
32. The China Global Investment Tracker. http://www.aei.org/china-global-investment-tracker/
33. Goodreau, S.M., Handcock, M.S., Hunter, D.R., Butts, C.T., Morris, M.: A statnet tutorial. J. Stat. Softw. **24**(9), 1 (2008)

The 2016 US Presidential Election and Its Chinese Audience

Jiahua Yue$^{(\boxtimes)}$, Yuke Li, and James Sundquist

Department of Political Science, Yale University,
Prospect St., New Haven 06520, USA
jiahua.yue@yale.edu

Abstract. Motivated by the question of how the public in an authoritarian political environment may perceive democratic elections, we analyze the underlying interests and sentiments of the Chinese audience regarding the 2016 US Presidential Election with the social media data collected from a large Chinese online community. We extract several latent topics of interest to the community from the text corpus by applying the unsupervised learning method of Latent Dirichlet Allocation (LDA), and explore the amount of interests received by each topic by applying the supervised learning methods, including the Bayesian Additive Regression Trees and the Bayesian LASSO model. Results reveal much more attentions paid by the audience to the sensational news, especially the controversies related to Hillary Clinton's email leakage and Donald Trump's anti-political-correctness and anti-globalization remarks, than to the substantive issues, e.g., regarding the candidates' policy agendas or the democratic process.

Keywords: Public opinion · Text mining · US Presidential Election · China

1 Introduction

The 2016 US Presidential Election is arguably the most dramatic event in that year that attracts global attention, and perhaps one of the most dramatic events in the US election history. The attention it has received even exceeds the "Obamamania" in the midst of the Great Recession in 2008 when President Obama articulated plans to fundamentally change his predecessor's unpopular unilateral policies and won a huge amount of admiration and support from the public. Two factors make the 2016 US Presidential Election one of the most peculiar races in the history of modern democracy: the first is the campaign strategy of the Republican party candidate, Donald Trump, which strongly leans toward, if not entirely rests on, searing and divisive rhetoric, unsubstantiated claims and controversial policy platforms. The second is the unprecedented amount of

We thank for kind suggestions from Joseph Chang, John Henderson and Wenhui Yang. All remaining errors are our own.

© Springer Nature Singapore Pte Ltd. 2017
X. Cheng et al. (Eds.): SMP 2017, CCIS 774, pp. 319–330, 2017.
https://doi.org/10.1007/978-981-10-6805-8_26

unverified information and unwarranted speculations that make all similar scenarios in the past pale in comparison, where the Democratic party candidate Hillary Clinton is the main target.

This article focuses on how foreign citizens under a sharply different political context - China in particular - view the 2016 US presidential election, given the exposure to a multitude of messages and information with mixed implications via the burgeoning social media. Specifically, we choose an online community that is predominantly composed of highly educated, young and urban citizens in China. A question motivating our study is that the most cautious observers would expect Sino-American relations to deteriorate sharply after Donald Trump began his presidency. When running as the presidential candidate, Trump had promised to get tough with China on trade, label it a currency manipulator, and explore the possibility of a nuclear-armed South Korea. Trump even proposed that the One China Policy, the bedrock of Sino-US relation since 1979, could be a bargaining chip in resolving disputes between the two countries. And shortly after being elected, Trump had a telephone conversation with Tsai Ing-wen, the leader of Taiwan, (certainly) without first consulting with the Chinese government. There are good reasons to expect that Trump's provocative stance would draw most attention from angry Chinese citizens, whose national pride is probably on the rise with China's growing comprehensive power and international status.

We report two parts of main results in Sect. 2. Our first part of the results show a somewhat surprising picture – many Chinese citizens do pay attention to a wide range of topics regarding the US 2016 presidential election, yet few topics directly pertains to the prospects of the Sino-US relations. Our second part of the results suggest that sensational stories, especially controversies related to Hillary Clinton, seem to be what the Chinese audience are most interested in.

In Sect. 3 we explain the data source, the first-hand social media data from Zhihu ("Do you know"), a Chinese question-and-answer (Q&A) website similar to Quora, and introduce our empirical strategy (details in the Online Appendix). Specifically, we use Latent Dirichlet Allocation (henceforth, LDA) to uncover the topics that have received interests among the Chinese audience, and Bayesian Additive Regression Trees (henceforth, BART) and the Bayesian LASSO model to further uncover their real political interests.

We will conclude by discussing the implications of our results – for instance, an implication can be drawn that citizens who are most attentive to democratic elections in other countries are nevertheless exposed to unverified information and sensational stories that went viral on the social media. We suggest that Chinese citizens may thereby become disillusioned with the effectiveness of the democratic process in general, as scholars have found a strong association between political cynicism and distrust of democratic institutions and ingrained beliefs in conspiracy theories [5,9,12]. Further research, ideally based on individual-level survey data, is still needed for establishing the explanation as a strict causal claim.

2   Results

2.1   Topic Analysis

We first present the top topics uncovered from the original document-term matrix of user-generated content using the unsupervised LDA procedure [1].[1] In Table 1, we report the eight primary topics associated with 15 top Chinese phrases and the corresponding English translation.

Table 1 shows the following. We find that changing K to other values in the range of 5 and 20 will produce similar results, and present those results in the Online Appendix.

1. One focal point of discussion is the US foreign policies and its relationship with China and the rest of the world (Topic 1). Yet somewhat surprisingly, specific issues pertaining to the Sino-US relationship, such as trade, exchange rate, and disputes over the South China Sea, are not very salient under this topic.
2. Much of the discussion are linked to controversies surrounding Hillary Clinton (Topic 5) as well as Trump's provocative platform built on anti-political-correctness and anti-globalization agenda (Topic 2 and Topic 7). Needless to say, those topics are highly contentious with considerable news coverage.
3. The Chinese audience pays attention to the presidential debates (Topic 4) and the electoral process, including the primary election, Bernie Sanders, the Establishment politicians (Topic 8) and comparison with Obama (Topic 6). In particular, keywords under Topic 4 suggest the Chinese audience seem to be more concerned with the outcome of presidential debates than the policy substances, and important economic issues in presidential debates, such as taxation, employment and investment on infrastructure, are not strongly associated with any of the eight topics.

Table 1. Topic model and associated keywords ($K = 8$)

Topic 1 Foreign Affair	Topic 2 Political Correctness	Topic 3 (Unclear Theme)	Topic 4 Presidential Debate	Topic 5 Hillary Controversy	Topic 6 Obama & Trump	Topic 7 Anti-elitism/Globalization	Topic 8 Electoral Procedure
美国 US	政治正确 Political Correctness	问题 Problem	特朗普 Trump	希拉里 Hillary	美国 US	精英 Elite	共和党 Republican
中国 China	穆斯林 Muslim	支持 Pro	希拉里 Hillary	媒体 Media	总统 President	社会 Society	民主党 Democrat
国家 Nation	白左 White-leftist	女性 Female	辩论 Debate	联邦调查局 FBI	奥巴马 Obama	民主 Democracy	大选 Election
经济 Economy	移民 Immigrant	观点 View	支持者 Supporter	邮件 Email	政治 Politics	世界 World	选举 Election
世界 World	白人 White	言论 Speech	希拉里* Hillary*	新闻 News	当选 Elected	国家 Nation	投票 Vote
上台 Inauguration	黑人 Black	反对 Con	攻击 Attack	调查 Investigation	竞选 Election	自由 Freedom	选民 Voter
日本 Japan	非法移民 Illegal Immigrant	大选 Election	主持人 Moderator	邮件门 Email Controversy	人民 People	人民 People	候选人 Candidate
利益 Interest	美国 US	影响 Influence	表现 Performance	事件 Event	伟大 Great	利益 Interest	民调 Poll
欧洲 Europe	华人 Chinese American	政治 Politics	演讲 Speech	团队 Team	国会 Congress	政治 Politics	桑德斯 Sanders
问题 Problem	同性恋 LGBT	评论 Comment	政策 Policy	视频 Video	政策 Policy	全球化 Globalization	支持 Pro
全球 World	歧视 Discrimination	讨论 Discussion	形象 Image	信息 Information	国家 Nation	西方 West	投票 Vote
俄罗斯 Russia	支持 Pro	选择 Choice	上台 Inauguration	报道 Report	民众 People	阶层 Class	代表 Represent
政策 Policy	反对 Con	立场 Stance	成功 Success	主流 Mainstream	政客 Politician	代表 Represent	建制 Establishment
战争 War	群体 Group	关注 Attention	节目 Show	维基解密 Wikileaks	床破 Trump*	人类 Human	情况 Situation
发展 Development		原因 Reason		公开 Publicize	政府 Government	资本 Capital	领先 Lead

*Informal abbreviations for Hillary and Trump in Chinese.

---

[1] To further mitigate the influence of extremely short answers, the LDA results presented are based on answers with more than ten political words (20, 637 in total). There is no substantive difference with empirical results by slightly increasing or decreasing the threshold.

To sum up, the Chinese audience pays attention to a multitude of topics related to the 2016 US Presidential Election. However, a majority of keywords derived from the latent topics are only remotely related to substantive political issues and the democratic process. Neither do those online discussions make extensive comparison on the fundamental political system of China and the United States, or show strong admiration on the democratic process.

In addition, we illustrate the output of predicted document topics from the LDA model with ideal point estimates by deriving the latent preference of texts from vectors of phrases and projecting the preference onto a unidimensional space of "ideology" [11]. Since it is difficult to accurately estimate the ideal points of short texts, we limit our analysis to answers with more than 100 phrases. We implement the Generalized Wordfish Model recently introduced by [8] developed under a Bayesian framework.

The distribution of ideal point estimate is displayed In Fig. 1 and is clustered by the primary topic predicted by the LDA model. Documents belonging to the same topic tend to have similar ideal point estimates. We may also tell the relative distance between topics from Fig. 1; for instance, ideal point estimates of Hillary Controversy (T5) are quite different from that of political correctness (T2) and anti-elitism/globalization (T7), which implies that these topics are only remotely related with each other.

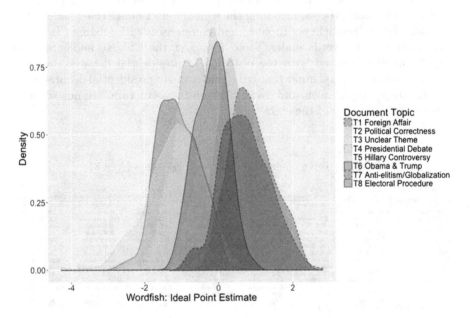

Fig. 1. Distribution of ideal point estimates, by topics

2.2   Estimation

In this section, we present results based on supervised learning, i.e. each observation is a pair constituting of an input vector (i.e., measuring latent political topics or raw text features) and a desired output value (i.e., measuring topic popularity). Our goal is two-fold: first, we examine which latent political topic is associated with the highest popularity among the audience as a benchmark; second, we directly select the most influential political phrases in predicting topic popularity.

For the first goal of predicting the popular topics, we conduct two basic tests. The first test is done by using the output from the LDA model as X, and regressing the number of upvotes (in log) an answer receives on them. The regression result is in Table 2. In Column (1), the input matrix X is the posterior probabilities of the 8 topics for each answer, and in Column (2), the input matrix X is a set of dummy variables indicating the most likely topic for each answer.

The regression results suggest that on average, answers featuring Hillary Controversy (Topic 5) receive more popularity among the audience than those

Table 2. Number of upvotes and latent topics

	Dependent variable: Log(Upvotes)	
	(1)	(2)
	Topic Prob	Topic Dummy
Foreign Affair (T1)	−0.748	−0.410*
	(0.254)	(0.047)
Political Correctness (T2)	1.522*	0.109
	(0.283)	(0.049)
Unclear Theme (T3)	−1.929*	−0.264*
	(0.352)	(0.054)
Presidential Debate (T4)	−1.612*	−0.161
	(0.366)	(0.050)
Hillary Controversy (T5)	3.498*	0.474*
	(0.295)	(0.054)
Obama & Trump (T6)	−2.694*	−0.331*
	(0.367)	(0.054)
Anti-elitism/Globalization (T7)	−2.111*	−0.472*
	(0.296)	(0.052)
Constant	2.817*	2.425*
	(0.192)	(0.035)
N	20637	20637
R-sq	0.032	0.023

Note: Robust standard errors in parenthesis. $^*p < 10^{-3}$
Electoral Procedure (T8) is the baseline comparison group.

of other topics. The topic of Political Correctness (T2) also receives significantly higher popularity than the baseline group. The main result remains unchanged with robust standard error clustered at the question level and additional controls, including the answer length and month fixed effects to model the first time the answer is posted. This result is also robust to alternative model specifications of count variables, such as the Poisson and the negative binomial regression models. Lastly, this result is robust to sample selection based on different length thresholds.

In the second test, we regress another objective indicator for the topic's popularity, the number of followers under a question, on two different sets of X with three experiments. In the first experiment, we still use the output from the LDA model above and aggregate them into topics at the question level. Similar to the test above, we regress the number of followers on the average probability of all the answers under each question belonging to each of the eight topics. Due to the space limitation, we relegate the regression results to the Online Appendix (Table A.6). Consistent with previous findings, the number of followers of a question is positively correlated with the relative weight of answers featuring Hillary Controversy (indicated by high topic probabilities). On the other hand, the coefficients for other topics are much smaller and less significant.

From Table A.6, we also find that the sheer volume of discussion, measured by the text lengths of a question, is a good predictor of the number of followers (see also Figs. A.1 and A.2 in the Online Appendix), capturing a large proportion of variance.

Motivated by this, for the next two experiments, we first have the variable of the number of followers residualized by the variable of text lengths and its squared term and control for the volume effects before variable selection using BART [3] and the sparse Bayesian regression [13]. We construct X as a set of dummy indicators that take the value of 1 if the proportion of a given political phrase under this question is larger than the median proportion of all the questions, which indicates the discussion focus under a posted question. The matrix sparsity of X (Dimension: $1,912 \times 1,436$) is 0.981.[2]

For the second goal of selecting the set of specific phrases related to the predicted topics, we use BART and the sparse Bayesian regression. Figure 2 shows the goodness of fit from BART when the number of trees to be grown in the sum-of-trees model is 150. In the Online Appendix (Fig. A.3), we also show the goodness of fit using different numbers of trees ($m = 25, 50, 100, 200, 250$). It can be observed that further increasing m makes little improvement in terms of data fitting beyond the threshold of 200; yet the iterative backfitting process becomes difficult for computers with small memories. When the value of m is small, e.g. 25 and 50, the predicted value is far less accurate and the percentage of coverage is 39.54% and 46.13% respectively. We emphasize here that since our main goal is to identify important text features, using small m will not jeopardize

---

[2] It should be noted that we use the LDA outputs instead of raw text features as the input X at the answer level (Table 2) due to the matrix sparsity and the latter's excessively high demand for computational power.

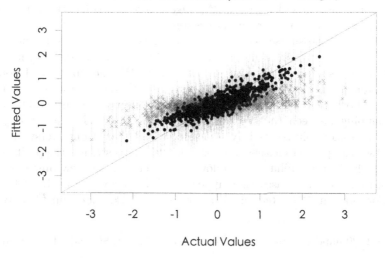

Fig. 2. BART: predicted value vs. fitted value

our inference, as BART always favors the most relevant predictors in the model. The following results are based on the model with $m = 150$.

We use the proportion of each predictor to all predictors chosen as a splitting rule in the BART model, to determine variable importance. The distribution of variable importance is shown in Fig. A.4 in the Online Appendix. And as anticipated, only a small number of political phrases are important and used in constructing the tree model. In particular, the numbers of variables whose proportion rates are over 0.010 and 0.005 are respectively 17 and 38. In Table 3, we

Table 3. Top 20 political phrases selected by BART

Y: Residualized log(No. Followers)

Abortion 堕胎 (−0.243), Update 更新 (0.226), FBI 联邦调查局 (0.265), Email 邮件 (0.209), Assange 阿桑奇 (0.276), Military 军事 (−0.220), Politics 政治 (−0.130), Insult 辱骂 (−0.232), Inauguration 上任 (0.127), Obama 奥观海 (0.105), Slave 奴隶 (−0.103), Trust 信任 (−0.050), Terror 恐怖 (0.163), Corporation 公司 (0.136), Reason 原因 (−0.094), Antipathy 反感 (−0.041), Videoclip 视频 (0.159), Hacker 黑客 (0.066), Development 发展 (−0.053), Romney 罗姆尼 (−0.114)

Note: Partial effect in the parenthesis.

present the twenty most important variables ordered by importance.[3] Consistent with previous findings from the LDA model, words and phrases associated with Hillary Controversy, e.g. FBI, email, Assange, hackers, predict stronger interests of citizens in discussions on related questions.

We then use the sparse Bayesian regression as an alternative to the BART model. The hyperparameters in the Bayesian LASSO model are tuned with cross-validation of a sample of observations ($n = 400$). We end up with 44 political phrases, and in Table 4 we present 20 phrases with the largest sizes in terms of effects, both positive and negative. We can observe a considerable degree of overlapping between the key phrases selected by BART and the Bayesian LASSO regression; phrases tightly associated with Hillary Controversy, e.g. FBI, email, Assange, predict citizens' interests in related questions – the positive sign indicates that more Zhihu users follow the update of the new answers under the question. Seven phrases that appear in Table 3 are not in Table 4: Trust 信任, Corporation 公司, Reason 原因, Antipathy 反感, Videoclip 视频, Hacker

Table 4. 20 important phrases selected by Bayesian LASSO and OLS estimation

Variable	LASSO		OLS	
	beta	t-statistics	beta	t-statistics
FBI 联邦调查局	0.325	6.010	0.361	6.147
Email 邮件	0.264	5.070	0.329	5.781
Update 更新	0.232	4.789	0.313	5.941
Obama 奥观海	0.221	3.060	0.352	3.910
Surveillance 监视	0.211	2.232	0.423	3.188
Shinzo Abe 安倍	0.175	2.102	0.369	3.081
Terror 恐怖	0.167	2.865	0.247	3.319
Assange 阿桑奇	0.155	2.659	0.239	3.174
Inauguration 上任	0.155	2.904	0.270	4.006
Pardon 赦免	0.113	1.527	0.276	2.134
Politics 政治	-0.102	3.504	-0.148	3.643
Development 发展	-0.108	2.910	-0.139	2.926
Division 分化	-0.111	2.113	-0.270	3.461
Sphere of Power 势力范围	-0.112	1.608	-0.334	2.810
Female 妇女	-0.156	2.860	-0.214	3.096
Slave 奴隶	-0.164	2.436	-0.297	3.245
Insult 辱骂	-0.254	3.122	-0.399	3.999
Military 军事	-0.254	4.709	-0.286	4.722
Trump-haters 川黑	-0.263	2.847	-0.451	3.884
Abortion 堕胎	-0.293	4.542	-0.368	5.086
N	1,912		1,912	

---

[3] A more delicate variable selection procedure is described in [2], which compares the variable's proportion rate to some thresholds obtained by permutation. This process is nevertheless computationally demanding.

黑客, Romney 罗姆尼. The Bayesian LASSO model actually selects six out of seven, which we omit for simplicity: Trust 信任 ($\beta = -0.089$), Corporation 公司 ($\beta = 0.106$), Antipathy 反感 ($\beta = -0.086$), Videoclip 视频 ($\beta = 0.108$), Hacker 黑客 ($\beta = 0.085$), and Romney 罗姆尼 ($\beta = -0.102$).

Lastly, as a robustness check, we also fit an unrestricted OLS model to the selected subset of 20 variables from the Bayesian LASSO model. The coefficients are statistically significant with larger magnitude and the same sign. It is because the LASSO model biases the estimates of the non-zero coefficients down towards zero. The R-squared of the OLS model is 0.186. In the Online Appendix, we show results using the LASSOplus model. It appears that LASSOplus selects a smaller subset of variables, but the positive effect of Hillary Controversy on citizens' interests still holds.

As a brief summary, we find a robust and strong association between latent text topics as well as text features pertaining to Hillary controversy and two measures of citizens' interests using OLS regression and other Bayesian-based supervised methods. Re-confirming our previous finding based on topic analysis, Chinese citizens do not show particularly strong interests in substantive policy proposals, nor do they pay considerable attention to those issues directly pertaining to the Sino-US relationship. Much like that in the United States, controversial information and sensational news dominate the public space in China [4].

3 Method

3.1 Data

As aforementioned, the data used in this article is collected from Zhihu, an emerging online community in China. It was launched on January 26, 2011, and the number of registered user was reportedly to have reached 17 millions as of May 2015, with 250 million monthly page views.[4] We collect all the answers under questions with three user-generated tags: Hillary, Trump, and the US presidential election. For the convenience of analysis, we remove questions with less than 5 answers; usually it indicates the question is weird, poorly organized, or similar to other questions that have been asked already. The time covered is between April 12th, 2015 (when Hillary announced her candidacy) and December 31st, 2016. Most questions and answers are posted in 2016, especially the days around the Presidential Election and other important events. Data collection is started and finished in early January, 2017.

We treat the texts as the bag of phrases because different from English and ancient Chinese texts, modern Chinese texts are primarily composed of two-

---

[4] Zhihu: Use Knowledge to Connect the World (in Chinese). Xinhuanet (http://news.xinhuanet.com/newmedia/2015-05/14/c_134238843.htm). 14 May 2015. Retrieved on 15 April 2017.

word and multi-word phrases.[5] And since Chinese is an ideographic language with no word or phrase delimiter, an important preprocessing step is segmentation, i.e. parsing a sentence into parts with substantive meanings. We use Jieba Parser (Project homepage: https://pypi.python.org/pypi/jieba/) which is available for multiple programing languages. We choose the default Mix Segmentation model that combines both the Maximum Probability Segmentation Model and the Hidden Markov Model to construct segmentation. Punctuations, numbers, stop words, and English are removed.[6] Some answers are composed mainly of pictures or links with few meaningful phrases, and we keep relatively long answers for analyses (at least 25 phrases except for stop words). In total, there are 26,917 answers under 1,912 questions. Other than the raw texts, we also collect two variables that measure citizens' interests in the latent political topics: the number of up-votes that an answer receives and the number of followers under a given question.

To reduce the demand for computational power, we create another dictionary of 1,436 phrases related to the presidential election and are also commonly seen in the cleaned corpus (frequency ≥ 50). We further remove around 6,000 answers with less than 10 political phrases. We include politicians' names and variants of informal abbreviations in Chinese (e.g. 希拉里, 希婆, 妖婆 for Hillary Clinton and 特朗普, 床破, 川皇 for Donald Trump),[7] partisanship and ideology (e.g. 白左, 白莲花, 自由派 for liberals, 茶党 for tea parties, 极右 for radical rightists), democratic process (e.g. 投票 for voting, 辩论 for debates), and controversies (e.g. 维基 and 维基解密 for Wikileaks).

3.2 Empirical Strategy

Due to limitation of space, we include details on the three methods used in our empirical analysis in the Online Appendix.

4 Discussion

To summarize, using original social media data, this paper shows that a large online community in China, composed mainly of educated and young urban citizens, displays strong interests in controversies and sensational stories instead of substantive political issues. Amid the top eight topics detected by the LDA

---

[5] For instance, 谏 (to advise someone earnestly) is a single-word phrase often used in ancient Chinese, and its counterparts commonly used in modern Chinese are 规劝 or 劝说.

[6] Some highly frequent English words, for instance Trump, Hillary, and LGBT, are converted to Chinese.

[7] How Chinese netizens call foreign politicians is less a strategy to avoid censorship than an expression of political preference. Comparatively, informal names that Chinese citizens use to call Hillary Clinton have stronger denigratory meaning in Chinese.

model, we find that the Chinese citizens pay more attention to Trump's anti-political-correctness platforms and controversies implicating Hillary Clinton. With the BART and the sparse Bayesian regression model, we directly select key political phrases to predict citizens' interests in latent political topics under different questions. Similarly, the appearance of political phrases related to Hillary Controversy predicts higher popularity measured by the number of the followers under the question.

The debate over the 2016 US presidential election and its global influence is far from settled. There are several implications of this paper.

First, this paper suggests that with accessibility to new sources of information, especially the social media, the Chinese audience may be inadvertently exposed to the negative sides of competitive elections, such as scandals, controversies, and other unverified information. A long-debated question in the Chinese politics is how the Chinese people view democracy as a potential alternative to the current one-party authoritarian system. Informed by this question, there is an emerging literature on the relationship between Chinese citizens' international knowledge of democratic countries, especially the United States, and domestic evaluations of the Chinese government [6, 7]. Citizens holding positive views of western democracies - even factually wrong - tend to hold negative view of the Chinese government, while exposure to negative information about democratic practice in other countries may undermine citizen's demand for domestic political reforms. Consistent with this theory, this paper suggests that when it comes to the 2016 US presidential election, Chinese citizens do not show admiration for open elections or discontent with China's current political system.

Some readers may argue that data from Zhihu, or any other social media based in China, does not accurately reflect the preference and interests of Chinese citizens. The Chinese government has invested heavily in official propaganda, promoting the ideal of the "China Model" and the importance of the strong Party Leadership in contrast with chaos and inefficiency that may accompany competitive elections. The lack of online discussions on the substance of democratic elections may result from a series of information control measures. This paper does not negate the government's role in shaping the focus of public discourse in China but we argue that such a role remains limited. The discussion of the current situation in China along with those of other advanced countries is common on the Chinese Internet, and generally perceived as posing little imminent threats, e.g., from collective actions, thus successfully avoiding being censored [10].

Second, besides the substantive implications, the methodological implication of the paper is on the use of text data in studying public opinion under authoritarian contexts. While maintaining an increasingly effective and sophisticated censorship apparatus on social media, the Chinese government also finds the Internet an efficient tool to interact with citizens and respond to people's demand. Consequently, there is a multitude of online communities for citizens to express political opinions that reflect their interests and preferences regarding a wide range of political issues. Due to many restrictions on conducting social surveys directly in China, large-scale text data collected directly from social media provides a comparatively more convenient way of studying public opinion, and complements other new quantitative methods, such as survey or lab experiments that have shown advantages in making strict causal claims.

References

1. Blei, D.M., Ng, A.Y., Jordan, M.I.: Latent Dirichlet allocation. J. Mach. Learn. Res. **3**, 993–1022 (2003)
2. Bleich, J., Kapelner, A., George, E.I., Jensen, S.T., et al.: Variable selection for BART: an application to gene regulation. Ann. Appl. Stat. **8**(3), 1750–1781 (2014)
3. Chipman, H.A., George, E.I., McCulloch, R.E., et al.: BART: Bayesian Additive Regression Trees. Ann. Appl. Stat. **4**(1), 266–298 (2010)
4. Calafiore, G.C., El Ghaoui, L., Preziosi, A., Russo, L.: Topic analysis in news via sparse learning: a case study on the 2016 US Presidential Elections. In: The 20th World Congress of International Federation of Automatic Control (2017, forthcoming)
5. Einstein, K.L., Glick, D.M.: Do I think BLS data are BS? The consequences of conspiracy theories. Polit. Behav. **37**(3), 679–701 (2015)
6. Huang, H.: International knowledge and domestic evaluations in a changing society: the case of China. Am. Polit. Sci. Rev. **109**(03), 613–634 (2015)
7. Huang, H., Yeh, Y.Y.: Information from abroad: foreign media, selective exposure, and political support in China. Br. J. Polit. Sci. (2016, forthcoming)
8. Imai, K., Lo, J., Olmsted, J.: Fast estimation of ideal points with massive data. Am. Polit. Sci. Rev. **110**(4), 631–656 (2016)
9. Jolley, D., Douglas, K.M.: The social consequences of conspiracism: exposure to conspiracy theories decreases intentions to engage in politics and to reduce one's carbon footprint. Br. J. Psychol. **105**(1), 35–56 (2014)
10. King, G., Pan, J., Roberts, M.E.: How censorship in China allows government criticism but silences collective expression. Am. Polit. Sci. Rev. **107**(02), 326–343 (2013)
11. Slapin, J.B., Proksch, S.O.: A scaling model for estimating time-series party positions from texts. Am. J. Polit. Sci. **52**(3), 705–722 (2008)
12. Swami, V., Chamorro-Premuzic, T., Furnham, A.: Unanswered questions: a preliminary investigation of personality and individual difference predictors of 9/11 conspiracist beliefs. Appl. Cogn. Psychol. **24**(6), 749–761 (2010)
13. Tibshirani, R.: Regression shrinkage and selection via the LASSO. J. Roy. Stat. Soc. Ser. B (Methodol.) **58**, 267–288 (1996)

Understanding the Pulse of the Online Video Viewing Behavior on Smart TVs

Tao Lian[✉], Zhumin Chen, Yujie Lin, and Jun Ma

School of Computer Science and Technology, Shandong University, Jinan, China
liantao1988@gmail.com, 1316975534@qq.com, {chenzhumin,majun}@sdu.edu.cn

Abstract. In recent years, millions of households have shifted from traditional TVs to smart TVs for the purpose of viewing online videos on
TV screens. In this paper, we examine a large-scale online video viewing
log on smart TVs over an extended period of time. Our aim is to understand the pulse of the collective behavior along the temporal dimension.
We identify eight interpretable daily patterns whose peak hours align
well to different dayparts. There also exists a holiday effect in the collective behavior. In addition, we detect three types of temporal habits
which characterize the differences between different households. Furthermore, we observe that the popularities of different video categories vary
depending on the dayparts. The obtained findings may provide guidance
on how to divide a day into several parts when developing time-aware
personalized video recommendation algorithms for smart TV viewers.

Keywords: Daily pattern · Online video viewing behavior · Smart TV

1 Introduction

In recent years, millions of households have shifted from traditional TVs to smart
TVs, which are equipped with Internet and interactive "Web 2.0" features and
hence can offer many more functions via apps than traditional TVs.[1] Many
people choose to purchase a smart TV and connect it to the Internet for the
purpose of viewing online videos on TV screens [19]. Online video service offers
a greater variety of content than live TV channels. Viewers can find interesting
videos on the Internet when they are willing to watch TV but there is nothing
interesting on live TV channels. In addition, it facilitates time shifting better
than live TV. Viewers can catch up on their favorite TV episodes that have
been missed when broadcast on TV channels. In short, online video service allows
smart TV viewers to watch whatever appeal to them at their convenience.

As we know, there are not much research investigating the online video viewing behavior on smart TVs yet, perhaps due to lack of open data. JuHaoKan[2],
a video content aggregation service platform for Hisense smart TVs, provides

---

[1] https://en.wikipedia.org/wiki/Smart_TV.
[2] http://www.juhaokan.org/.

© Springer Nature Singapore Pte Ltd. 2017
X. Cheng et al. (Eds.): SMP 2017, CCIS 774, pp. 331–342, 2017.
https://doi.org/10.1007/978-981-10-6805-8_27

us with a large-scale detailed online video viewing log on smart TVs over an extended period of time, which enables us to gain understanding of the collective behavior. Time—particularly our daily and weekly cycles of free and busy time—influences every aspect of our lives. In broadcast programming, dayparting is a common practice which divides a day into several parts and broadcast different types of programs at different parts of the day based on the usage patterns of the audience. Interestingly, some studies [1,13] demonstrated the existence of dayparts in the behavior of Internet users—the usage levels, audience compositions, and types of accessed content differ by daypart. Therefore, we want to understand the pulse of the online video viewing behavior on smart TVs along the temporal dimension.

It is reasonable to hypothesize that there exist some temporal patterns at the crowd level. One reason is that different people get used to watching TV in different time periods of the day, since most people have a regular yet different daily routine on most days. People with a day job have to work during the daytime, thus on workdays they can only watch TV after work (e.g., in the evening or late night). Students have to attend school during the daytime, thus on weekdays they can only watch TV after school (e.g., in the early fringe or evening). People who are often free at home, such as the elderly, the unemployed, full-time mothers and preschool children, may watch TV in the morning or lazy afternoon. Besides, different people usually prefer videos of different categories/genres. The other reason is that the household structure varies from household to household. That is to say, different families may be comprised of one, two, or three kinds of people mentioned above. If a household is comprised of a young couple who both have a day job, it is unlikely to observe any video viewing record for this household during the daytime on workdays. However, if a household includes people who are often free at home, there probably be quite a few video viewing records during the daytime on workdays.

To understand the pulse of the online video viewing behavior on smart TVs, we first obtain a set of 24-dimensional daily data points by measuring the amount of time per hour spent in watching online videos on smart TV by each household on each day. Next we identify typical daily patterns by applying the K-means algorithm on those daily data points. Then the temporal habit of a household is reflected by the K-dimensional cluster membership vector of the daily data points involving it. By further applying the clustering algorithm on the K-dimensional cluster membership vectors of all households, we can identify typical types of temporal habits. At last, we examine the popularity variations of different video categories over 24 h of the day.

The key findings include: (i) We identify eight interpretable daily patterns whose peak hours align well to different dayparts. (ii) There exists a holiday effect in the online video viewing behavior on smart TVs. That is to say, viewers tend to spend more time in watching online videos on smart TVs during the daytime on holidays than on workdays. (iii) We identify three types of temporal habits. Compared to the average, some households are more likely to watch online videos on smart TVs in the evening; some households tend to do that during the daytime; others rarely use the online video service on smart TVs. (iv) The most popular video categories are animation, movie, TV drama, followed by sports, children's program, and variety show. But their popularities vary depending on the dayparts.

2   Related Work

There were some qualitative studies of TV viewing behavior based on small-scale interviews, surveys and diaries [11,17]. They identified several contextual factors characterizing typical viewing situations at home, among which time is an important factor. In this paper, we employ standard data mining methods to perform quantitative analyses along the temporal dimension on a large-scale online video viewing log on smart TVs.

Note that a (smart) TV is shared by multiple users in a household. Temporal information is an important contextual factor for distinguishing and identifying different users in a household [2,5,9]. However, in real situations, users are reluctant to login with different accounts when watching TV. It is difficult to obtain such ground truth. What is observed on a smart TV is the mixed behavior of multiple users in the same household. Even though they can not be explicitly told apart, since they usually exhibit different temporal behavior, time-aware recommender systems [3] can still be helpful for improving personalized recommendation performance in this scenario, without the need to explicitly identify individual users within a household.

Some interesting studies were also conducted along the temporal dimension in other domains. For example, it was demonstrated that the behavior of Internet users differ by daypart [1,13]. Wu et al. [18] evaluated the sleep quality of microblog users based on the timestamps of posted microblogs. Ren et al. [14] categorized user queries into different types according to their search volume time series.

3   Methodology

3.1   Data Processing

On Hisense smart TVs, viewers can stream a variety of videos from the video content aggregation service platform JuHaoKan. At the same time, the platform gathers their detailed video viewing behavior in log files, including when a smart TV starts to play a video, which video is played, and when it stops playing the video, etc. We examined the online video viewing log over the period from 2015-12-21 to 2016-04-24. Each video viewing record r can be represented as a five-tuple $(h_r, d_r, v_r, s_r, e_r)$, where h_r denotes the smart TV which set off the record—we interpret it as a household since a smart TV is shared by multiple users in the same household, d_r denotes the date on which the record occurred, v_r denotes the video being played, s_r denotes the time when the smart TV started to play the video, and e_r denotes the time when it stopped playing the video. If a video viewing record cut across two days (i.e., the smart TV started to play the video before midnight and stopped the next day), we split it into two records. Let $\mathcal{R}$ be the set of all video viewing records. In total, there are $389\,564\,260$ records involving $4\,615\,220$ households and $80\,712$ videos.

In order to understand the pulse of the online video viewing behavior on smart TVs along the temporal dimension, we first measure the amount of time per hour that a household spent in watching online videos on smart TV on each day. For each household, we generate a daily data point for each day since the date when the household watched an online video for the first time, rather than average over all days to yield a single data point. For each pair (h, d), let $\mathcal{R}^{(h,d)} = \{r \in \mathcal{R} \mid h_r = h \wedge d_r = d\}$ denote the subset of video viewing records by the household h on the day d. Then the amount of time that the household h spent in watching online videos during each hour on the day d can be summarized as a 24-dimensional daily data point $x^{(h,d)} \in [0,1]^{24}$, where $x_m^{(h,d)} \in [0,1]$ is computed by summing up the overlap between the time span (i.e., $[s_r, e_r]$) of each video viewing record $r \in \mathcal{R}^{(h,d)}$ and the time slot of the mth hour. If the household h did not watch any online videos on the day d, then $x^{(h,d)} = \mathbf{0}$. We sampled $10\,000$ relatively active households in our analysis and discarded 10 abnormal ones for whom there were an unusually large number of video viewing records per day on many days. The total number of daily data points in $\mathcal{X} = \{x^{(h,d)}\}$ is $1\,199\,954$, among which 14.45% are $\mathbf{0}$.

3.2 Clustering Problem

Our goal is to uncover typical daily patterns from the set of daily data points $\mathcal{X} = \{x^{(h,d)}\}$, where $x^{(h,d)} \in [0,1]^{24}$ represents the variation of the amount of time that the household h spent in watching online videos on smart TV over the 24 h of the day d. It is in nature an unsupervised task, thus we resort to clustering techniques, which can automatically identify the unknown structures in a collection of data points by grouping them into several meaningful clusters such that the data points in a cluster are similar to one another but are dissimilar to the data points in the other clusters.

K-means is one of the most widely used clustering algorithms due to its simplicity, efficiency, and empirical success [6]. It partitions the data points into K disjoint clusters $\mathcal{C} = \{\mathcal{C}_1, \ldots, \mathcal{C}_K\}$. Each cluster $\mathcal{C}_k$ is characterized by its centroid μ_k, which is randomly initialized at the beginning. Then, K-means iteratively optimizes the objective function (1) by alternating between the two steps: (i) Each data point is assigned to the cluster whose centroid is the nearest to it, i.e., $x^{(h,d)} \in \mathcal{C}_{k^\star}$, where $k^\star = \arg\min_k \left\| x^{(h,d)} - \mu_k \right\|_2^2$. (ii) The centroid of each cluster is updated to be the mean of the data points currently assigned to it, i.e., $\mu_k = \frac{1}{|\mathcal{C}_k|} \sum_{x^{(h,d)} \in \mathcal{C}_k} x^{(h,d)}$. Thus, the centroid of each cluster can be thought of as the representative of the data points in the cluster. In our setting, we treat it as a typical daily pattern of the online video viewing behavior on smart TVs (Sect. 4.1).

$$J = \sum_{k=1}^{K} \sum_{x^{(h,d)} \in \mathcal{C}_k} \left\| x^{(h,d)} - \mu_k \right\|_2^2. \tag{1}$$

3.3 Clustering Tendency

Given a set of data points, before applying any clustering algorithm, we need to assess whether the data has a clustering tendency. Although clustering algorithms can always partition the data into multiple groups in any case, forcing unstructured data into clusters could lead to erroneous conclusions about the underlying data organization.

The Hopkins statistic [4] is a simple and intuitive measure of clustering tendency that compares the real data set with a set of artificial data points distributed uniformly in the same data space. If the data set is arranged in tight clusters, then on average the distance from a real data point to its nearest real data point will be much smaller than the distance from an artificial data point to its nearest real data point, so the Hopkins statistic will be much larger than 0.5, approaching 1. However, if the data set is no more clustered compared with uniformly distributed artificial data points, the Hopkins statistic will be approximately 0.5. We computed the Hopkins statistic 10 times with different samples of artificial data points in the space $[0, 1]^{24}$. The average value is 0.91, and the standard deviation is 9.8×10^{-4}.

However, a set of uniformly distributed artificial data points is a relatively weak competitor. Lawson and Jurs [8] proposed to compare the set of real data points with a set of artificial data points which not only lies in the same space as the real data points, but also has identical individual univariate distributions— not multivariate distribution—to those of the real data points rather than uniform distributions. Specifically, each dimension of an artificial data point is sampled from the empirical distribution of the values in the corresponding dimension of the real data points. Again, we repeated the procedure 10 times. The average value is 0.62, and the standard deviation is 3.2×10^{-3}. Thus, we conclude that our data set has a clear clustering tendency.

3.4 Cluster Membership

Recall that for each household, we generate a 24-dimensional data point for each day since the date that the household used the online video service for the first time. Let $\mathcal{X}^{(s)} = \{ x^{(h,d)} \in \mathcal{X} \mid h = s \}$ denote the subset of daily data points involving the household s. Once K-means partitions $\mathcal{X}$ into K clusters, every daily data point in $\mathcal{X}^{(s)}$ is assigned to one of the K clusters, but they may belong to different clusters. That is to say, the behavior of the household s on different days may be similar to different daily patterns. To gain deeper insights into the household's habit, we analyze the cluster membership of $\mathcal{X}^{(s)}$, which can be represented by a K-dimensional vector $\theta^{(s)}$. Each component,

$$\theta_k^{(s)} = \frac{\left| \mathcal{X}^{(s)} \cap \mathcal{C}_k \right|}{\left| \mathcal{X}^{(s)} \right|}, \tag{2}$$

is the fraction of $\mathcal{X}^{(s)}$ assigned to the cluster $\mathcal{C}_k$. We can think of $\theta_k^{(s)}$ as the possibility of the household s to follow the daily pattern corresponding to the cluster centroid μ_k. And $\boldsymbol{\theta}^{(s)}$ encodes the temporal habit of the household s. By further applying the clustering algorithm on these K-dimensional vectors for all households, we can obtain typical types of temporal habits (Sect. 4.3).

4   Results

4.1   Daily Patterns

Number of Daily Patterns. As we adopt the K-means algorithm to cluster the set of daily data points $\mathcal{X} = \left\{ \boldsymbol{x}^{(h,d)} \right\}$ and treat each cluster centroid as a typical daily pattern, we first need to determine the number of clusters present in the data. A commonly used method [16] is to try different numbers and inspect the variation of the objective value (1) with respect to the number of clusters K. We tried different values of K in the range $[2, 32]$. By further examining the resulting daily patterns from the perspective of discriminability and interpretability, we decided to set K to 8. The eight daily patterns, i.e., $\{\mu_k\}$ when $K = 8$, are plotted in Fig. 1. In each sub-figure, the horizontal axis represents the 24 h of the day, and the vertical axis represents the average amount of time per hour spent in watching online videos on smart TV.

Interpretation of Daily Patterns. As shown in Fig. 1, the eight daily patterns are discernible. Note that the peak hours of different daily patterns occur in different time slots, which inspires us to interpret them by referring to the television dayparts. In broadcast programming, dayparting[3] is a common practice which divides a day into several parts based on the usage patterns of the audience. We divide a day into eight parts as listed in Table 1 according to the industrial practice [10, Chap. 4] and our own daily viewing habits. Surprisingly, the peak hours of different daily patterns except for Figs. 1g and h align well to different dayparts. Thus, each pattern is given a name based on its peak hours.

   Note that these daily patterns should be interpreted at the crowd level rather than at the individual level. The daily behavior of a household on a certain day may not be exactly the same as any of the eight daily patterns, but only roughly similar to one of them. Since each daily pattern corresponds to the centroid of one cluster, which is the mean of the data points belonging to it, the subtle differences between the data points in the same cluster average out whereas the commonalities stand out.

4.2   Cluster Sizes

Now we analyze the cluster sizes to understand the population's online video viewing habits on smart TVs. The second column of Table 2 lists the distribution

---

[3] https://en.wikipedia.org/wiki/Dayparting.

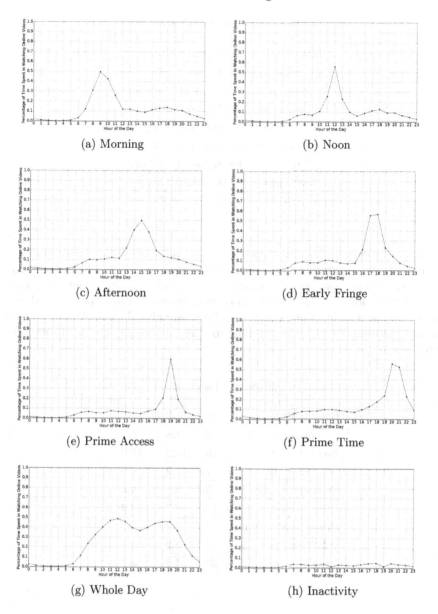

(a) Morning

(b) Noon

(c) Afternoon

(d) Early Fringe

(e) Prime Access

(f) Prime Time

(g) Whole Day

(h) Inactivity

Fig. 1. Daily patterns

of $\mathcal{X}$ among the eight clusters. A key observation is that the cluster "Whole Day" is the smallest, containing only 5.1% of the data points, while the cluster "Inactivity" is the largest, which contains 38.5% of the data points, including those (14.45%) that are **0**. It indicates that users only spend much time in watching online videos on smart TVs on a few days, while on many days they

Table 1. Television dayparts in China

Dayparts	Time period
Morning	06:00 a.m.–11:00 a.m.
Noon	11:00 a.m.–01:00 p.m.
Afternoon	01:00 p.m.–04:00 p.m.
Early fringe	04:00 p.m.–07:00 p.m.
Prime access	07:00 p.m.–08:00 p.m.
Prime time	08:00 p.m.–11:00 p.m.
Late night	11:00 p.m.–01:00 a.m.
Overnight	01:00 a.m.–06:00 a.m.

rarely or never use the online video service on smart TVs. Possible reasons include: (i) As reported by Nielsen [12], most users appear to be supplementing, rather than replacing, live TV programs with online videos. They still watch live TV programs. (ii) Nowadays most people own a smart phone, and most households own at least one computer. There are abundant choices of pastimes besides watching TV, such as listening to music, playing games, and surfing the Internet.

Table 2. Cluster sizes

Daily patterns	$\% \mathcal{X}$	$\% \mathcal{X}_{\text{holi}}$	$\% \mathcal{X}_{\text{work}}$
Morning (Fig. 1a)	8.5	11.8	6.0
Noon (Fig. 1b)	10.0	10.2	10.0
Afternoon (Fig. 1c)	7.8	10.2	6.0
Early fringe (Fig. 1d)	9.4	8.8	9.8
Prime access (Fig. 1e)	11.9	10.3	13.2
Prime time (Fig. 1f)	8.7	9.2	8.3
Whole day (Fig. 1g)	5.1	8.2	2.9
Inactivity (Fig. 1h)	38.5	31.4	43.8

Holiday Effect. Next we make a distinction between holidays and workdays, since the amount of time spent in watching online videos on smart TVs greatly depends on whether the users are free at home. Holidays include those official public holidays in China[4] in the period from 2015-12-21 to 2016-04-24. In addition, all students in China have a winter vacation lasting about four weeks (from 2016-01-25 to 2016-02-21) around the Spring Festival. Besides, all weekends are also included in the holidays. All the other days are considered as workdays.

---

[4] english.gov.cn/services/2015/12/11/content_281475252239869.htm.

The set of daily data points $\mathcal{X}$ is split into two subsets $\mathcal{X}_{\mathrm{holi}}$ and $\mathcal{X}_{\mathrm{work}}$, where $|\mathcal{X}_{\mathrm{holi}}| = 510\,278$ and $|\mathcal{X}_{\mathrm{work}}| = 689\,676$. The distributions of $\mathcal{X}_{\mathrm{holi}}$ and $\mathcal{X}_{\mathrm{work}}$ among the eight clusters are shown in the last two columns of Table 2. There is a clear difference between these two distributions. The percentage of daily data points in $\mathcal{X}_{\mathrm{holi}}$ belonging to the cluster "Inactivity" is much lower than that of daily data points in $\mathcal{X}_{\mathrm{work}}$, while the percentages of daily data points in $\mathcal{X}_{\mathrm{holi}}$ belonging to the cluster "Morning", "Afternoon" and "Whole Day" are much higher than those of daily data points in $\mathcal{X}_{\mathrm{work}}$. Therefore, smart TV viewers tend to spend more time in watching online videos during the daytime on holidays than on workdays. A Chi-squared test [15, Sect. 4.3] confirms that the observed holiday effect is not due to chance.

4.3   Types of Temporal Habits

As mentioned in Sect. 1, since different families are comprised of different kinds of people, thus the daily behavior of different households may have different possibilities to follow the eight daily patterns. For example, if a household is comprised of a young couple who both have a day job, it is unlikely to observe many video viewing records for this household during the daytime on workdays. However, if a household includes people who are often free at home, there probably be quite a few video viewing records during the daytime on workdays. In other words, different households may have different types of temporal habits.

Now we arrange the K-dimensional vector $\boldsymbol{\theta}^{(s)}$ (Sect. 3.4) of all households into a matrix $\boldsymbol{\Theta}$. By further applying the clustering algorithm on the normalized matrix $\tilde{\boldsymbol{\Theta}}^{5}$, we can obtain clusters of households. After trying different numbers, we obtained three clusters. The corresponding cluster centroids are plotted in Figs. 2a, c and e, and household examples of the clusters are presented in Figs. 2b, d and f. We can observe that different clusters have disparate possibilities to follow those daily patterns. They exhibit different types of temporal habits. Compared with the average, some households are more likely to watch online videos on smart TVs in the evening (Figs. 2a and b); some households tend to do that during the daytime (Figs. 2c and d); others rarely use the online video service on smart TVs (Figs. 2e and f).

4.4   Dynamics of Video Categories

Each video in the watch log is assigned to one of 16 video categories: animation, movie, TV drama, sports, children's program, variety show, music, news, lifestyle, education, documentary, entertainment, autos, info, short film, and others. Further insights may be gained by investigating dynamics of video categories over the 24 h of the day.

We first break down the total number of views on workdays by category, and then break down the percentage of views received by each category by

---

[5] $\boldsymbol{\Theta}$ is normalized to $\tilde{\boldsymbol{\Theta}}$, where $\tilde{\theta}_k^{(s)} = \frac{\theta_k^{(s)} - \bar{\theta}_k}{\sigma_k}$, i.e., each dimension is subtracted by its mean and divided by its standard deviation.

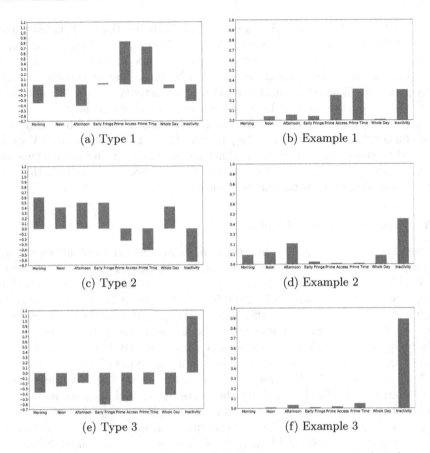

(a) Type 1   (b) Example 1

(c) Type 2   (d) Example 2

(e) Type 3   (f) Example 3

Fig. 2. Types of temporal habits (left) and household examples (right)

hour of the day. Figure 3a illustrates the percentage of views received by each category in each hour on workdays. To facilitate understanding, Fig. 3b shows the composition of views in each hour on workdays—the total for each bar adds up to 100%. Taking Figs. 3a and 3b together, we can make the following observations[6]: (i) The most popular video categories are animation, movie, TV drama, followed by sports, children's program, and variety show (Fig. 3a). Together they account for approximately 90% of views in each hour of the day (Fig. 3b). (ii) Though animation is the most popular video category during the daytime, it is less popular than movie and TV drama during the late night and overnight, i.e., 10:00 p.m.–6:00 a.m. Perhaps because pre-school children and students usually go to sleep early in the evening (Figs. 3a and b). (iii) The highest volume occurs in the hour 6:00 p.m.–7:00 p.m. for animation, sports, and children's program, while the highest volume occurs in the hour 7:00 p.m–8:00 p.m. for movie, TV

---

[6] We repeat the analysis on holidays, the results are very similar with minor differences. Due to space limitations, we omit the figures here.

drama, and variety show (Fig. 3a). (iv) The percentages of views for animation, sports and children's program dip slightly in the period 1:00 p.m.–4:00 p.m.

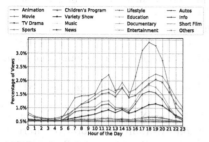

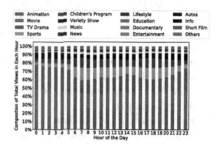

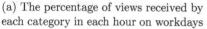

(a) The percentage of views received by each category in each hour on workdays

(b) The composition of views in each hour on workdays

Fig. 3. Popularity variations of different video categories

5 Conclusion and Future Work

In this paper, we perform extensive analyses on a large-scale online video viewing log on smart TVs with the aim of understanding the pulse of the collective behavior. By clustering the daily behavior on many days by a large number of households, we identify eight interpretable daily patterns whose peak hours align well to different dayparts. We also verify that there exists a holiday effect in the collective behavior. In addition, by analyzing the relationship between each household and the eight daily patterns, we identify three types of temporal habits, which characterize the difference between households. Finally, we observe that the popularities of different video categories differ by daypart. In the future, we plan to explore time-aware video recommendation algorithms on smart TVs, such as tensor factorization [7] and profile splitting [20]. Both of them need to discretize the temporal dimension, the findings in this paper may provide guidance on how to divide a day into several parts.

Acknowledgments. This work is supported by the Natural Science Foundation of China (61672322, 61672324), the Natural Science Foundation of Shandong Province (2016ZRE27468) and the Fundamental Research Funds of Shandong University. We also thank Hisense for providing us with a large-scale watch log on smart TVs.

References

1. Beyers, H.: Dayparting online: living up to its potential? Int. J. Media Manag. **6**(1–2), 67–73 (2004)
2. Campos, P.G., Bellogin, A., Díez, F., Cantador, I.: Time feature selection for identifying active household members. In: Proceedings of the 21st ACM International Conference on Information and Knowledge Management, pp. 2311–2314. ACM (2012)

3. Campos, P.G., Díez, F., Cantador, I.: Time-aware recommender systems: a comprehensive survey and analysis of existing evaluation protocols. User Model. User Adap. Interact. **24**(1), 67–119 (2014)
4. Hopkins, B., Skellam, J.G.: A new method for determining the type of distribution of plant individuals. Ann. Bot. **18**(2), 213–227 (1954)
5. Iguchi, K., Hijikata, Y., Nishida, S.: Individualizing user profile from viewing logs of several people for TV program recommendation. In: Proceedings of the 9th International Conference on Ubiquitous Information Management and Communication, pp. 61:1–61:8. ACM (2015)
6. Jain, A.K.: Data clustering: 50 years beyond K-means. Pattern Recogn. Lett. **31**(8), 651–666 (2010)
7. Karatzoglou, A., Amatriain, X., Baltrunas, L., Oliver, N.: Multiverse recommendation: N-dimensional tensor factorization for context-aware collaborative filtering. In: Proceedings of the Fourth ACM Conference on Recommender Systems, pp. 79–86. ACM (2010)
8. Lawson, R.G., Jurs, P.C.: New index for clustering tendency and its application to chemical problems. J. Chem. Inf. Comput. Sci. **30**(1), 36–41 (1990)
9. Lesaege, C., Schnitzler, F., Lambert, A., Vigouroux, J.R.: Time-aware user identification with topic models. In: IEEE 16th International Conference on Data Mining, pp. 997–1002 (2016)
10. Lin, Y.: Repeat viewing in China: an expansion of determinants of program choice. Master's thesis, Ohio University (2008)
11. Mercer, K., May, A., Mitchel, V.: Designing for video: investigating the contextual cues within viewing situations. Pers. Ubiquit. Comput. **18**(3), 723–735 (2014)
12. Nielsen: Video on demand: How worldwide viewing habits are changing in the evolving media landscape (2016). http://www.nielsen.com/us/en/insights/reports/2016/video-on-demand.html
13. Online Publishers Association: The OPA white papers: The existence and characteristics of dayparts on the internet (2003)
14. Ren, P., Chen, Z., Ma, J., Zhang, Z., Si, L., Wang, S.: Detecting temporal patterns of user queries. J. Assoc. Inf. Sci. Technol. **68**(1), 113–128 (2017)
15. Shasha, D., Wilson, M.: Statistics is Easy!, 2nd edn. Synthesis Lectures on Mathematics and Statistics, vol. 3, issue 1, pp. 1–174 (2010)
16. Sugar, C.A., Lenert, L.A., Olshen, R.A.: An application of cluster analysis to health services research: empirically defined health states for depression from the SF-12. Tech. report, Stanford University (1999)
17. Vanattenhoven, J., Geerts, D.: Contextual aspects of typical viewing situations: a new perspective for recommending television and video content. Pers. Ubiquit. Comput. **19**(5), 761–779 (2015)
18. Wu, K., Ma, J., Chen, Z., Ren, P.: Sleep quality evaluation of active microblog users. In: Cheng, R., Cui, B., Zhang, Z., Cai, R., Xu, J. (eds.) APWeb 2015. LNCS, vol. 9313, pp. 178–189. Springer, Cham (2015). doi:10.1007/978-3-319-25255-1_15
19. YuMe: Emerging growth opportunities for connected TV and advertisers (2012). www.yume.com/whitepaper/whitepaper-emerging-growth
20. Zheng, Y., Burke, R., Mobasher, B.: Splitting approaches for context-aware recommendation: an empirical study. In: Proceedings of the 29th Annual ACM Symposium on Applied Computing, pp. 274–279. ACM (2014)

Hierarchical Community Detection Based on Multi Degrees of Distance Space and Submodularity Optimization

Shu Zhao[1,2,3], Chengjin Yu[1,2,3], and Yanping Zhang[1,2,3(✉)]

[1] Key Laboratory of Intelligent Computing and Signal Processing of Ministry
of Education, Anhui University, Hefei 230601, China
zhangyp2@gmail.com
[2] Center of Information Support and Assurance Technology, Anhui University,
Hefei 230601, China
[3] School of Computer Science and Technology, Anhui University,
Hefei 230601, China

Abstract. Detecting hierarchical community is crucial to analyze complex networks. In this paper, we propose a model for Hierarchical Community Detection based on Multi Degrees of distance space and submodularity function optimization (MD-HCD). First, an original network is divided into many communities under one degree of distance space. Then each community in original network is regarded as a super node, so those super nodes and corresponding edges construct a quotient network. And the same method is used to identify communities in quotient network. During hierarchical process, target function holds the property of submodularity, so that a result with $[1 - 1/e]$ approximation is guaranteed. Experiments reveal the benefits of the multi degree of distance space. The proposed method generally detects a hierarchical community structure that includes three layers and has a stable performance in terms of modularity compared with many other main stream algorithms.

Keywords: Hierarchical community structure · Quotient network · Multi degrees of distance · Submodularity optimization

1 Introduction

Detecting hierarchical community structures in social networks is a very important task in social network analysis, which has attracted a lot of attention. Accordingly, hierarchical community structure refers to that a community contains small communities, and the small communities further contain smaller communities based on the similarities or strength [1–3]. Mining such hierarchical structure is of great significance for us to analyze and understand networks.

Traditionally, hierarchical clustering techniques are often used to address hierarchical community problem [4]. Their main idea is computing similarity or strength between nodes in the network, and nodes with larger similarity tend to same community while nodes with smaller similarity tend to different communities. They can be classified in two categories include divisive methods and agglomerative methods.

© Springer Nature Singapore Pte Ltd. 2017
X. Cheng et al. (Eds.): SMP 2017, CCIS 774, pp. 343–354, 2017.
https://doi.org/10.1007/978-981-10-6805-8_28

The most famous divisive algorithm is GN that proposed by Girvan and Newman [1]. It computes the betweenness of all edges and deletes the edge with largest betweenness every step. Some other works are also presented based on different centrality [5, 6]. However, divisive techniques have been rarely used in the past duo to its huge time complexity. Most researches concentrate on agglomerative methods. In this way, modularity [7] is the most popular criterion to measure the quality of community detection. And modularity gain is often used as a criterion for hierarchical aggregation. First modularity optimization algorithm is NFA [7]. It achieves maximum growth of modularity for each merging based on greedy techniques. Then Clauset and Newman proposed an improved method named CNM [8], which has lower time complexity. Many other strategies are also used to optimize modularity, include spectral clustering [9], simulated annealing strategy [10], combining content with links algorithms [11] and extremal optimization [12]. But detecting the highest modularity value is proven to be NP-hard [13]. It is impossible to find the optimal solution in a limited time. Previous algorithms are all based on approximate optimization but without the proof that its approximation is valid. Besides modularity, similarity between objects is also an important aggregation strategy [14–17]. And larger communities can be obtained by merging small communities based on a similarity measure. However, for most hierarchical community detection methods, whether divisive methods or agglomerative methods, they only divide a community into two small communities, or merge two small communities into a large one in each layer. They design heuristics to generate hierarchical community, but lack of deeper consideration from real networks.

Recently, to better address hierarchical community detection, many efforts have been done based on the perspective of influence of entities. [18, 20, 21] identify the core-periphery structure in networks. Those works divide the network into two layers and consider the core is more important. But such two-level structure is too simple to reveal sophisticated social behaviors and interactive relationships among nodes in social networks. A more basic research is proposed by Christakis [22], who gives the concepts of strong connections and weak connections. Connections within three degrees of distance can lead to behavior, while connections over three degrees of distance only can lead to pass information. Therefore, with the increase of degree distance, it might exhibit different property among entities, which infers us to achieve a hierarchical understanding of whole network through multi degrees of distance.

Working with these discussions, we propose a model for hierarchical community detection based on multi degrees of distance space and submodularity optimization (MD-HCD). In such work, hierarchical community detection problem is formulated as general community detection problem under multi degrees of distance. More specifically, an original network is divided into many communities. Then a corresponding quotient network can be constructed, and same method is used to identify communities in such quotient network. The results of proof and experiments indicate that our model can achieve a result with $[1 - 1/e]$ approximation [23] and have a stable performance in terms of modularity. Besides, MD-HCD detects hierarchical community under multi degrees of distance and allows us to get a novel understanding of hierarchical structure under the view of degree distance.

The remainder of this paper is organized as follows. We will give problem statement and preliminaries in Sect. 2. The complete description of MD-HCD will be introduced in Sect. 3. Section 4 will analyze experimental results. Finally, we will conclude the paper with future works in Sect. 5.

2 Preliminaries and Problem Statement

In this section, we first introduce the preliminaries in our work. We then formally state our hierarchical community detection problem as a generalization of the community detection problem under different degree space.

Generally, a network is typically modeled as an undirected graph $G = (V, E)$ with each node $x \in V$ to represent an entity, and the edge $e(x, y) \in E$ to reflect the relationships between x and y. And community detection is finding a partition $C = \{c_x \subseteq V, \cup\, c_y = V$ and $c_x \cap c_y = \emptyset\}$ to maximize the utility function $F(C)$. In this work, we consider the hierarchical community problem as generalization of the community detection problem under multi degrees of distance space. After one degree of distance space is further considered, a new community structure is detected so that a new layer is further formed. Specifically, for a node x, we call $t_x^i = \{e(x, y)|\forall y \in V\}$ one degree of distance space of x. For a network G, we call $T_1 = \bigcup_{x \in V} t_x^1$ one degree of distance space of G. Then $t_x^2 = \{e(x, y), e(y, z)|\forall y, z \in V\}$ is two degree of distance space of x, and $T_2 = \bigcup_{x \in V} t_x^2$ is two degrees of distance space of G. Similarly, we have multi degrees of distance space T_i of G. And during the hierarchical process, if a community c_x is detected under T_i, the community c_x will be maintained or be incorporated into a large community under T_{i+1}.

Based on above discussion, we model the multi degree of distance space as a series of quotient networks, $G_1, G_2, \ldots, G_k$. We assume that the original network G_0 is divided into many communities $C^1 = \{c_1, c_2, \ldots, c_s\}$ under one degree of distance space $[T]_1$ of G_0. Then each community in original network is regarded as a super node so that those super nodes and corresponding edges construct a quotient network $G_1 = (V^1, E^1), \{V^1 = \{c_x|c_x \in C^0\}\}$. We then detect community structure C^2 under one degree of distance space $[T]_1$ of G_1, which approximately equivalents to the communities detection under two degrees of distance space in original network. Finally, we define the quotient network $G_i = (V^i, E^i), \{V^i = \{c_x|c_x \in C^{i-1}\}\}$ and the one degree of distance space $[T]_i = \bigcup_{x \in V^i} [t]_x^1$ of G_i.

Our goal is detecting hierarchical community structure $C^T = \{C^1, C^2, \ldots, C^k\}$ based on community detection under multi degrees of distance space. On a quotient network G_i, we try to find a best community structure C^i to maximize utility function $F_G(C)$. More formally, we define the above task as the hierarchical community detection.

Definition 1 Hierarchical community detection (HCD). Given a network G, the task of HCD is finding a hierarchical community structure $C^T = \{C^1, C^2, \ldots, C^k\}$, such that:

$$C^{i+1} = \arg\max_{C||[T]_i} F_G(C). \tag{1}$$

The key steps of MD-HCD to solve HCD are constructing quotient network and find a good community structure under one degree of distance space of such quotient network. Both two steps are flexible. For example, [16, 17] proposed different methods to address previous step. For the later step, we can treat the different quotient networks G_i independent and solve them as community detection problem for each G_i by algorithms such as [3, 7, 8].

3 Hierarchical Community Detection Based on Multi Degree of Distance Space and Submodularity Optimization (MD-HCD)

3.1 Model

Based on the above idea, we propose MD-HCD model for hierarchical community detection problem. Figure 1 is graphical representation of MD-HCD model. Algorithm 1 describes the main steps of MD-HCD.

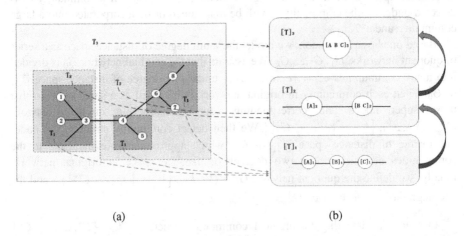

(a)   (b)

Fig. 1. Graphical representation of MD-HCD model. (a) donates the original network. In (b), A = {1, 2, 3}, B = {4, 5} and C = {6, 7, 8} are small communities that identified under one degree of distance space on original network.

As described in Algorithm 1, our model has three main steps. First, we should get one degree of distance space from original network or quotient network. Specifically, we get one degree of distance space from i-th quotient network, which is approximately regarded as i degrees of distance space of original network. Step 2 is flexible, and many methods can be adopted, such as simple greedy technique mentioned in [7, 8]. In step 3, we will use weighted edges to address the problem. Next, we detail the step 1 and step 2 in Algorithm 2, and then construct the quotient network.

Algorithm 1. MD-HCD model.

Input: A network $G = (V, E)$, objective function $F(C)$;
Output: Hierarchical community $C^T = \{C^1, C^2, ..., C^k\}$;
Initialize: $i = 0, G_i = G, C^0 = \{x_1, x_2, ..., x_n\}$;
Repeat
Step 1: Get $[T]_i$ from G_i;
Step 2: $C^{i+1} = \arg\max_{C|[T]_i} F_G(C)$;
Step 3: Construct G_{i+1} according to C^{i+1} and $[T]_i$;
Until *Convergence*;
Return: $C^T = \{C^1, C^2, ..., C^k\}$;

As Algorithm 2 shown, line 1 first computes the similarities among nodes. But if only there is an edge between two nodes, a similarity would be computed. And similarity is compute by Eqs. (2) or (3). Then, in lines 3 and 4, we would select a biggest similarity value from similarity set, and a node would be divided into a community with its neighborhoods if their similarity value is not less than the selected similarity value. We would handle the overlapping nodes based on a membership function $Belongness(x, c_{x_{cur}})$ as shown in Eq. (4). At the end, while the $\Delta F(C) < 0$ ($\Delta F(C)$ is computed based on [7]), the algorithm stops. In Algorithm 2, we should compute similarity both on original network and quotient networks. First, we use the Jaccard [19] index to compute the similarity on original network. Jaccard index is shown as Eq. (2).

$$Sim_1(x, y) = \frac{|\Gamma(x) \cap \Gamma(y)|}{|\Gamma(x) \cup \Gamma(y)|}. \tag{2}$$

Where $\Gamma(x)$ donates the neighborhoods of x, and $|\cdot|$ donates the number of $\cdot$.

The properties of quotient network are different. Since a super node represents a community, we give following similarity index, which is simple but effective.

$$Sim_2(x, y) = \frac{\tau(x, y)}{|x| * |y|} \tag{3}$$

As x is community, $|x|$ is the number of nodes in x. $\tau(x, y)$ is the number of edges between x and y. In this way, the information of super nodes is considered. In fact, such $Sim_2(x, y)$ value is also the weight of our super edges. Compared with unweight edges, the weighted edges contain more information.

In the end, we give a membership function to handle the overlapping nodes. Equation (4) is easy to understand. But it must be explained that $|c_{x_{cur}}|$ is the number of super nodes when detects on quotient network.

$$Belongness(x, c_{x_{cur}}) = \frac{\sum_{y \in c_{x_{cur}}} Sim(x, y)}{|c_{x_{cur}}|}. \tag{4}$$

Algorithm 2. Community detection under one degree of distance space

---
Input: A network $G = (V, E)$, objective function $F(C)$;
Output: Community structure $C = \{c_1, c_2, \ldots, c_l\}$;
Initialize: $C = \{c_1, c_2, \ldots, c_n\}, \{c_1 = x_1, c_2 = x_2, \ldots, c_n = x_n\}$;
01: $\forall e(x, y) \in E, s_{xy} = Sim(x, y)$ and $s_{xy} \rightarrow S$;
02: while $\triangle F(C) \geq 0$ do
03: $s^* = \max(S), S = S - s^*$;
04: $\forall x \in V, \exists y \in V$, if $s_{xy} \geq s^*$,
05: $V = V - x, c_x = c_x \cup y$, end if;
06: $\forall c_{x_p}, c_{x_q} \in C$, let $c^* = c_{x_p} \cap c_{x_q}$, if $c^* \neq \varnothing$,
07: let $\forall x \in c^*$, let $cur = \arg\min_{cur} Belongness(x, c_{x_{cur}})$,
08: $c_{x_{cur}} = c_{x_{cur}} - x$;
09: end if;
10: end while;

3.2 The Submodularity of Target Function

We will prove that the utility function $F(C)$ is monotonic and submodular. In this section, we first illustrate the property of $F(C)$ on original network. And then prove that the $F(C)$ holds submodularity during our hierarchical process.

First, we only consider the property of target function on original network. We define a function $f(c_x)$ to evaluate the quality of a specific community c_x. Then we have $F(C) = \sum_{c_x \in C} f(c_x), \{c_x, c_y \in C \text{ and } c_x \cap c_y = \emptyset\}$. Therefore, if an i degrees of distance space T_i is given of network G, we can get corresponding $F_{T_i}(C) = \sum_{c_x^i \in C^i} f_{t_x^i}(c_x^i)$ where c_x^i is a specific community when i degrees of distance space is given of node x. We need the lemma 1 to characterize the properties of $f(c_x)$.

Lemma 1. For $\forall x \in V$, c_x^i, t_x^i and $i, j = 1, 2, 3, \ldots, \{j > i\}$, we have,
Monotonicity:

$$f_{t_x^i}(c_x^i) \leq f_{t_x^{i+1}}(c_x^{i+1}). \tag{5}$$

Submodularity:

$$f_{t_x^i \cup t_x^j}\left(c_x^{\max\{i,j\}}\right) - f_{t_x^i}\left(c_x^i\right) \ge f_{t_x^{i+1} \cup t_x^j}\left(c_x^{\max\{(i+1),j\}}\right) - f_{t_x^{i+1}}\left(c_x^{i+1}\right). \tag{6}$$

Proof. In Eq. (5), monotonicity can be easily proved, since $c_x^{i+1} = c_x^i \cup V^*$. V^* is a subset of nodes that can be reached by x under t_x^{i+1} and it is selected based on a simple greedy strategy to maximize $f(c_x)$. In (6), since $j > i$, we have $t_x^i \cup t_x^j = t_x^j$. Then, Eq. (6) can be rewritten as Eq. (7).

$$f_{t_x^j}\left(c_x^j\right) - f_{t_x^i}\left(c_x^i\right) \ge f_{t_x^j}\left(c_x^j\right) - f_{t_x^{i+1}}\left(c_x^{i+1}\right). \tag{7}$$

According to Eq. (5) in Lemma 1, it is easy to prove that Eq. (7) is workable. Based on above discussion, we can conclude that $F_{T_i}(C)$ holds submodularity on original network.

Then, we consider the property of target function on quotient network. Give a $G_i = (V^i, E^i)$, and a specific node $v \in V^i$. In fact, v represents a community $c_x^{i-1} \in C^{i-1}$. The diameter of communities in C^{i-1} ranges from 0 to 2^{i-1}. For convenience, we assume that all nodes in community c_x^{i-1} that can be reached by x under 2^{i-1} degrees of distance space t_x^{i-1}. We have $F_{[T]_i}(C) = \sum_{c_x^i \in C^i} f_{[t]_x^i}\left(c_x^i\right)$, where $c_x^i \in C^i$ is a community that is detected under one degree of distance space $[t]_x^i$ of x on G_i.

Lemma 2 For $\forall x \in V$, c_x^i, $[t]_x^i$ and $i,j = 1, 2, 3, \dots, \{j > i\}$, we have,
Monotonicity:

$$f_{[t]_x^i}\left(c_x^i\right) \le f_{[t]_x^{i+1}}\left(c_x^{i+1}\right). \tag{8}$$

Submodularity:

$$f_{[t]_x^i \cup [t]_x^j}\left(c_x^{\max\{i,j\}}\right) - f_{[t]_x^i}\left(c_x^i\right) \ge f_{[t]_x^{i+1} \cup [t]_x^j}\left(c_x^{\max\{(i+1),j\}}\right) - f_{[t]_x^{i+1}}\left(c_x^{i+1}\right). \tag{9}$$

Proof. In Eq. (8), it is similar to Eq. (5). Based on a greedy strategy, a subset of nodes (communities) are selected and added to c_x^i. So, $f_{[t]_x^i}\left(c_x^i\right)$ is monotonic. According to (6), we have $f_{[t]_x^i \cup [t]_x^j}\left(c_x^{\max\{i,j\}}\right) - f_{[t]_x^i}\left(c_x^i\right) \ge f_{[t]_x^{i+1} \cup t_x^k}\left(c_x^{\max\{(i+1),k\}}\right) - f_{[t]_x^{i+1}}\left(c_x^{i+1}\right), \{k > 2^i\}$, since that only the communities with diameter is less than $(k - 2^i)$ can be all reached under t_x^k. So, the proving of Eq. (9) can be translated into Eq. (10).

$$f_{[t]_x^i \cup t_x^j}\left(c_x^{\max\{i,k\}}\right) - f_{[t]_x^i}\left(c_x^i\right) \ge f_{[t]_x^{i+1} \cup t_x^k}\left(c_x^{\max\{(i+1),k\}}\right) - f_{[t]_x^{i+1}}\left(c_x^{i+1}\right). \tag{10}$$

If $k \leq 2^i$, we have $t_x^k = [t]_x^i \cup t_x^k$ and $[t]_x^{i+1} = [t]_x^{i+1} \cup t_x^k$, otherwise, we have $t_x^k = [t]_x^i \cup t_x^k$ and $t_x^k = [t]_x^{i+1} \cup t_x^k$. According to Eq. (8) in lemma 2, it is easy to prove that Eq. (10) is workable.

At the end, we can conclude that $F_{[T]_i}(C)$ holds submodularity on quotient network. So during the hierarchical process, target function holds the property of submodularity. According to [23], a result with $[1 - 1/e]$ approximation is guaranteed.

4 Experiments

In this section, in order to evaluate performance of the proposed MD-HCD approach, we do tests on real-world networks and compare it to several mainstream community detection algorithms.

BGLL [3] is one of well-known modularity based algorithm. It allows for hierarchical community detection and has lower time complexity.

Infomap [25] envisions community detection problem as a coding problem, and aims at finding the optimal partitions based on minimum description length principle.

LPA [26] is one of the fastest algorithms, which can identify communities in large network.

4.1 Metric

Modularity is the most popular internal measure that could evaluate the quality of communities produced by different algorithms. The modularity is defined as follow.

$$Q = \frac{1}{2m} * \sum_{i,j} \left[A_{i,j} - \frac{k_i * k_j}{2m} \right] \delta(C_i, C_j). \tag{11}$$

In Eq. (11), $A_{i,j}$ is the element of adjacency matrix of network, and the value of $A_{i,j}$ is 1 or 0 when there is a link between i and j or not. k_i is degree number of node i. Finally, the value of $\delta(C_i, C_j)$ is 1 when i and j belong to same community, otherwise, the value is 0. In general, the Q-value ranges from 0.3 to 0.7. And the greater of the value, the better the community structure quality.

4.2 Data Sets

We test our MD-HCD model, and compared algorithms on five networks [24] that are widely used by other researchers for evaluating community detection algorithms. Table 1 lists the features of these networks.

Table 1. Information of data sets. n and m donate the number of nodes and edges of networks, respectively. d-avg is the average of all nodes' degree. L is the average of path length. The r is the assortativity coefficient which is the Pearson correlation coefficient of degree between pairs of linked nodes. Positive values of r indicate a correlation between nodes of similar degree, while negative values indicate relationships between nodes of different degree. In general, r lies between -1 and 1.

Networks	n	m	d-avg	L	r
Karate	34	78	4.6	2.7	−0.211
Dolphin	62	159	5.1	3.4	0.306
Football	115	613	10.7	2.3	0.936
Netscience	1589	2742	3.45	4.0	0.439
Power	4941	6594	2.67	19.0	0.226

4.3   Experimental Results and Analysis

In this section, we test our MD-HCD approach on given data sets, and record the results of the experiment.

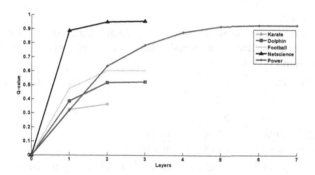

Fig. 2. Change of Q-value during hierarchical process

Firstly, Fig. 2 exhibits the change of Q-value during hierarchical process. It is easy to find that Q-value rise gradually and eventually tend to be stable on all networks. MD-HCD detects a hierarchical community structure that includes three layers on Dolphin, Football and Netscience. In first layer, an acceptable community structure can be identified with Q-value greater than 0.3 under one degree distance space. And with the increase of degree distance, the best community structure can be obtained under three degrees distance space. It shows that when we consider 'my friend' [23] in a community, an acceptable community structure can be obtained, when we consider 'the friends of my friends' friends" [23] in a community, the optimal community structure can be obtained. For Karate network, MD-HCD detects best community structure under two degree distance. In fact, a few core nodes are almost connected to all nodes with assortativity coefficient $r = -0.211$ in Karate network, which is not easy for degree of distance based hierarchical analysis. The Power network is a sparse network. Though

MD-HCD can detect a competitive community structure under three degree distance space, it would consider further degrees of distance space to approximate optimal community structure duo to the value of average of path length is 19.0. In Table 2, we show the change of quotient networks' size during hierarchical process. G_i is a quotient network each super node represents a community detected in G_{i-1}. And each edge represents interaction between super nodes. We can see that the size of quotient network is reduced effectively compared with the size of original network.

In the end, we compare MD-HCD with other algorithms in terms of Q-value as shown in Table 3. Our method has the highest Q-value on Football and Netscience. On Dolphin network, MD-HCD has a similar result with Infomap, which is obviously better than the other two algorithms. Alike, MD-HCD achieves a results with the value of Q is 0.931 on Power network, which is slightly less than BGLL with the value of Q is 0.935, but much higher than other algorithms. On Karate network, MD-HCD achieves a result with 88% approximation of the optimal result detected via BGLL. In the last row of Table 3 is average Q-value of four algorithms. We can see that the MD-HCD performs the best among these algorithms. On the whole, our algorithm performs best on Football and Netscience, or has an approximate result compared with best performing algorithms on other networks. We can conclude that the proposed method has a stable performance.

Table 2. Change of quotient networks' size during hierarchical process. For number 'a,b' in every cell, a donates the number of nodes and b donates the number of edges in G_i. And '——' donates the algorithm stopped.

Networks	G_1	G_2	G_3	G_4	G_5	G_6	G_7
Karate	6,6	2,1	——	——	——	——	——
Dolphin	26,41	6,11	4,6	——	——	——	——
Football	21,98	11,50	10,41	——	——	——	——
Netscience	485,113	413,26	403,12	——	——	——	——
Power	2814,3863	1063,1877	462,937	177,382	53,110	34,74	32,70

Table 3. Comparison of Q-value by different algorithms on real-world networks

Networks	BGLL	Infomap	LPA	MD-HCD
Karate	**0.411**	0.402	0.384	0.363
Dolphin	0.418	**0.525**	0.501	0.522
Football	0.603	0.601	0.581	**0.604**
Netscience	0.950	0.931	0.943	**0.954**
Power	**0.934**	0.829	0.605	0.931
Average	0.663	0.658	0.601	**0.675**

5   Conclusion

In this paper, we propose a multi degrees of distance space and submodularity optimization based hierarchical community detection (MD-HCD) model, which allows us to get a novel understanding of hierarchical structure under the view of degree distance and achieves a result where the $[1 - 1/e]$ approximation is guaranteed. The experimental results show that the proposed method often detects a hierarchical community structure that includes three layers and an optimal community structure can be obtained when we consider 'the friends of my friends' friends" in a community. Compared with other main stream algorithms, MD-HCD has a stable performance in terms of Q-value. In the future, we will pay more attention to the relationship between hierarchical community structure and topology structure.

Acknowledgements. This work was partially supported by National Natural Science Foundation of China (Grants #61402006, #61602003 and #61673020), National High Technology Research and Development Program (863 Plan) (Grant #2015AA124102), National Defense Science and technology innovation special zone project (Grant 17-H863-01-ZT-005-007-03), the Provincial Natural Science Foundation of Anhui Province (Grants #1508085MF113 and #1708085QF156), Scientific Research Foundation for the Returned Overseas Chinese Scholars, State Education Ministry (Forty-ninth batch) and the Recruitment Project of Anhui University for Academic and Technology Leader.

References

1. Girvan, M., Newman, M.E.J.: Community structure in social and biological networks. Proc. Natl. Acad. Sci. U.S.A. **99**(12), 7821–7826 (2002)
2. Yang, B., Di, J., Liu, J., et al.: Hierarchical community detection with applications to real-world network analysis. Data Knowl. Eng. **83**(90), 20–38 (2013)
3. Blondel, V.D., Guillaume, J.L.R., Lambiotte, L.E.: Fast unfolding of communities in large networks. J. Stat. Mech. Theor. Exp. **2008**(10), P10008 (2008)
4. Yin, C., Zhu, S., Chen, H., et al.: A method for community detection of complex networks based on hierarchical clustering. Int. J. Distrib. Sens. Netw. **11**(6), 849140 (2015)
5. Guohui, D., Huimin, S., Chunlong, F., Yan, S.: Community detection algorithm of the large-scale complex networks based on random walk. In: Song, S., Tong, Y. (eds.) WAIM 2016. LNCS, vol. 9998, pp. 269–282. Springer, Cham (2016). doi:10.1007/978-3-319-47121-1_23
6. Bae, S.H., Halperin, D., West, J.D., et al.: Scalable and efficient flow-based community detection for large-scale graph analysis. ACM Trans. Knowl. Discov. Data **11**(3), 32 (2017). ACM
7. Newman, M.E.J.: Detecting community structure in networks. Eur. Phys. J. B Condens. Matter Complex Syst. **38**(2), 321–330 (2004)
8. Clauset, A., Newman, M.E.J., Moore, C.: Finding community structure in very large networks. Phys. Rev. E Stat. Nonlinear Soft Matter Phys. **70**(2), 066111 (2004)
9. Li, Y., He, K., Bindel, D., et al.: Overlapping community detection via local spectral clustering. arXiv preprint arXiv:1509.07996 (2015)
10. Wang, F., Chen, J.: A community detection combining simulated annealing and greedy method. Proc. Nat. Acad. Sci. **103**(23), 8577–8582 (2006)

11. Yang, T., Jin, R., Chi, Y., et al.: Combining link and content for community detection. In: Alhajj, R., Rokne, J. (eds.) Encyclopedia of Social Network Analysis and Mining, pp. 190–201. Springer, New York (2014). doi:10.1007/978-1-4614-6170-8_214

12. Suciu, M., Lung, R.I., Gaskó, N.: Mixing network extremal optimization for community structure detection. In: Ochoa, G., Chicano, F. (eds.) EvoCOP 2015. LNCS, vol. 9026, pp. 126–137. Springer, Cham (2015). doi:10.1007/978-3-319-16468-7_11

13. Fortunato, S., Castellano, C.: Community structure in graphs. In: Meyers, R.A. (ed.) Computational Complexity, pp. 490–512. Springer, Heidelberg (2012). doi:10.1007/978-1-4614-1800-9_33

14. Shen, H., Cheng, X., Cai, K., et al.: Detect overlapping and hierarchical community structure in networks. Physica A Stat. Mech. Appl. 388(8), 1706–1712 (2009)

15. Zhao, S., Ke, W., Chen, J., et al.: Tolerance granulation based community detection algorithm. Tsinghua Sci. Technol. 20(6), 620–626 (2015)

16. Shang, R., Luo, S., Li, Y., et al.: Large-scale community detection based on node membership grade and sub-communities integration. Physica A Stat. Mech. Appl. 428, 279–294 (2015)

17. Soundarajan, S., Hopcroft, J.E.: Use of local group information to identify communities in networks. ACM Trans. Know. Discov. Data 9(3), 1–27 (2015)

18. Bing-Bing, X., et al.: A unified method of detecting core-periphery structure and community structure in networks. Preprint arXiv:1612.01704 (2016)

19. Niwattanakul, S., Singthongchai, J., Naenudorn, E., et al.: Using of Jaccard coefficient for keywords similarity. In: Proceedings of the International MultiConference of Engineers and Computer Scientists, vol. I. Hong Kong (2013)

20. Borgatti, S.P., Everett, M.G.: Models of core/peripherystructures. Soc. Netw. 21(4), 375–395 (2000)

21. Wang, L., et al.: Detecting community kernels in large social networks. In: IEEE, International Conference on Data Mining, pp. 784–793. IEEE Computer Society (2011)

22. Walker, S.K.: Connected: the surprising power of our social networks and how they shape our lives. J. Fam. Theor. Rev. 3(3), 220–224 (2011)

23. Fisher, M.L., Nemhauser, G.L., Wolsey, L.A.: An analysis of approximations for maximizing submodular set functions. Math. Program. 14, 265–294 (1978)

24. Zhang, X., et al.: Efficient community detection based on label propagation with belonging coefficient and edge probability. In: Li, Y., Xiang, G., Lin, H., Wang, M. (eds.) Chinese National Conference on Social Media Processing, pp. 54–72. Springer, Singapore (2016). doi:10.1007/978-981-10-2993-6_5

25. Rosvall, M., Bergstrom, C.T.: Maps of random walks on complex networks reveal community structure. Proc. Nat. Acad. Sci. U.S.A. 4(105), 1,118–1,123 (2008)

26. Raghavan, U.N., Albert, R., Kumara, S.: Near linear time algorithm to detect community structures in large-scale networks. Phys. Rev. E Stat. Nonlinear Soft. Matter Phys. 76, 036106 (2013)

Author Index

Chang, Biao 77
Chen, Enhong 77
Chen, Guang 206
Chen, Zhumin 331

Ding, Xiao 40
Du, Jiachen 29

Feng, Chong 116
Feng, Shi 218, 283

Gao, Qinghong 29
Gui, Lin 29
Guo, Yuhang 91

Hao, Yi-Jing 3
Hong, Yu 141
Hu, Xuexian 129
Huang, Heyan 3, 16, 91
Huang, Zhenhua 53

Jian, Ping 91
Jin, Tianyuan 77

Li, Dayu 166
Li, Deyu 232
Li, Fang 256
Li, Luying 181
Li, Shoushan 153, 193
Li, Yang 166
Li, Yuke 319
Li, Zhongyang 40
Lian, Tao 331
Lin, Hongfei 181
Lin, Hua-Kang 16
Lin, Yuan 181
Lin, Yujie 331
Liu, Jinglian 283
Liu, Mengyi 141
Liu, Pei 104
Liu, Ting 40, 244
Liu, Wenfen 129
Lu, Zhonglei 129
Luo, Zhunchen 116

Ma, Jun 331
Mao, Xian-Ling 3, 16

Pei, Yuxia 116

Qin, Bing 244

Shi, Xuewen 91
Su, Tingxuan 53
Sundquist, James 319

Tang, Jian 141
Tang, Xijin 65
Tang, Yi-Kun 3, 91

Wang, Binyu 129
Wang, Chu 218
Wang, Daling 218, 283
Wang, Dan 16
Wang, Suge 166, 232
Wang, Xinbo 206
Wang, Zhefeng 77
Wang, Zhenyu 53
Wei, Xiaochi 91

Xu, Bo 181
Xu, Cheng 3
Xu, Jian 193
Xu, Kan 181
Xu, Nuo 65
Xu, Ruifeng 29
Xu, Shuaishuai 296
Xu, Tong 77

Yao, Jianmin 141
Ye, Zhe 256
Yi, Chengqi 53
Yin, Hao 193
Yu, Chengjin 343
Yuan, Jianhua 244
Yue, Jiahua 319

Zhan, Zhiqiang 104
Zhang, Chunhong 104
Zhang, Le 77
Zhang, Lun 271
Zhang, Weiming 104
Zhang, Xinzhi 308
Zhang, Yanping 343
Zhang, Yifei 218, 283
Zhang, Zike 296
Zhao, Chuanjun 232
Zhao, He 116

Zhao, Sendong 40
Zhao, Shu 343
Zhao, Weiji 283
Zhao, Yanyan 244
Zhao, Zhishan 29
Zhou, Guodong 153, 193
Zhou, Yanfang 129
Zhou, Yinzuo 296
Zhu, Suyang 153
Zhu, Yingbo 53
Zhuang, Benhui 104

Printed in the United States
By Bookmasters